MW00465011

Electronics Portable
Handbook

Electronics Portable Handbook

Stan Gibilisco

McGraw-Hill

New York   San Francisco   Washington, D.C.   Auckland   Bogotá
Caracas   Lisbon   London   Madrid   Mexico City   Milan
Montreal   New Delhi   San Juan   Singapore
Sydney   Tokyo   Toronto

Library of Congress Cataloging-in-Publication Data

Stan Gibilisco.
    Electronics portable handbook / Stan Gibilisco, editor-in-chief.
       p.   cm.
    Includes Index.
    ISBN 0-07-134415-2
    1. Electronics Handbooks, manuals, etc. I. Gibilisco, Stan.
TK7825.E373 1999
621.381—dc21
                                  99-27804
                                       CIP

McGraw-Hill

*A Division of The **McGraw·Hill** Companies*

1 2 3 4 5 6 7 8 9 0 DOC/DOC 9 0 4 3 2 1 0 9

ISBN 0-07-134415-2

*The sponsoring editor for this book was Scott Grillo, the editing supervisor was
Nancy Young, and the production supervisor was Pamela A. Pelton. It was set
in Century Schoolbook per the MHT 5 x 8 design by Michele Pridmore and Paul
Scozzari of McGraw-Hill's Hightstown, N.J., Professional Book Group composition unit.*

Printed and bound by R. R. Donnelley & Sons Company.

McGraw-Hill books are available at special quantity discounts to use as pre-
miums and sales promotions, or for use in corporate training programs. For
more information, please write to the Director of Special Sales, McGraw-Hill,
11 West 19th Street, New York, NY 10011. Or contact your local bookstore.

*To Samuel, Tim, and Tony
from Uncle Stan*

Contents

Preface

This is a handy, easy-to-carry general reference for electronics technicians, engineers, hobbyists, and students. Some information is provided in the fields of mathematics, physics, and chemistry as applicable to electronics. Lists of electronics abbreviations and schematic symbols are also included.

Every effort has been made to arrange this book in a logical manner, and to portray the information in concise but understandable terms. Suggestions for future editions are welcome. I can be reached by e-mail from links to my Web site at http://members.aol.com/stangib.

Stan Gibilisco

Acknowledgments

Illustrations in this book were generated with CorelDRAW. Some clip art is courtesy of Corel Corporation, 1600 Carling Avenue, Ottawa, Ontario, Canada K1Z 8R7.

Acknowledgments

Direct Current

Direct current (dc) is a flow of electrical charge carriers that always takes place in the same direction. This is what distinguishes direct current from *alternating current* (ac). The current need not always have the same magnitude, but if it is to be defined as dc, the direction of the charge-carrier flow must never reverse.

The Nature of DC

Figure 1.1 illustrates four waveform graphs of current versus time. The graphs in Fig. 1.1A, B, and C depict dc because the current always flows in the same direction, even though the amplitude (intensity) might change with time. The rendition in Fig. 1.1D is not dc because the direction of current flow does not remain constant.

Current

Current is a measure of the rate at which electrical charge carriers, usually electrons, flow. A current of 1 *ampere* (1 A) represents 1 *coulomb* (6.24×10^{18}), 1 C of charge carriers per second past a given point. Often, current is specified in

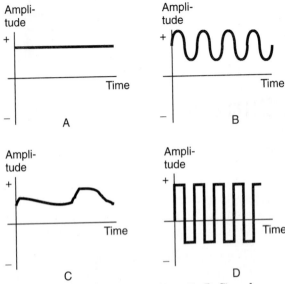

Figure 1.1 Examples of dc waveforms (A, B, C), and a non-dc waveform (D).

terms of *milliamperes,* abbreviated mA, where 1 mA = 0.001 A. You will also sometimes hear of *microamperes* (μA), where 1 μA = 10^{-6} A = 0.001 mA. And it is increasingly common to hear about *nanoamperes* (nA), where 1 nA = 0.001 μA = 10^{-9} A.

A current of a few milliamperes will give you a shock, 50 mA will jolt you severely, and 100 mA can cause death if it flows through your chest cavity. In some circuits extremely large currents flow. This might happen through a metal bar placed directly at the output of a massive electric generator. The resistance is extremely low in this case, and the generator is capable of driving huge amounts of charge.

In some semiconductor electronic devices, such as microcomputers, a few nanoamperes will suffice for the execution of complicated electronic processes. Some electronic clocks draw so little current that their batteries last as long as they would if left on the shelf.

Resistance

Resistance is the opposition that a circuit offers to the flow
of electric current. The standard unit of resistance is the
ohm (Ω). Other common resistance units include the *kil-
ohm* (abbreviated kΩ), where 1 kΩ = 1000 Ω, and the
megohm (abbreviated MΩ), where 1 MΩ = 1000 K = 10^6 Ω.

When current flows through a resistive material, there is
always a potential difference (voltage) across the resistive
object. The larger the current through the resistance, the
greater the voltage across it. In general, this voltage is
directly proportional to the current, provided the resistance
does not change.

Electrical circuits always have some resistance. When
certain metals are chilled to extremely low temperatures,
they lose practically all of their resistance, but they never
become theoretically perfect conductors. There is no theoretically infinite resistance, either, although resistance can
be infinite in a practical sense. Even air conducts to some
extent, although the effect is usually so small that it can be
ignored. In some electronic applications, materials are
selected on the basis of how nearly infinite their resistance
is. These materials make good electric insulators and good
dielectrics for capacitors.

In a practical electronic circuit, the resistance of a particular component might vary depending on the conditions
under which it is operated. A transistor, for example, might
have extremely high resistance some of the time and very
low resistance at other times. This high/low fluctuation can
be made to take place thousands, millions, or billions of
times each second.

Electromotive force

An *electromotive force* (EMF) of 1 *volt* (1 V), across a resistance of 1 Ω, will drive an electric current of 1 A through
that resistance. This is how the volt is generally defined.
But it is possible to have an EMF without a flow of current. An example occurs just before a lightning stroke

occurs. It is true of a battery when there is nothing connected to it. Charge carriers will move only if a conductive path is provided.

Even a large EMF might not drive much current. A good example is your body after walking around on a carpet. Although the notion of thousands of volts building up on your body sounds dangerous, there are not many coulombs of electrons involved. Therefore, not many electrons will flow through your finger, in relative terms, when you touch a grounded object. The shock you receive will be small.

If there are many coulombs of available electric charge carriers, a moderate EMF such as 100 V can drive a lethal current through the human body. This is why it is dangerous to repair some electronic devices, particularly those using vacuum tubes, when powered-up. The power supply of such a system can pump an unlimited number of coulombs of charge through your body. If the conductive path is through the heart, electrocution can occur.

Sources of dc

Typical sources of dc include power supplies, electrochemical cells and batteries, and photovoltaic cells and panels. The intensity, or amplitude, of a direct current might fluctuate with time, and this fluctuation might be periodic. In some such cases the dc has an ac component superimposed on it (as in Fig. 1.1B). An example of this is the output of a photovoltaic cell that receives a modulated-light communications signal. A source of dc is sometimes called a *dc generator.*

Batteries and various other sources of dc produce a constant voltage. This is called *pure dc* and can be represented by a straight, horizontal line on a graph of voltage versus time (as in Fig. 1.1A). The peak and effective values are the same. The peak-to-peak value is zero because the instantaneous amplitude never changes. In some instances the value of a dc voltage pulsates or oscillates rapidly with time, in a manner similar to the changes in an ac wave. The unfiltered output of a half-wave or a full-wave rectifier, for example, is *pulsating dc.*

Ohm's Law

Most dc circuits can be reduced to three major components: a voltage source, a set of conductors, and a resistance. The voltage, or EMF source, is called E, the current in the conductor is called I, and the resistance is called R. The interdependence between current, voltage, and resistance is one of the most fundamental rules of electrical circuits. It is called *Ohm's Law,* named after the scientist who supposedly first expressed it. Three formulas denote this law:

$$E = IR$$

$$I = \frac{E}{R}$$

$$R = \frac{E}{I}$$

You need only remember the first equation to derive the others. The easiest way to remember it is to learn the abbreviations E for EMF, or voltage, I for current, and R for resistance and then remember that they appear in alphabetical order with the equals sign after the E. You can also imagine them in a triangle, as shown in Fig. 1.2. The letters proceed in alphabetical order (E, I, R) from top to bottom and left to right.

If the initial quantities are given in units other than volts, amperes, and ohms, you must convert to these units and then calculate. After that, you can convert the units back again to whatever you like. For example, if you get 13,500,000 Ω as a calculated resistance, you might prefer to say that it is 13.5 MΩ.

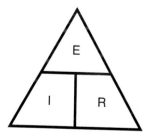

Figure 1.2 The Ohm's Law triangle can serve as a memory aid.

Current calculations

The first way to use Ohm's Law is to find current values in dc circuits. To find the current, you must know the voltage and the resistance or be able to deduce them.

Refer to the schematic diagram of Fig. 1.3. It consists of a variable dc generator, a voltmeter, some wire, an ammeter, and a calibrated, wide-range potentiometer. Suppose the dc generator produces 10 V and that the potentiometer is set to a value of 10 Ω. Then the current can be found by the formula $I = E/R$. Plug in the values for E and R; they are both 10, because the units were given in volts and ohms. Then $I = 10/10 = 1$ A.

Voltage calculations

The second use of Ohm's Law is to find unknown voltages when the current and the resistance are known. Suppose the potentiometer (Fig. 1.3) is set to 100 Ω, and the measured current is 10 mA. Use the formula $E = IR$ to find the voltage. First, convert the current to amperes: 10 mA = 0.01 A. Then multiply: $E = 0.01 \times 100 = 1$ V.

If the potentiometer is set to a value of 157 kΩ and the current reading is 17 mA, you must convert both the resistance and the current values to their proper units. A resistance of 157 kΩ is 157,000 Ω; a current of 17 mA is 0.017 A. Then $E = IR = 0.017 \times 157,000 = 2669$ V = 2.669 kV. You might want to round this off to 2.67 kV.

Resistance calculations

Ohm's Law can be used to find a resistance between two points in a dc circuit, when the voltage and the current are known. Suppose both the voltmeter and ammeter scales in Fig. 1.3 are visible but that the potentiometer is uncalibrated. If the voltmeter reads 24 V and the ammeter shows 3.0 A, the value of the potentiometer can be found via the formula $R = E/I$: $R = 24/3.0 = 8.0$ Ω.

If the current is 18 mA and the voltage is 229 mV, convert these values to amperes and volts. This gives $I = 0.018$ A

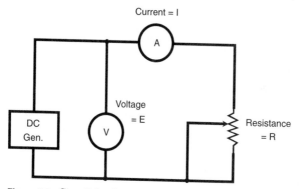

Figure 1.3 Circuit for demonstrating Ohm's Law.

and $E = 0.229$ V. Then plug into the equation $R = E/I = 0.229/0.018 = 13\ \Omega$.

Power calculations

You can calculate the power in a dc circuit as shown in Fig. 1.3 by the formula $P = EI$, or the product of the voltage in volts and the current in amperes. You might not be given the voltage directly, but you can calculate it if you know the current and the resistance. If you know I and R but do not know E, you can get the power P by means of the formula $P = (IR)I = I^2R$. You can also get the power if you are not given the current directly. Suppose you know only the voltage and the resistance. Then $I = E/R$; therefore, $P = E(E/R) = E^2/R$.

Stated all together, these power formulas are

$$P = EI = I^2R = \frac{E^2}{R}$$

Suppose that the voltmeter in Fig. 1.3 reads 12 V and the ammeter shows 50 mA. Use the formula $P = EI$ to find the power dissipated in the resistor. First, convert the current to amperes, getting $I = 0.050$ A. Then $P = EI = 12 \times 0.050 = 0.60$ W.

If the resistance is 999 Ω and the voltage source delivers 3 V, use the formula $P = E^2/R = 3 \times 3/999 = 9/999 = 0.009$ W = 9 mW.

If the resistance is 0.47 Ω and the current is 680 mA, use the formula $P = I^2R$, after converting to ohms and amperes. Then $P = 0.680 \times 0.680 \times 0.47 = 0.22$ W.

Resistive Networks

Combinations of resistances are used in dc circuits to regulate current, control voltage, and in general set the operating parameters as desired.

Resistances in series

When you place resistances in series, their ohmic values simply add together to get the total resistance. This is intuitive and is simple to remember.

Suppose the following resistances are hooked up in series with each other: 112, 470, and 680 Ω. The total resistance of the series combination is found by adding the values, getting a total of $112 + 470 + 680 = 1262$ Ω.

Resistances in parallel

When resistances are placed in parallel, they behave differently than they do in series. In general, if you have a resistor of a certain value and you place other resistors in parallel with it, the overall resistance will decrease.

One way to look at resistances in parallel is to consider them as conductances instead. In parallel, conductances add, just as resistances add in series. If you change all the ohmic values to siemens (the standard unit of conductance, abbreviated S), you can add these figures up and convert the final answer back to ohms.

The symbol for conductance is G. This figure, in siemens, is related to the resistance R, in ohms, by the formulas

$$G = \frac{1}{R}$$

$$R = \frac{1}{G}$$

To find the net resistance of a set of three or more resistors in parallel, follow this procedure:

- Convert all resistance values to ohms.
- Find the reciprocal of each resistance; these are the respective conductances in siemens.
- Add up the individual conductance values; this is the net conductance of the combination.
- Take the reciprocal of the net conductance; this is the net resistance, in ohms, of the parallel combination.

Consider five resistors in parallel. Call them R_1 through R_5, and call the total resistance R as shown in Fig. 1.4. Let $R_1 = 100$ Ω, $R_2 = 200$ Ω, $R_3 = 300$ Ω, $R_4 = 400$ Ω, and $R_5 = 500$ Ω. Converting the resistances to conductance values, you get $G_1 = 1/100 = 0.01$ S, $G_2 = 1/200 = 0.005$ S, $G_3 = 1/300 = 0.00333$ S, $G_4 = 1/400 = 0.0025$ S, and $G_5 = 1/500 = 0.002$ S. Adding these gives $G = 0.01 + 0.005 + 0.00333 + 0.0025 + 0.002 = 0.0228$ S. The total resistance is therefore $R = 1/G = 1/0.0228 = 43.8$ Ω.

When there are several resistances in parallel and their values are all equal, the total resistance is equal to the resistance of any one component, divided by the number of components.

Division of power

When combinations of resistances are hooked up to a source of voltage, they will draw current. You can determine how much current they will take by calculating the total resistance of the combination and then considering the network as a single resistor.

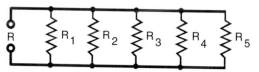

Figure 1.4   Five resistances in parallel, R_1 through R_5, give a total resistance R as discussed in the text.

If the resistances in the network all have the same ohmic value, the power from the source will be evenly distributed among the resistances, whether they are hooked up in series or in parallel. If the resistances in the network do not all have identical ohmic values, they divide up the power unevenly. In that case, the power must be calculated by determining either the current through each resistor or the voltage across each and using the dc power formulas, $P = I^2R$ or $P = E^2/R$.

Resistances in series-parallel networks

Sets of resistors, all having identical ohmic values, can be connected together in parallel sets of series networks or in series sets of parallel networks. By doing this, the total power-handling capacity of the resistance can be greatly increased over that of a single resistor.

Sometimes, the total resistance of a series-parallel network is the same as the value of any one of the resistors. This is always true if the components are identical and are in a network called an *n-by-n matrix*. That means, when n is a whole number, there are n parallel sets of n resistors in series (Fig. 1.5A), or else there are n series sets of n resistors in parallel (Fig. 1.5B). Either arrangement will give exactly the same results in practice. Engineers and technicians sometimes use this to advantage to get resistors with large power-handling capacity. Each resistor should have the same rating, say 1 watt, or 1 W. Then the combination of n-by-n resistors will have n^2 times that of a single resistor. A 3-by-3 series-parallel matrix of 2-W resistors can handle $3^2 \times 2 = 9 \times 2 = 18$ W, for example. A 10×10 array of 1-W resistors can take 100 W.

Nonsymmetrical series-parallel networks, made up from identical resistors, will increase the power-handling capability. But in these cases, the total resistance will not be the same as the value of the single resistors. The overall power-handling capacity will always be multiplied by the total number of resistors, whether the network is symmetrical or not, provided all the resistors are the same. In engineering

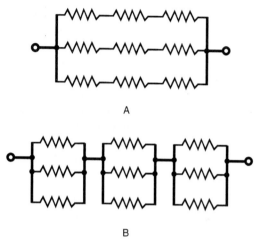

A

B

Figure 1.5 Series-parallel combinations. (A) Sets of
series resistors are connected in parallel; (B) sets of
parallel resistors are connected in series.

work, cases sometimes arise where nonsymmetrical networks
fit the need.

Currents through series resistances

In a series dc circuit, the current at any given point is the
same as the current at any other point. This is true no matter what the components actually are and regardless of
whether or not they all have the same resistance.

If the components have different resistances in a series dc
circuit, some of them consume more power than others. In
case one of the components shorts out, the current through
the whole chain will increase, because the overall resistance
of the string will go down. If a component opens, the current
drops to zero at every point.

Voltages across series resistances

In a series circuit, the voltage is divided up among the components. The sum total of the potential differences across
each resistance is equal to the power supply voltage. This is

always true, no matter how large or how small the resistances and whether or not they are all of the same value. This also holds for 60-hertz (Hz) utility ac circuits almost all the time.

The voltage across any resistor in a series combination is equal to the product of the current and the resistance. Remember that you must use volts, ohms, and amperes when making calculations. To find the current in the circuit, I, you need to know the total resistance and the supply voltage. Then $I = E/R$. First find the current in the whole circuit; then find the voltage across any particular resistor.

Voltages across parallel resistances

In a parallel circuit, the voltage across each component is always the same and is always equal to the power supply or battery voltage. The current drawn by each component depends only on the resistance of that particular device. In this sense, the components in a parallel-wired circuit work independently, as opposed to the series-wired circuit in which they all interact.

If any one branch of a parallel circuit is taken away, the conditions in the other branches will remain the same. If new branches are added, assuming the power supply can handle the load, conditions in previously existing branches will not be affected.

Currents through parallel resistances

Refer to the schematic diagram of Fig. 1.6. The resistors are called R_n. The total parallel resistance in the circuit is R. The battery voltage is E. The current in branch n, containing resistance R_n, is measured by ammeter A and is called I_n.

The sum of all the I_n's in the circuit is equal to the total current I drawn from the source. That is, the current is divided up in the parallel circuit, similarly to the way that voltage is divided up in a series circuit.

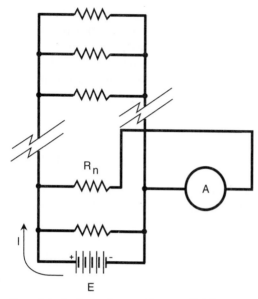

Figure 1.6 Analysis of current in a parallel dc circuit.

Power distribution in series circuits

When calculating the power dissipated by a particular resistor R_n in a circuit containing n resistors in series, find the current I that the circuit is carrying. Then it is easy to calculate the power P_n, based on the formula $P_n = I^2 R_n$.

The total power dissipated in a series circuit is equal to the sum of the wattages dissipated in each resistor. In this way, the distribution of power in a series circuit is like the distribution of the voltage.

Power distribution in parallel circuits

When resistances are wired in parallel, they each consume power according to the same formula, $P = I^2 R$. But the current is not the same in each resistance. An easier method to find the power P_n, dissipated by resistor R_n, is by using the formula $P_n = E^2/R_n$, where E is the voltage of the power supply. This voltage is the same across every resistor in a parallel circuit.

In a parallel circuit, the total power consumed is equal to the sum of the wattages dissipated by the individual resistances. In this respect, the parallel circuit acts like the series circuit. All the power must be accounted for. Power can never appear from nothing or disappear into nowhere.

Kirchhoff's Laws

Kirchhoff's First Law is that the current going into any point in a dc circuit is the same as the current going out. This is true no matter how many branches lead into or out of the point (Fig. 1.7A). A qualitative way of saying this is that current can never appear from nothing or disappear into nowhere.

Kirchhoff's Second Law is that the sum of all the voltages, as you go around a dc circuit from some fixed point and return there from the opposite direction, and taking polarity into account, is always zero (Fig. 1.7B). Stated another way, voltage can never arise from nothing or disappear into nowhere.

Voltage dividers

Resistances in series produce ratios of voltages, and these ratios can be tailored to meet certain needs. When designing voltage divider networks, the resistance values should be as small as possible, without causing too much current drain on the power supply. In practice the optimum values depend on the nature of the circuit being designed. The voltage divider "fixes" the intermediate voltages most effectively when the resistance values are as small as the current-delivering capability of the power supply will allow.

Figure 1.8 illustrates the principle of voltage division. The individual resistances are R_1, R_2, R_3, ..., R_n. The total resistance is $R = R_1 + R_2 + R_3 + \cdots + R_n$. The supply voltage is E, and the current in the circuit is therefore $I = E/R$. At the various points P_1, P_2, P_3, ..., P_n, the voltages are E_1, E_2, E_3, ..., E_n. The last voltage, E_n, is the same as the power supply

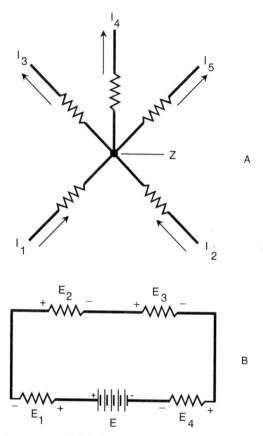

Figure 1.7 (A) Kirchhoff's First Law. The current entering point Z is equal to the current leaving point Z. In this case, $I_1 + I_2 = I_3 + I_4 + I_5$. (B) Kirchhoff's Second Law. The sum of the voltages $E + E_1 + E_2 + E_3 + E_4 = 0$, taking polarity into account.

voltage, E. All the other voltages are less than E, so $E_1 < E_2 < E_3 < \cdots < E_n = E$.

The voltages at the various points increase according to the sum total of the resistances up to each point, in proportion to the total resistance, multiplied by the supply voltage. The voltage E_1 is equal to $E(R_1/R)$. The voltage E_2 is equal to $E(R_1 + R_2)/R$. The voltage $E_3 = E(R_1 + R_2 + R_3)/R$, and so on.

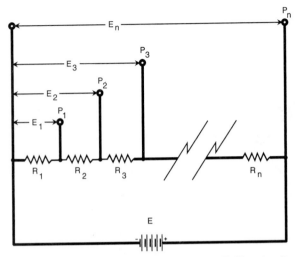

Figure 1.8 General arrangement for voltage divider circuit.

DC Magnetism

Whenever the atoms in a material are aligned, a *magnetic field* exists. A magnetic field can also be caused by the motion of electric charge carriers, either in a wire or in free space.

Magnetic flux

Physicists consider magnetic fields to be made up of *flux lines.* The intensity of the field is determined according to the number of flux lines passing through a certain cross-sectional area, such as a square centimeter (cm^2) or a square meter (m^2).

A magnetic field is considered to begin at the north magnetic pole and to terminate at the south magnetic pole. The lines of flux in a magnetic field always connect the two poles. The greatest *flux density,* or magnetic field strength, around a bar magnet is near the poles, where the lines converge. Around a current-carrying wire, the greatest field strength is near the wire.

Magnetic field strength

The overall magnitude of a magnetic field is measured in *webers,* abbreviated Wb. A smaller unit, the *maxwell* (Mx), is used if a magnetic field is weak. One weber is equal to 10^8 Mx. Conversely, $1 \text{ Mx} = 10^{-8}$ Wb.

The flux density is a more useful expression for magnetic effects than the overall quantity of magnetism. A flux density of 1 *tesla* (1 T) is equal to 1 Wb/m². A flux density of 1 *gauss* (1 G) is equal to 1 Mx/cm². The gauss is 10^{-4} T. Conversely, 1 T is 10^4 G. With electromagnets, another unit is employed: the *ampere-turn* (At). A wire, bent into a circle and carrying 1 A of current, will produce 1 At of magneto-motive force. The *gilbert* (Gb) is also sometimes used to express magnetomotive force. To get ampere-turns from gilberts, multiply by 1.26; to get gilberts from ampere-turns, divide by 1.26.

Ampere's Law

The intensity of the magnetic field surrounding a conductor is directly proportional to the current in the conductor. For a straight wire, the magnetic flux takes the form of concentric cylinders centered on the wire. A cross-sectional view of this situation is shown in Fig. 1.9. The wire is perpendicular to the page, so its axis appears as a point.

Physicists define electric current as flowing from the positive pole to the negative (plus to minus). With this in mind, suppose the current depicted in the figure is flowing out of the page toward you. According to *Ampere's Law,* the direction of the magnetic flux is counterclockwise in this situation.

Ampere's Law is sometimes called the *right-hand rule.* If you hold your right hand with the thumb pointing out straight and the fingers curled, and then point your thumb in the direction of current flow in a straight wire, your fingers will curl in the direction of the magnetic flux. Similarly, if you orient your right hand so your fingers curl in the direction of the magnetic flux, your thumb will point in the direction of the current.

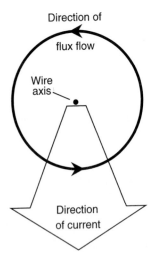

Figure 1.9 Ampere's Law for a current-carrying conductor.

Permeability

Permeability is a measure of the extent to which a material concentrates magnetic flux. A material that increases magnetic flux density is called *ferromagnetic*. These substances can be "magnetized." Iron and nickel are examples. Some alloys are more ferromagnetic than pure iron or pure nickel. A substance that dilates, or dilutes, magnetic flux is called *diamagnetic*. Wax, dry wood, bismuth, and silver are substances that decrease magnetic flux density compared with air or a vacuum.

Permeability is measured on a scale relative to a vacuum, or free space. Free space is assigned permeability 1. The permeability of pure air is approximately equal to 1. If you place an iron core in a coil, the flux density increases by a factor ranging from 60 to 8000, depending on the purity of the iron. Thus, the permeability of iron can be anywhere from 60 (impure) to as much as 8000 (highly refined). If you use certain alloys as the core material in electromagnets, you can increase the flux density, and therefore the local strength of the field, by as much as 1,000,000 times. The permeability factors of some common materials are shown in Table 1.1.

Usually, diamagnetic substances are used to keep magnetic objects apart while minimizing the interaction

TABLE 1.1 Permeability Figures for Some Common Materials

Substance	Permeability (approximate)
Aluminum	Slightly more than 1
Bismuth	Slightly less than 1
Cobalt	60–70
Ferrite	100–3000
Free space	1
Iron	60–100
Iron, refined	3000–8000
Nickel	50–60
Permalloy	3000–30,000
Silver	Slightly less than 1
Steel	300–600
Super permalloys	100,000 to 1,000,000
Wax	Slightly less than 1
Wood, dry	Slightly less than 1

between them. Nonferromagnetic metals, such as copper and aluminum, have the useful property that they conduct electric current well, but magnetic current poorly. They can be used for electrostatic shielding, a means of allowing magnetic fields to pass through while blocking electric fields.

Relay

A *relay* (Fig. 1.10) makes use of an electromagnet to allow remote-control switching. The movable lever, called the *armature,* is held to one side by a spring when there is no current flowing through the electromagnet. Under these conditions, terminal X is connected to Y, but not to Z. When a sufficient current is applied, the armature is pulled over to the other side. This disconnects terminal X from terminal Y and connects X to Z.

A normally closed relay completes the circuit when there is no current flowing in its electromagnet and breaks the circuit when current flows. A normally open relay is just the opposite. Some relays have several sets of contacts. Some are designed to remain in one state (either with current or without) for a long time, whereas others are built to switch several times per second.

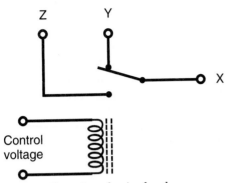

Figure 1.10 Operation of a simple relay.

Magnetic tape

Magnetic tape is used for home entertainment, especially high-fidelity (hi-fi) music and home video. The tape consists of millions of particles of iron oxide, attached to a flexible, nonmagnetic strip. A fluctuating magnetic field, produced by the recording head, polarizes these particles. As the field changes in strength next to the recording head, the tape passes by at a constant, controlled speed. This produces regions in which the iron-oxide particles are polarized in either direction.

When the tape is run at the same speed through the recorder in the playback mode, the magnetic fields around the individual particles cause a fluctuating field that is detected by the pickup head. This field has the same pattern of variations as the original field from the recording head.

The data on a magnetic tape can be distorted or erased by external magnetic fields. Extreme heat can also result in loss of data and possible physical damage to the tape.

Magnetic disk

The principle of the *magnetic disk,* on the micro scale, is the same as that of the magnetic tape. The information is stored in digital form; that is, there are only two different ways that the particles are magnetized. This results in almost perfect, error-free storage.

On a larger scale, the disk works differently than tape because of the difference in geometry. On a tape, the information is spread out over a long span, and some bits of data are far away from others. But on a disk, no 2 bits are farther apart than the diameter of the disk. This means that data can be stored and retrieved much more quickly onto, or from, a disk than is possible with a tape.

The same precautions should be observed when handling and storing magnetic disks as are necessary with magnetic tape.

Alternating Current

In ac, the direction of current, or the polarity of voltage, reverses at regular intervals. The instantaneous absolute magnitude usually changes because of this polarity reversal, except in a theoretically perfect ac square wave, where the instantaneous absolute magnitude does not change even though the polarity does.

Frequency and Waveform

In a *periodic ac wave,* the function of magnitude versus time repeats endlessly. The length of time between one iteration, or *cycle,* and the next is called the *period* of the wave. This is illustrated in Fig. 2.1 for a simple ac wave. The period can, in theory, range anywhere from a miniscule fraction of 1 second (1 s) to many centuries. Radio-frequency currents reverse polarity millions or billions of times a second. The charged particles held captive by the magnetic field of the sun reverse their direction over a period of years. Period, when measured in seconds, is denoted by T.

The *frequency,* denoted f, of a wave is the reciprocal of the period. That is, $f = 1/T$ and $T = 1/f$. The standard unit of frequency is the *hertz,* which represents one complete ac cycle

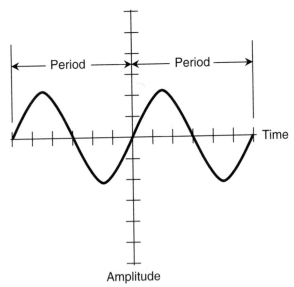

Figure 2.1 A sine wave. The period of the length of time required for one cycle to be completed.

per second. Higher frequencies are given in *kilohertz* (kHz), *megahertz* (MHz) or *gigahertz* (GHz). The relationships are

$$1 \text{ kHz} = 1000 \text{ Hz}$$

$$1 \text{ MHz} = 1000 \text{ kHz} = 10^6 \text{ Hz}$$

$$1 \text{ GHz} = 1000 \text{ MHz} = 10^9 \text{ Hz}$$

Sometimes an even larger unit, the *terahertz* (THz), is used. This is a trillion (10^{12}) hertz. Electrical currents generally do not attain such frequencies, although electromagnetic radiation can.

Some ac waves have only one frequency. These are *pure sine waves*. Often, there are components, known as *harmonics*, at whole-number multiples of the *fundamental frequency*. There might also be components at unrelated frequencies. Some waves are extremely complex, consisting of hundreds, thousands, or even infinitely many different component frequencies.

Sine wave

Sometimes, ac has a sine-wave, or sinusoidal, nature. The waveform in Fig. 2.1 is a sine wave. Any ac wave that consists of a single frequency has a perfect sine-wave shape. Any perfect sine-wave current contains only one component frequency.

In practice, a wave might be so close to a sine wave that it looks exactly like the sine function on an oscilloscope, when in reality there are traces of other frequencies present. Imperfections are often too small to see. Utility ac in the United States has an almost perfect sine-wave shape, with a frequency of 60 Hz.

Square wave

On an oscilloscope, a perfect *square wave* looks like a pair of parallel, dotted lines, one having positive polarity and the other having negative polarity (Fig. 2.2A). In reality, the transitions can often be seen as vertical lines (Fig. 2.2B).

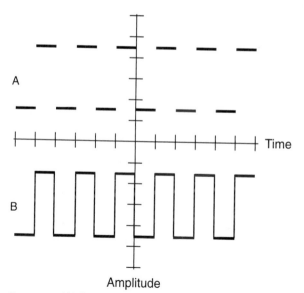

Amplitude

Figure 2.2 (A) A theoretically perfect square wave; (B) the more common rendition.

A square wave might have equal negative and positive peaks. Then the absolute amplitude of the wave is constant at a certain voltage, current, or power level. Half of the time the amplitude is $+x$ volts, amperes, or watts and the other half of the time it is $-x$.

Some square waves are asymmetrical, with the positive and negative magnitudes differing. If the length of time for which the amplitude is positive differs from the length of time the amplitude is negative, the wave is not truly square but is described by the more general term *rectangular wave*.

Sawtooth waves

Some ac waves reverse their polarity at constant, but not instantaneous, rates. The slope of the amplitude-versus-time line indicates how fast the magnitude is changing. Such waves are called *sawtooth waves* because of their appearance.

Sawtooth waves are generated by certain electronic test devices. These waves provide ideal signals for control purposes. Integrated circuits can be wired so that they produce sawtooth waves that have an exact desired shape.

In Fig. 2.3, one form of sawtooth wave is shown. The positive-going slope (rise) is extremely steep, as with a square wave, but the negative-going slope (fall or decay) is gradual. The period of the wave is the time between points at identical positions on two successive pulses.

Another form of sawtooth wave is just the opposite, with a gradual positive-going slope and a vertical negative-going transition. This type of wave is sometimes called a *ramp* (Fig. 2.4). This waveform is used for scanning in television (TV) sets and oscilloscopes.

Sawtooth waves can have rise and decay slopes in an infinite number of different combinations. One example is shown in Fig. 2.5. In this case, the positive-going slope is the same as the negative-going slope. This is a *triangular wave*.

Complex and irregular waves

Figure 2.6 shows an example of a *complex wave*. There is a definite period and therefore a specific and measurable

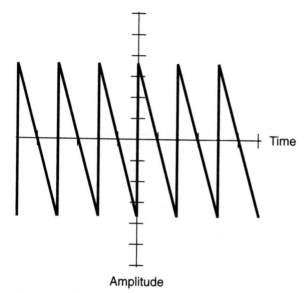

Figure 2.3   A fast-rise, slow-decay sawtooth wave.

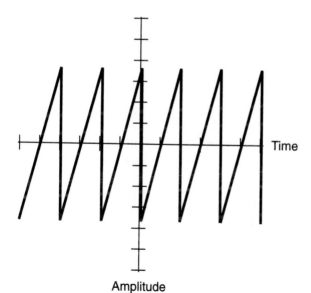

Figure 2.4   A slow-rise, fast-decay sawtooth wave, also called a ramp wave.

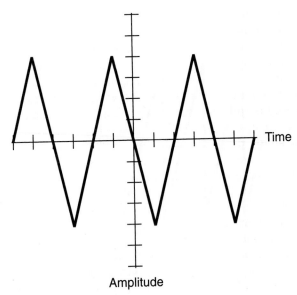

Figure 2.5 A triangular wave.

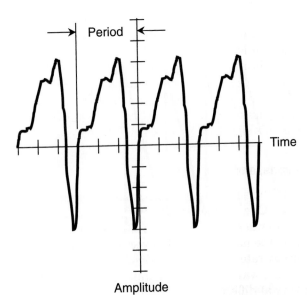

Figure 2.6 A complex waveform. Four complete cycles are shown.

frequency. The period is the time between two points on succeeding wave repetitions.

With some waves, it can be difficult, or almost impossible, to determine the period because the wave has two or more components that are nearly the same magnitude. When this happens, the frequency spectrum of the wave is multifaceted. The energy is split up among two or more frequencies.

An ac sine wave, as displayed on a spectrum analyzer, appears as a single pip (Fig. 2.7A). This means that all of the energy in the wave is concentrated at one frequency. Many ac waves contain harmonics along with the fundamental frequency. In general, if a wave has a frequency equal to n times the fundamental, that wave is the nth harmonic. In the illustration of Fig. 2.7B, a wave is shown along with several harmonics, as it would look on the display screen of a spectrum analyzer. The frequency spectra of square waves and sawtooth waves contain harmonic energy in addition to the fundamental. The wave shape depends on the amount of energy in the harmonics and the way in which this energy is distributed among the harmonic frequencies.

Irregular waves can have practically any frequency distribution. An example is shown in Fig. 2.8. This is a spectral display of an amplitude-modulated voice radio signal. Much of the energy is concentrated at the center of the pattern, at the frequency shown by the vertical line. There is also plenty of energy near, but not exactly at, the center frequency. On an oscilloscope, this signal would look like a fuzzy sine wave.

Degrees, Radians, and Amplitude

Engineers break the ac cycle down into small parts for analysis and reference. One complete cycle corresponds to the revolution of a vector once around a circle in the coordinate plane. For a sine wave, the magnitude of the vector is constant, and the instantaneous rate of revolution (angular velocity) is also constant. For other waveforms, the magnitude and/or angular velocity can vary at different points in the cycle.

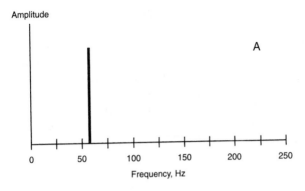

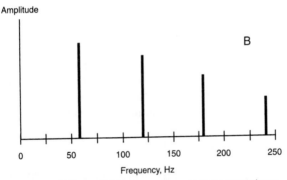

Figure 2.7 (A) Pure 60-Hz sine wave as seen on a spectrum-analyzer display; (B) 60-Hz wave containing harmonic energy.

Degrees versus radians

One method of specifying the details of an ac cycle is to divide it into 360 equal *degrees of phase*. The value 0° is assigned to the point in the cycle where the magnitude is 0 and positive-going. The same point on the next cycle is given the value 360°. Then halfway through the cycle is 180°; a quarter cycle is 90°, and so on.

The other method of specifying phase is to divide the cycle into 2π (approximately 6.28) equal parts called *radians of phase*. This is approximately the number of radii of a circle that can be laid end to end around the circumference. A radian is equal to approximately 57.3°. This unit of phase is common among physicists.

Sometimes, the frequency of an ac wave is measured in *radians per second,* rather than in hertz. Because there are 2π (about 6.28) radians in a complete cycle of 360°, the angular frequency of a wave, in radians per second, is roughly equal to 6.28 times the frequency in hertz.

Amplitude

Amplitude is the magnitude, level, or intensity of an ac wave, described either for a specific instant in time or gen-

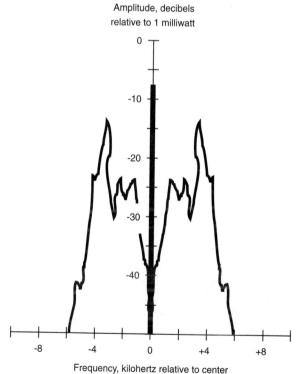

Figure 2.8 Amplitude-modulated radio signal as seen on a spectrum-analyzer display.

erally over a span of time. Depending on the quantity being measured, amplitude might be specified in amperes (for current), volts (for voltage), or watts (for power). It can also be expressed logarithmically in *decibels*.

The *instantaneous amplitude* is the amplitude at some precise moment in time. This constantly changes. The manner in which it varies depends on the waveform. Instantaneous amplitudes are represented by individual points on the wave curves.

The *peak (pk) amplitude* is the maximum extent, either positive or negative, that the instantaneous amplitude attains. In many waves, the positive and negative peak amplitudes are the same, but sometimes they differ. Figure 2.9 is an example of a wave in which the positive peak amplitude is the same as the negative peak amplitude. Figure 2.10 is an illustration of a wave that has different positive and negative peak amplitudes.

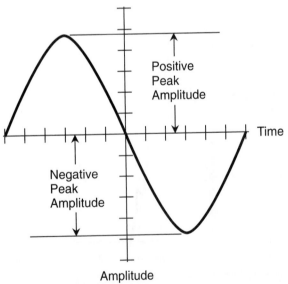

Figure 2.9 Positive and negative peak amplitudes. In this case they are equal.

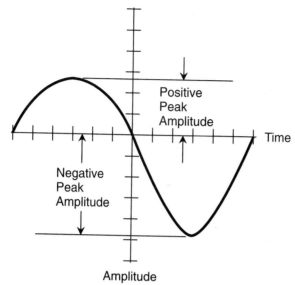

Figure 2.10 A wave in which the positive and negative peak amplitudes differ.

The *peak-to-peak (pk-pk) amplitude* of a wave is the total difference between the positive peak amplitude and the negative peak amplitude (Fig. 2.11). Another way of saying this is that the pk-pk amplitude is equal to the positive peak amplitude plus the negative peak amplitude. In many waves, the pk-pk amplitude is twice the peak amplitude, but this is not always true.

Often, it is necessary to express the effective level of an ac wave. This is the voltage, current, or power that a dc source would be required to produce for the same general effect. The most common specification is the *root-mean-square (rms) amplitude*. For a perfect sine wave, the rms value is equal to 0.707 times the peak value, or 0.354 times the pk-pk value. For a perfect square wave, the rms value is the same as the peak value. For sawtooth and irregular waves, the relationship between rms and peak depends on the shape of the wave, but the rms amplitude can never be greater than the peak amplitude.

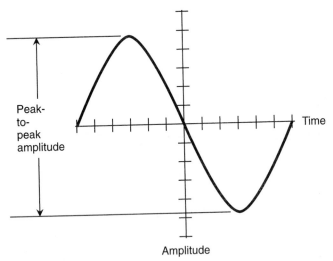

Figure 2.11 Peak-to-peak amplitude.

Composite ac/dc

An ac wave can have a dc component superimposed on it. If the dc component exceeds the peak value of the ac wave, fluctuating, or pulsating, dc is the result. This would happen, for example, if a 200-V dc source were connected in series with the 117-V ac from a standard utility outlet. Pulsating dc would appear, with an average value of 200 V, but with instantaneous values much higher and lower. The waveform in this case is illustrated by Fig. 2.12.

Hybrid ac/dc is not often generated deliberately. But such waveforms are sometimes observed at certain points in electronic circuitry.

Phase Relationships

When two sine waves have the same frequency, they can behave much differently if their cycles begin at different times. Whether or not the *phase difference,* often called the *phase angle* and specified in degrees, matters depends on the nature of the circuit.

Voltage

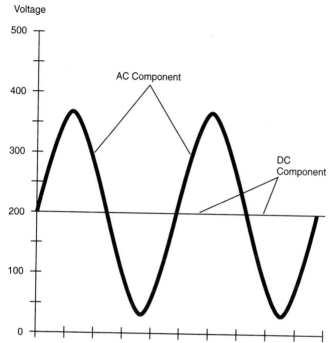

Figure 2.12   Composite ac/dc wave resulting from 117 volts rms ac in series with +200 V dc.

Phase coincidence

Phase angle can have meaning only when two waves have identical frequencies. If the frequencies differ, the relative phase constantly changes. In the following discussions of phase angle, assume that the two waves always have identical frequencies.

Phase coincidence means that two waves begin at exactly the same moment. This is shown in Fig. 2.13 for two waves having different amplitudes. The phase difference in this case is 0°. If two sine waves are in phase coincidence, the peak amplitude of the resultant wave, which is also a sine wave, is equal to the sum of the peak amplitudes of the two composite waves. The phase of the resultant wave is the same as that of the composite waves.

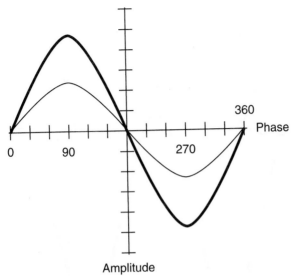
Amplitude

Figure 2.13 Two sine waves in phase coincidence.

Phase opposition

When two waves begin exactly ½ cycle, or 180°, apart, they are said to be in *phase opposition*. This is illustrated by Fig. 2.14. In this situation, engineers sometimes say that the waves are 180° out of phase.

If two sine waves have the same amplitude and are in phase opposition, they completely cancel each other, because the instantaneous amplitudes of the two waves are equal and opposite at every moment in time. If two sine waves have different amplitudes and are in phase opposition, the peak value of the resultant wave, which is a sine wave, is equal to the difference between the peak amplitudes of the two composite waves. The phase of the resultant wave is the same as the phase of the stronger of the two composite waves.

A sine wave has the unique property that, if its phase is shifted by 180°, the resultant wave is the same as turning the original wave "upside-down." Not all waveforms have this property.

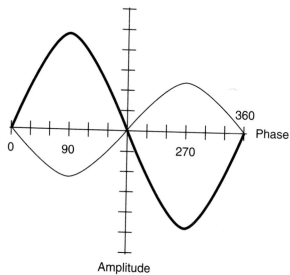

Amplitude

Figure 2.14   Two sine waves in phase opposition.

Leading phase

Two waves can differ in phase by any amount from 0° (in phase), through 180° (phase opposition), to 360° (in phase again).

Suppose there are two sine waves, X and Y, with identical frequencies. If wave X begins a fraction of a cycle earlier than wave Y, wave X is said to be leading wave Y in phase. For this to be true, X must begin its cycle less than 180° before Y. Figure 2.15 shows wave X leading wave Y by 90° of phase. The difference could be anything greater than 0°, up to 180°.

Note that if wave X (the thinner curve) is leading wave Y (the thicker curve), wave X is somewhat to the left of wave Y. In a time line, the left is earlier and the right is later.

Lagging phase

Suppose that wave X begins its cycle more than 180°, but less than 360°, ahead of wave Y. In this situation, it is easier to imagine that wave X starts its cycle later than wave Y,

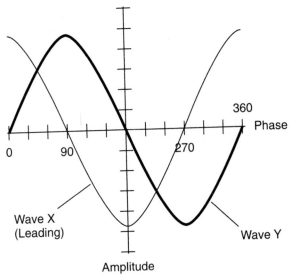

Figure 2.15 Wave X leads wave Y by 90°.

by some value between 0 and 180°. Then wave X is not leading but instead is lagging wave Y. Figure 2.16 shows wave X lagging wave Y by 90°.

If two waves have the same frequency and different phase, how do you know that one wave is really leading the other by some small part of a cycle, instead of lagging by a cycle and a fraction, or by a few hundred, thousand, million, or billion cycles and a fraction? The answer lies in the real-life effects. By convention, phase differences are expressed as values between 0 and 180°, either lagging or leading. Sometimes this range is given as −180 to +180° (lagging to leading).

Note that if wave X (the thinner curve in Fig. 2.16) is lagging wave Y (the thicker curve), wave X is somewhat to the right of wave Y.

Vector diagrams

If a sine wave X is leading a sine wave Y by some number of degrees, the two waves can be drawn as vectors, with vector

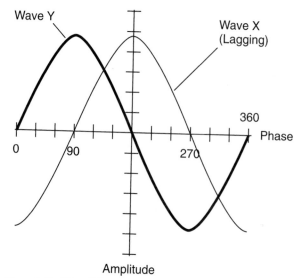

Wave Y

Wave X
(Lagging)

360
Phase

0   90   270

Amplitude

Figure 2.16   Wave X lags wave Y by 90°.

X being that number of degrees counterclockwise from vector Y. If wave X lags Y by some number of degrees, X will be clockwise from Y by that amount.

If two waves are in phase, their vectors overlap (line up). If they are in phase opposition, they point in exactly opposite directions.

Figure 2.17 shows four phase relationships between waves X and Y. Wave X always has twice the amplitude of wave Y, so vector X is always twice as long as vector Y. In Fig. 2.17A, X is in phase with Y. In Fig. 2.17B, X leads Y by 90°. In Fig. 2.17C, X and Y are 180° opposite in phase; in Fig. 2.17D, X lags Y by 90°. In all cases, the vectors rotate counterclockwise at the rate of one complete circle per wave cycle.

Utility Power Transmission

Electrical energy undergoes various transformations from the point of origin to the end users. The initial source consists of potential or kinetic energy in some nonelectrical form, such as falling or flowing water (*hydroelectric energy*),

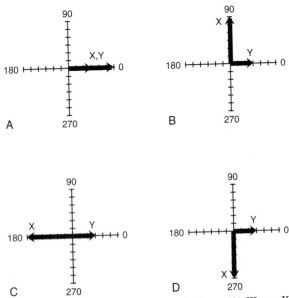

Figure 2.17 Vector representations of phase. (A) Waves X and Y are in phase; (B) X leads Y by 90°; (C) X and Y are in phase opposition; (D) X lags Y by 90°.

coal or oil (*fossil-fuel energy*), radioactive substances (*nuclear energy*), moving air (*wind energy*), light or heat from the sun (*solar energy*), or heat from the earth's interior (*geothermal energy*).

Generating plants

In fossil-fuel, nuclear, geothermal, and some solar electric power-generating systems, heat is used to boil water, producing steam under pressure to drive *turbines*. These turbines produce the rotational force (torque) necessary to drive large *electric generators*. The greater the power demand becomes, the more torque is required to turn the generator shaft at the required speed.

In a *photovoltaic energy-generating system,* sunlight is converted into dc electricity by semiconductor devices. This dc must be changed to ac for use by appliances. This is done by a *power inverter.* Electrochemical batteries are necessary if a *stand-alone photovoltaic system* is to provide useful energy at night. Photovoltaic systems cannot generate much power and are therefore used mainly in homes and small businesses. These systems produce no waste products other than chemicals that must be discarded when storage batteries wear out. Photovoltaics can be used in conjunction with existing utilities, without storage batteries, to supplement the total available energy supply to all consumers in a power grid. This is an *interactive photovoltaic system.*

In a *hydroelectric power plant,* the movement of water (waterfalls, tides, river currents) drives turbines that turn the generator shafts. In a *wind-driven electric power plant,* moving air operates devices similar to windmills, producing torque that turns the generator shafts. These power plants do not pollute directly, but large dams can disrupt ecosystems, have adverse affects on agricultural and economic interests downriver, and displace people upriver by flooding their land. Many people regard arrays of windmill-like structures as eyesores, but to generate significant power, hundreds or even thousands of the devices must be arrayed.

Whatever source is used to generate the electricity at a large power plant, the output is ac on the order of several hundred kilovolts and in some cases more than a megavolt.

High-tension lines

When electric power is transmitted over wires for long distances, power is lost because of *ohmic resistance* in the conductors. This loss can be minimized in two ways. First, efforts can be made to keep the wire resistance to a minimum. This involves using large-diameter wires made from metal that has excellent conductivity and by routing the power lines in such a way as to keep their lengths as short as possible. Second, the highest possible voltage can be used. Long-distance, high-voltage power lines are called *high-tension lines.*

The reason that high voltage minimizes power-transmission line loss is apparent upon analysis of equations denoting the relationship among power, current, voltage, and resistance. One equation can be stated as

$$P_{\text{loss}} = I^2R$$

where P_{loss} is the power (in watts) lost in the line as heat, I is the line current (in amperes), and R is the line resistance (in ohms). For a given span of line, the value of R is a constant. The value of P_{loss} thus depends on the current in the line. This current depends on the load, that is, on the collective demand of the end users.

Power-transmission line current also depends on line voltage. Current is inversely proportional to voltage at a given fixed power load. That is,

$$I = \frac{P_{\text{load}}}{E}$$

where I is the line current (in amperes), P_{load} is the total power demanded by the end users (in watts), and E is the line voltage (in volts). By making a substitution from this equation into the previous one, the following formula is obtained:

$$P_{\text{loss}} = \left(\frac{P_{\text{load}}}{E}\right)^2 R$$

$$= (P_{\text{load}})^2 \frac{R}{E^2}$$

For any given fixed values of P_{load} and R, doubling the line voltage E will cut the power loss P_{loss} to one-quarter (25 percent) of its previous value. Multiply the line voltage by 10, and in theory the line loss will drop to 1/100 (1 percent) of its former value. Thus it makes sense to generate as high a voltage as possible for efficient long-distance electric power transmission.

The loss in a power line can be reduced further by using dc, rather than ac, for long-distance transmission. Direct currents do not produce electromagnetic fields as alternat-

ing currents do, so a secondary source of power-line ineffi-
ciency—*electromagnetic radiation loss*—is eliminated by the
use of dc. However, long-distance dc power transmission is
a difficult and costly scheme to implement. It necessitates
high-power, high-voltage rectifiers at a generating plant
and power inverters at distribution stations where high-tension lines branch out into lower-voltage, local lines.

Transformers

The standard utility voltages are 117 and 234 V rms. These
lower voltages are obtained from higher voltages via *transformers.*

Step-down transformers reduce the voltage of high-tension lines (100,000 V or more) down to a few thousand volts
for distribution within municipalities. These transformers
are physically large because they must carry significant
power. Several of them might be placed in a building or a
fenced-off area. The outputs of these transformers are fed to
power lines that run along city streets.

Smaller transformers, usually mounted on utility poles
or underground, step the municipal voltage down to 234 V
rms for distribution to individual homes and businesses.
The 234-V electricity is provided in three phases to the distribution boxes in each house, apartment, or business.
Some utility outlets are supplied directly with this *three-
phase ac.* Large appliances such as electric stoves, ovens,
and laundry machines employ this power. Smaller wall outlets and lamp fixtures are provided with single-phase, 117-
V rms ac electricity.

3

Impedance

Impedance is the opposition that a component or circuit offers to ac. Impedance is a two-dimensional quantity, consisting of two independent components, *resistance* and *reactance*. Resistance in ac circuits is the same as its dc counterpart and is expressed as a positive ohmic quantity. Reactance can be inductive (nonnegative ohmic) or capacitive (nonpositive ohmic).

Inductive Reactance

Resistance is a *scalar quantity* because it can be expressed on a one-dimensional scale. Given a certain dc voltage, the current decreases as the resistance increases, in accordance with Ohm's Law. The same law holds for ac through a resistance. But in an inductance or capacitance, the situation is more complicated.

Coils and current

If you wind a length of wire into a coil and connect it to a source of dc, the wire will draw current $I = E/R$, where I is the current, E is the dc voltage, and R is the resistance of the wire. The coil will get hot as energy is dissipated in the resistance of the wire. If the voltage of the power supply is increased, the wire in the coil will get hotter.

Suppose you change the voltage source, connected across the coil, from dc to ac. You vary the frequency from a few hertz to many megahertz. The coil has a certain *inductive reactance* (X_L), so it takes some time for current to establish itself in the coil. A point will be reached, as the ac frequency increases, at which the current cannot get fully established in the coil before the polarity of the voltage reverses. As the frequency is raised, this effect becomes more pronounced. Eventually, if you keep increasing the frequency, the coil will not even come near establishing a current with each cycle. Hardly any effective current will flow through it.

The X_L of an inductor can vary from zero (a short circuit) to a few ohms (a small coil) to kilohms or megohms (large coils). Like pure resistance, inductive reactance affects the current in an ac circuit. But unlike pure resistance, reactance changes with frequency. This affects the way the current flows with respect to the voltage.

X_L versus frequency

If the frequency of an ac source is given (in hertz) as f, and the inductance of a coil is given (in henrys) as L, then the inductive reactance (in ohms), X_L, is given by

$$X_L = 2\pi fL \approx 6.28 fL$$

This same formula applies if the frequency, f, is in kilohertz and the inductance, L, is in millihenrys (mH). It also applies if f is in megahertz and L is in microhenrys (μH).

Inductive reactance increases linearly with increasing ac frequency. Inductive reactance also increases linearly with inductance. The value of X_L is directly proportional to f; X_L is also directly proportional to L. These relationships are graphed, in relative form, in Fig. 3.1.

Points in the *RL* plane

Inductive reactance can be plotted along a half line, as is resistance. In a circuit containing both resistance and inductance, the characteristics become two-dimensional. You can orient the resistance and reactance half lines perpendicular to each other to make a quarter-plane coordinate system. In this scheme, *RL* combinations form *complex impedances.* Each point corresponds to one unique impedance, and each *RL* impedance value corresponds to one unique point. Impedances on this *RL plane* are written in the form $R + jX_L$, where R is the resistance and X_L is the inductive reactance. (See p. 59–60 for an explanation of j.)

If you have a resistance $R = 5\ \Omega$, the complex impedance is $5 + j0$, and is at the point (5,0) on the plane. If you have a pure inductive reactance, such as $X_L = 3\ \Omega$, the complex impedance is $0 + j3$ and is at the point (0,3) on the *RL* plane. Often, resistance and inductive reactance exist together. This gives impedance values such as $2 + j3$ or $4 + j1.5$. These are shown as points in Fig. 3.2.

Vectors in the *RL* plane

Engineers commonly represent points in the *RL* plane as *vectors.* This gives each point a specific and unique *magnitude* and *direction.*

In Fig. 3.2, various points are shown. Another way to think of these points is to draw lines from the origin out to them. Then the points become rays, each having a certain magnitude and a direction (angle counterclockwise from the R axis). These rays, going out to the points, are *complex impedance vectors* (Fig. 3.3).

Current and Voltage in *RL* Circuits

Inductance stores electrical energy as a magnetic field. When a voltage is placed across a coil, it takes awhile for the current to build up to full value. Thus, when ac is placed across a coil, the current lags the voltage in phase.

Inductance and resistance

Suppose that you place an ac voltage across a low-loss coil, with a frequency high enough so that X_L is much larger than

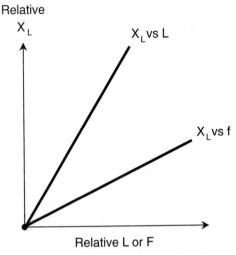

Figure 3.1 Inductive reactance is directly proportional to inductance and also to frequency.

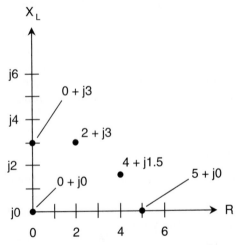

Figure 3.2 Some points in the *RL* impedance plane.

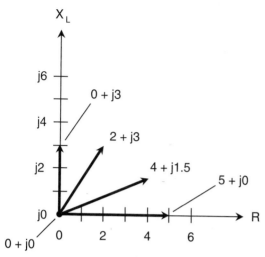

Figure 3.3 Some vectors in the *RL* impedance plane.

the dc resistance *R*. In this situation, the current lags the voltage by almost 90° (1/4 cycle). When X_L is huge compared with the value of *R* in a circuit, the *RL*-plane vector points almost straight up along the X_L axis. Its angle is almost 90° from the *R* axis.

When the resistance in an *RL circuit* is significant compared with the inductive reactance, the current lags the voltage by less than 90° (Fig. 3.4). If *R* is small compared with X_L, the current lag is almost 90°; as *R* gets relatively larger, the lag decreases. When *R* is many times greater than X_L, the vector lies almost on the *R* axis, going "east," or to the right. The *RL* phase angle, ϕ_{RL}, is nearly zero. When the complex impedance becomes a pure resistance, the current comes into phase with the voltage.

Calculating *RL* phase angle

You can use a ruler that has centimeter and millimeter markings and a protractor to calculate *RL* phase angle. Draw a horizontal line a little more than 100 mm long,

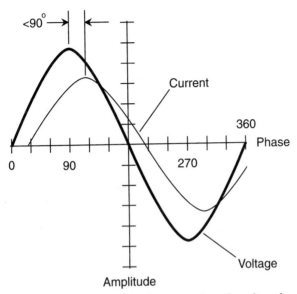

Figure 3.4 In an *RL* circuit, the current lags the voltage by less than 90°.

going from left to right. Construct a vertical line off the left end of this first line, going vertically upward. Make this line at least 100 mm long. The horizontal line is the *R* axis. The vertical line is the X_L axis.

If you know the values of X_L and *R*, divide them down or multiply them up so that they are both between 0 and 100. For example, if $X_L = 680 \ \Omega$ and $R = 840 \ \Omega$, divide them both by 10 to get $X_L = 68$ and $R = 84$. Plot these points by making hash marks on the axes. The *R* mark will be 84 mm to the right of the origin. The X_L mark will be 68 mm up from the origin. Draw a line connecting the hash marks, as shown in Fig. 3.5. This line will form a right triangle with the two axes. Measure the angle between the slanted line and the *R* axis. This is ϕ_{RL}.

The actual vector, $R + jX_L$, is found by constructing a rectangle using the origin and hash marks as three of the four vertices and drawing horizontal and vertical lines to complete the figure. The vector is the diagonal of this rectangle,

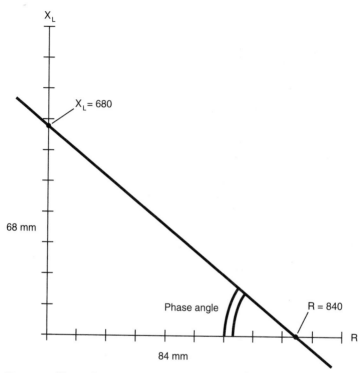

Figure 3.5 Pictorial method of finding *RL* phase angle.

as shown in Fig. 3.6. Then ϕ_{RL} is the angle between this vector and the *R* axis.

A more exact value for ϕ_{RL}, especially if it is near 0 or 90°, making pictorial constructions difficult, can be found using a calculator that has inverse trigonometry functions. The angle is the arctangent of the ratio of inductive reactance to resistance:

$$\phi_{RL} = \tan^{-1} \frac{X_L}{R}$$

Capacitive Reactance

Inductive reactance has its counterpart in the form of *capacitive reactance*. This, too, can be represented as a ray,

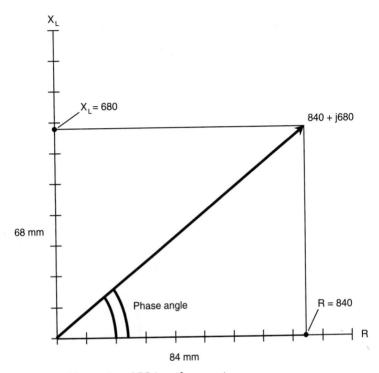

Figure 3.6   Illustration of *RL* impedance vector.

starting at the same zero point as inductive reactance but running off in the opposite direction, having negative ohmic values.

Capacitors and current

Imagine two gigantic, flat, parallel metal plates, both of which are excellent electrical conductors. If supplied with dc, the plates will become electrically charged and will reach a potential difference equal to the dc source voltage. The current, once the plates are charged, will be zero. If a dielectric material, such as glass, is placed between the plates, their mutual voltage will not change, although the charging time will increase.

Suppose the source is changed from dc to ac. Imagine you can adjust the frequency of this ac from a few hertz to many megahertz. At first, the voltage between the plates will follow almost exactly along as the ac polarity reverses; the set of plates will act as an open circuit. As the frequency increases, the charge will not get well established with each cycle. When the frequency becomes extremely high, the set of plates will behave like a short circuit.

The opposition that the set of plates offers to ac is capacitive reactance (X_C). It varies with frequency. It is measured in ohms, like inductive reactance (X_L) and resistance (R). But X_C is, by convention, assigned negative values rather than positive ones. The value of X_C increases negatively as the frequency goes down.

Sometimes X_C is talked about in terms of its absolute value, with the minus sign removed. But in complex impedance calculations, X_C is always considered a nonpositive quantity. This prevents it from getting confused with inductive reactance X_L, which is always nonnegative.

X_C versus frequency

Capacitive reactance behaves, in many ways, like a mirror image of inductive reactance. In another sense, X_C is an extension of X_L into negative values.

If the frequency of an ac source is given (in hertz) as f, and the value of a capacitor is given (in farads) as C, then the capacitive reactance (in ohms), X_C, is given by

$$X_C = - \frac{1}{2\pi f C} \approx - \frac{1}{6.28 f C}$$

This same formula applies if the frequency f is in megahertz and the capacitance C is in microfarads.

Capacitive reactance varies inversely with the frequency. The function X_C versus f appears as a curve when graphed, and this curve "blows up" as the frequency nears zero. Capacitive reactance also varies inversely with the actual value of capacitance, given a fixed frequency. Therefore, the function of X_C versus C also appears as a curve that "blows up" as the capacitance approaches zero. Relative graphs of these functions are shown in Fig. 3.7.

Points in the *RC* plane

In a circuit containing resistance and capacitive reactance, the characteristics are two-dimensional, in a way that is analogous to the situation with the *RL* plane. The resistance and the capacitive-reactance rays can be placed end to end at right angles to make the *RC plane* (Fig. 3.8). Resistance is plotted horizontally, with increasing values toward the right. Capacitive reactance is plotted vertically, with increasingly negative values downward. Each point corresponds to one and only one impedance. Conversely, each specific impedance coincides with one and only one point on the plane.

Impedances that contain resistance and capacitance are written in the form $R - jX_C$. If the resistance is pure, say $R = 3\ \Omega$, the complex impedance is $3 - j0$, and this corre-

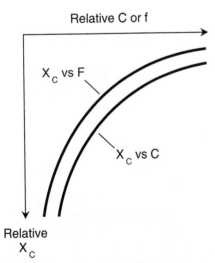

Relative C or f

X_C vs F

X_C vs C

Relative X_C

Figure 3.7 Capacitive reactance is negatively, and inversely, proportional to frequency (f) and also to capacitance (C).

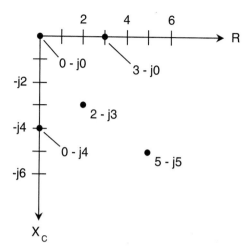

Figure 3.8 Some points in the *RC* impedance plane.

sponds to the point (3,0) on the *RC* plane. If you have a pure capacitive reactance, say $X_C = -4 \ \Omega$, the complex impedance is $0 - j4$, and this is at the point $(0,-4)$ on the *RC* plane. Often, resistance and capacitive reactance both exist, resulting in complex impedances such as $2 - j3$ and $5 - j5$.

Vectors in the *RC* plane

Figure 3.8 shows several different impedance points. Each one is represented by a certain distance to the right of the origin (0,0) and a certain displacement downward.

Impedance points in the *RC* plane can be rendered as vectors, just as can be done in the *RL* plane. The magnitude of a vector is the distance of the point from the origin; the direction is the angle measured clockwise from the resistance (*R*) line and is specified in negative degrees. The vectors for the points of Fig. 3.8 are depicted in Fig. 3.9.

Current and Voltage in *RC* Circuits

Capacitance stores energy in the form of an electric field. When an ac voltage source is placed across a capacitor, the

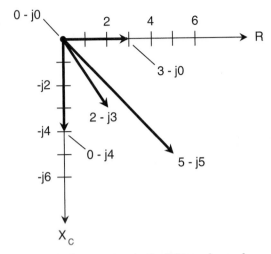

Figure 3.9 Some vectors in the *RC* impedance plane.

voltage lags the current in phase. The phase difference can range from zero to a quarter of a cycle (90°).

Capacitance and resistance

When the resistance in an *RC circuit* is significant compared with the capacitive reactance, the current leads the voltage by something less than 90° (Fig. 3.10). If *R* is small compared with X_C, the difference is almost a quarter of a cycle. As *R* gets larger, or as X_C becomes smaller, the phase difference decreases. The value of *R* might increase relative to X_C because resistance is deliberately put into a circuit. Or, it might happen because the frequency becomes so low that X_C rises to a value comparable with the leakage resistance of the capacitor. In either case, the situation can be represented by a resistor and capacitor in series.

As the resistance in an *RC* circuit gets large compared with the capacitive reactance, the *RC* phase angle becomes smaller. The same thing happens if the value of X_C gets small compared with the value of *R*. Note that when we call X_C "large," we mean large negatively. When

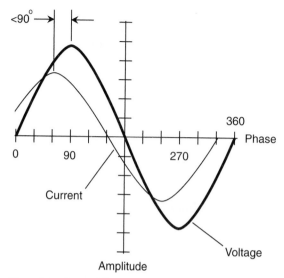

Figure 3.10 In an RC circuit, the current leads the voltage by less than 90°.

we say X_C is "small," we mean that it is close to zero, or small negatively.

When R is many times larger than X_C, the vector in the RC plane lies almost along the R axis. Then the RC phase angle will be nearly zero. Ultimately, if the capacitive reactance gets small enough, the circuit will act as a pure resistance, and the current will come into phase with the voltage.

Calculating *RC* phase angle

If you know the ratio X_C/R in an RC circuit, you can find the RC *phase angle*. You can use a protractor and a ruler to find phase angles for RC circuits, just as with RL circuits, as long as the angles are not too close to 0 or 90°.

First, draw a line somewhat longer than 100 mm, going from left to right on the paper. Then, use the protractor to construct a line going somewhat more than 100 mm vertically downward, starting at the left end of the horizontal line. The horizontal line is the R axis of the RC plane. The line going down is the X_C axis.

If you know the values of X_C and R, divide or multiply them by a constant, chosen to make both values fall between -100 and 100. For example, if $X_C = -3800$ Ω and $R = 7400$ Ω, divide them both by 100, getting -38 and 74. Plot these points on the lines as hash marks. The X_C mark goes 38 mm down from the origin. The R mark goes 74 mm to the right of the origin. Draw a line connecting the hash marks, as shown in Fig. 3.11. Measure the angle between the slanted line and the R axis. This angle will be between 0 and 90°. Multiply by -1 to get the RC phase angle.

The complex impedance vector is diagrammed by constructing a rectangle using the origin and the hash marks, making new perpendicular lines to complete the figure. The vector is the diagonal of this rectangle, running out from the

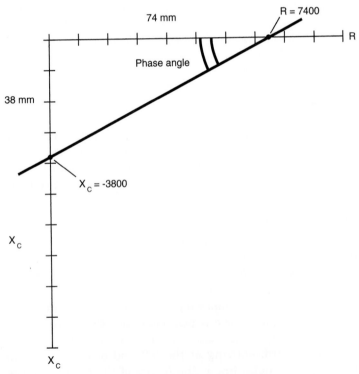

Figure 3.11 Pictorial method of finding RC phase angle.

origin (Fig. 3.12). The phase angle is the angle between the R axis and this vector, multiplied by -1.

The more accurate way to find RC phase angles is to use trigonometry. Determine the ratio X_C/R and enter it into a calculator. This ratio will be a negative number or zero, because X_C is always negative or zero and R is always positive. Find the arctangent (arctan or $\tan^{-1}$) of this number. This is the RC phase angle ϕ_{RC}. Mathematically,

$$\phi_{RC} = \tan^{-1}\frac{X_C}{R}$$

Imaginary Numbers

In expressions of impedance such as $4 + j7$ and $45 - j83$, the lowercase j represents the positive square root of -1. The

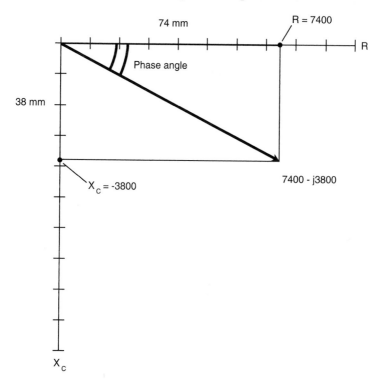

Figure 3.12   Illustration of RC impedance vector.

entire set of imaginary numbers derives from this single unit.

The unit imaginary number j can be multiplied by any real number, getting an infinitude of imaginary numbers, forming an *imaginary number line.* This is a duplicate of the *real number line* and is generally placed at a right angle to the real number line to form the *complex impedance plane,* also called the *RX plane* (Fig. 3.13).

Complex numbers

When you add a real number and an imaginary number, you have a *complex number.* Real numbers are one dimensional. Imaginary numbers are also one dimensional. Complex numbers need two dimensions to be defined.

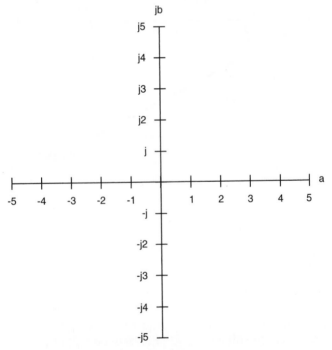

Figure 3.13 The complex number plane.

Adding complex numbers requires adding the real parts and the complex parts separately. The sum of $4 + j7$ and $45 - j83$ is therefore $(4 + 45) + j(7 - 83) = 49 + j(-76) = 49 - j76$.

Subtracting complex numbers works similarly. The difference $(4 + j7) - (45 - j83)$ is found by multiplying the second complex number by -1 and then adding the result, getting $(4 + j7) + [-1(45 - j83)] = (4 + j7) + (-45 + j83) = -41 + j90$. Remember that subtraction can always be performed as an addition, and you will minimize the risk of getting signs confused.

The general formula for the sum of two complex numbers $(a + jb)$ and $(c + jd)$ is

$$(a + jb) + (c + jd) = (a + c) + j(b + d)$$

The product of the same two complex numbers is given by

$$(a + jb)(c + jd) = (ac - bd) + j(ad + bc)$$

Complex number vectors

Complex numbers can be represented as vectors in the complex plane. This gives each complex number a unique magnitude and direction. The magnitude is the distance of the point $a + jb$ from the origin $0 + j0$. The direction is the angle of the vector, measured counterclockwise from the $+a$ axis. This is shown in Fig. 3.14.

The *absolute value* of a complex number $a + jb$ is the length, or magnitude, of its vector in the complex plane, measured from the origin $(0,0)$ to the point (a,b). In the case of a pure real number $a + j0$, the absolute value is the number a if it is positive and the additive inverse $-a$ if it is negative. In the case of a pure imaginary number $0 + jb$, the absolute value is equal to b when b is positive and to $-b$ if b is negative.

If a complex number is neither pure real or pure imaginary, the absolute value can be found using the *pythagorean theorem* from plane geometry. First, square both a and b.

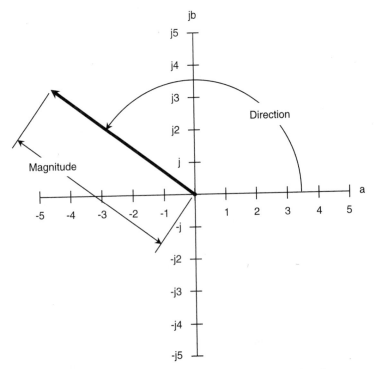

Figure 3.14 Magnitude and direction of a vector in the complex plane.

Then add them. Finally, take the square root. This is the length of the vector $a + jb$, as shown in Fig. 3.15.

The *RX* plane

Recall the plane for R and X_L. This is the same as the upper-right quadrant of the complex-number plane shown in Fig. 3.13. Similarly, the plane for R and X_C is the same as the lower-right quadrant of the complex number plane. Resistances are represented by nonnegative real numbers. Reactances, whether inductive (positive) or capacitive (negative), correspond to imaginary numbers.

To construct the *RX plane*, remove the upper-left and lower-left quadrants of the complex-number plane, obtaining a half plane, as shown in Fig. 3.16.

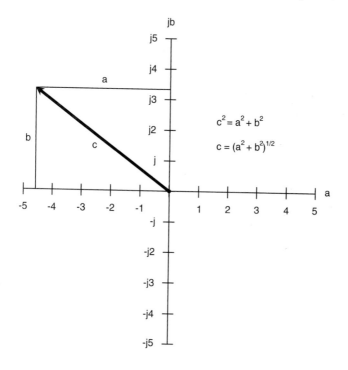

Figure 3.15 Calculation of absolute value, or vector length.

Impedance vectors

Any impedance $R + jX$ can be represented by a complex number of the form $a + jb$. (Let $R = a$ and $X = b$.) The impedance vector changes magnitude and direction as R and/or X are varied.

Sometimes, the capital letter Z is used in place of the word *impedance* in general discussions, as in "$Z = 8\ \Omega$." Such an expression, if no specific complex impedance is given, can refer to $8 + j0$, $0 + j8$, $0 - j8$, or any value on a half circle of points in the RX plane that are at distance eight units away from $0 + j0$. This is shown in Fig. 3.17. In theory, there are infinitely many complex impedances with $Z = 8\ \Omega$.

If you're not specifically told what complex impedance is meant when a single-number ohmic figure is quoted, the number usually refers to a *nonreactive impedance*. That

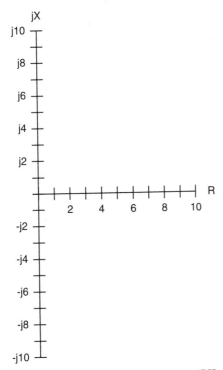

Figure 3.16 The complex impedance (RX) plane.

means the imaginary, or reactive, factor is zero. This is also called a *resistive impedance* or a *pure resistance*.

Impedances in series

Given two impedances $Z_1 = R_1 + jX_1$ and $Z_2 = R_2 + jX_2$ connected in series, the net impedance Z is their vector sum, given by

$$Z = (R_1 + R_2) + j(X_1 + X_2)$$

For complex impedances in parallel, the process of finding the net impedance is more complicated and is outlined later in this chapter.

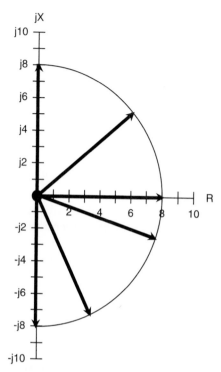

Figure 3.17 Some vectors representing an absolute-value impedance of 8 Ω.

Characteristic Impedance

There is one property that is often (somewhat inaccurately) called an "impedance": *characteristic impedance* or *surge impedance*. It is abbreviated Z_0 and is a specification of transmission lines. It can always be expressed as a positive real number.

Transmission lines

When it is necessary to get energy or signals from one place to another, a transmission line is required. These almost always take either of two forms, coaxial or parallel wire. Examples of transmission lines include the "ribbon" or "twin lead" that goes from a TV antenna to the receiver, the cable running from a hi-fi amplifier to the speakers, and high-tension wires that carry electricity over long distances.

The Z_0 of a parallel-wire transmission line depends on the diameter of the wires, on the spacing between the wires, and on the nature of the insulating material separating the wires. In general, Z_0 increases as the wire diameter gets smaller and decreases as the wire diameter gets larger, all other things being equal.

In a coaxial line, the thicker the center conductor, the lower the Z_0 if the shield stays the same size. If the center conductor stays the same size and the shield tubing increases in diameter, the Z_0 increases. In general, the Z_0 increases as the spacing between wires, or between the center conductor and the shield, increases. Solid dielectrics such as polyethylene reduce the Z_0 of a transmission line, compared with an air dielectric.

Standing waves

Standing waves are usually present in antenna radiators. The pattern of standing waves on a half-wave resonant radiator, fed at the center, is shown in Fig. 3.18A. The pattern of standing waves on a full-wave resonant radiator, fed at the center, is shown in Fig. 3.18B. The pattern of standing waves on a full-wave resonant radiator, fed at one end, is shown in Fig. 3.18C.

Some antenna radiators do not exhibit standing waves; instead, the current and voltage are uniform all along the length of the radiating element. This is the case in antennas that have a terminating resistor with a value equal to the Z_0 of the radiator (about 600 Ω). Some long-wire and rhombic antennas employ such resistors to obtain a unidirectional pattern.

On an *RF* transmission line terminated in an impedance that differs from the Z_0 of the line, a nonuniform distribution of current and voltage exists. The greater the *impedance mismatch,* the greater the nonuniformity. The ratio of the maximum voltage to the minimum voltage, or

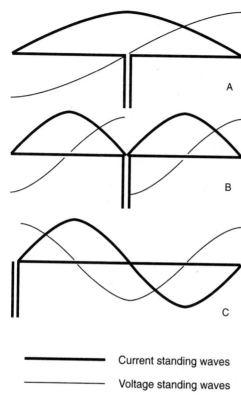

Current standing waves

Voltage standing waves

Figure 3.18 (A) Standing waves on a center-fed half-wave antenna; (B) on a center-fed full-wave antenna; (C) on an end-fed full-wave antenna.

the maximum current to the minimum current, is called the *standing-wave ratio* (SWR) on the line. The SWR is 1:1 when the current and voltage are in the same proportions everywhere along the line. The SWR can be 1:1 only when a transmission line is terminated in a nonreactive (purely resistive) load having the same ohmic value as the Z_0 of the line.

In the case of a short circuit, open circuit, pure inductance, or pure capacitance at the load end of the line, the SWR is theoretically infinite, because the current and the voltage fall to zero at certain points and rise to high values at other points. In practice, line losses prevent the SWR from becoming infinite, but it can reach 40:1 or more.

An extremely large SWR can cause significant loss in a transmission line. In any line, the loss is smallest when the SWR is 1:1. If the SWR is not 1:1, the line loss increases. This additional loss is called *SWR loss, impedance-mismatch loss,* or *feed-line mismatch loss.* Figure 3.19 shows the feed-line mismatch loss that occurs for various values of matched-line loss and SWR. These relations are independent of the physical construction of the line.

Admittance

Admittance is a measure of the ease with which a medium carries ac. It is the ac counterpart of dc conductance. Admittance is a complex quantity, as is impedance.

AC conductance

In an ac circuit, electrical conductance works the same way as it does in a dc circuit. Conductance is symbolized by the capital letter G. The relationship between conductance and resistance is simple:

$$G = \frac{1}{R}$$

The unit of conductance is the *siemens,* sometimes called the *mho.* The larger the value of conductance, the smaller the resistance, and the more current will flow. Conversely, the smaller the value of G, the greater the value of R, and the less current will flow.

Susceptance

Susceptance is symbolized by the capital letter B. It is the reciprocal of ac reactance:

$$B = \frac{1}{X}$$

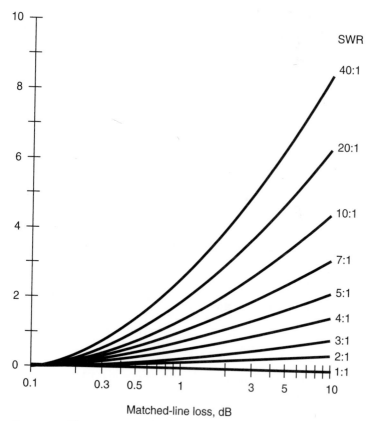

Figure 3.19 Extra loss produced by impedance mismatches on an RF transmission line.

Susceptance can be either capacitive or inductive. These quantities are symbolized as B_C and B_L, respectively. Therefore,

$$B_C = \frac{1}{X_C}$$

$$B_L = \frac{1}{X_L}$$

The expression of susceptance requires the j operator, as does the expression of reactance. The j operator behaves strangely in quotients. The multiplicative inverse (reciprocal) of j is also its additive inverse (negative); that is, $1/j = -j$. Therefore, when calculating susceptance in terms of reactance, the sign changes.

The formula for *capacitive susceptance* is

$$B_C = 2\pi fC \approx 6.28fC$$

This resembles the formula for inductive reactance. The formula for *inductive susceptance* is similar to that for capacitive reactance:

$$B_L = \frac{-1}{2\pi fL} \approx \frac{-1}{6.28fL}$$

Complex conductance and complex susceptance combine to form *complex admittance,* symbolized by the capital letter Y.

Complex admittance

Admittance is the complex composite of conductance and susceptance:

$$Y = G + jB$$

The j factor might be negative, so there are times you will write $Y = G - jB$.

Admittance, rather than impedance, is best for working with parallel ac circuits. Resistance and reactance combine in messy fashion in parallel circuits, but conductance (G) and susceptance (B) simply add together, yielding admittance (Y). The situation is similar to the behavior of resistances in parallel when you work with dc.

The *GB* plane

Admittance can be depicted on a plane that looks like the complex impedance (RX) plane. Actually, it is a half plane, because there is ordinarily no such thing as negative conductance. Conductance is plotted along the horizontal, or G, axis on this coordinate half plane, and susceptance is plotted along the B axis (Fig. 3.20).

The center, or *origin,* of the *GB plane* represents that point at which there is no conduction for dc or ac. In the *RX* plane, the origin represents a perfect short circuit; in the *GB* plane it corresponds to a perfect open circuit. The open circuit in the *RX* plane is represented by points infinitely far from the origin in all directions. In the *GB* plane, these points represent a short circuit.

Vector representation of admittance

Complex admittances can be shown as vectors, just as can complex impedances. In Fig. 3.21, the points from Fig. 3.20

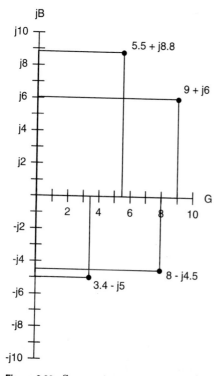

Figure 3.20 Some points in the *GB* plane and their components on the G and B axes.

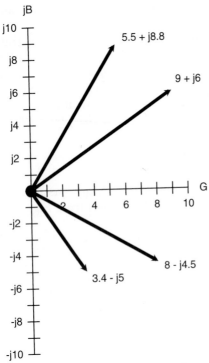

Figure 3.21 Some vectors in the *GB* plane.

are rendered as vectors. Longer vectors indicate, in general, greater flow of current, and shorter ones indicate less current.

Vectors pointing generally "northeast," or upward and to the right, represent conductances and capacitances in parallel. Vectors pointing in a more or less "southeasterly" direction, or downward and to the right, represent conductances and inductances in parallel.

Admittances in parallel

Given two admittances $Y_1 = G_1 + jB_1$ and $Y_2 = G_2 + jB_2$ connected in parallel, the net admittance Y is their vector sum, given by

$$Y = (G_1 + G_2) + j\,(B_1 + B_2)$$

When it is necessary to find the net impedance of two components in parallel, first convert each resistance to conductance and each reactance to susceptance. Combine these to
get the admittances. Use the above formula to find the net
admittance. Then convert the net conductance back to resistance and the net susceptance back to reactance. The resulting composite is the net impedance of the components in
parallel.

4

Digital Basics

A signal or quantity is *digital* when it can attain only a finite number of levels or values. This is in contrast to *analog* signals or quantities that vary over a continuous range of levels or values. A simple analog waveform is shown in Fig. 4.1A. The amplitude varies continuously from instant to instant. Figure 4.1B is a digital approximation of the analog waveform in Fig. 4.1A.

Numbering Systems

People are used to dealing with the *decimal number system,* which has 10 unique digits, but machines use schemes that have some power of 2 unique digits—most often 2 (2^1), 8 (2^3), or 16 (2^4).

Decimal

The *decimal number system* is also called *modulo 10, base 10,* or *radix 10.* Digits are representable by the set {0, 1, 2, 3, 4, 5, 6, 7, 8, 9}. The digit just to the left of the decimal point is multiplied by 10^0, or 1. The next digit to the left is multiplied by 10^1, or 10. The power of 10 increases as you move further to the left. The first digit to the right of the

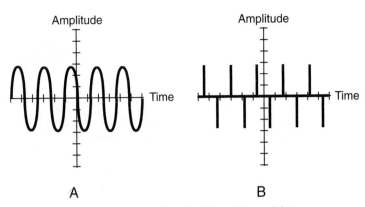

Figure 4.1 (A) An analog wave; (B) a digital rendition of this wave.

decimal point is multiplied by a factor of 10^{-1}, or 1/10. The next digit to the right is multiplied by 10^{-2}, or 1/100. This continues as you go further to the right. Once the process of multiplying each digit is completed, the resulting values are added. This is what is represented when you write a decimal number. For example,

$$2704.53816 = 2\times10^3 + 7\times10^2 + 0\times10^1 + 4\times10^0 + 5\times10^{-1} \\ + 3\times10^{-2} + 8\times10^{-3} + 1\times10^{-4} + 6\times10^{-5}$$

Binary

The *binary number system* is a method of expressing numbers using only the digits 0 and 1. It is sometimes called *base 2, radix 2,* or *modulo 2.* The digit immediately to the left of the radix point is the "ones" digit. The next digit to the left is a "twos" digit; after that comes the "fours" digit. Moving further to the left, the digits represent 8, 16, 32, 64, and so on, doubling every time. To the right of the radix point, the value of each digit is cut in half again, that is, $\frac{1}{2}$, $\frac{1}{4}$, $\frac{1}{8}$, $\frac{1}{16}$, $\frac{1}{32}$, $\frac{1}{64}$, and so on.

Consider an example using the decimal number 94:

$$94 = (4\times10^0) + (9\times10^1)$$

In the binary number system the breakdown is

$$1011110 = (0 \times 2^0) + (1 \times 2^1) + (1 \times 2^2) + (1 \times 2^3) + (1 \times 2^4) +$$
$$(0 \times 2^5) + (1 \times 2^6)$$

When you work with a computer or calculator, you give it a decimal number that is converted into binary form. The computer or calculator does its operations with zeros and ones. When the process is complete, the machine converts the result back into decimal form for display.

In a communications system, binary numbers represent alphanumeric characters, shades of color, frequencies of sound, and other variable quantities.

Octal

Another scheme, sometimes used in computer programming, is the *octal number system,* so named because it has eight symbols, or 2^3. Every digit is an element of the set {0, 1, 2, 3, 4, 5, 6, 7}.

Hexadecimal

Yet another numbering scheme, also used in computer work, is the *hexadecimal number system,* so named because it has 16 symbols, or 2^4. These digits are the usual 0 through 9 plus six more, represented by A through F, the first six letters of the alphabet. The digit set thus becomes {0, 1, 2, 3, 4, 5, 6, 7, 8, 9, A, B, C, D, E, F}.

Colors can be represented by six-digit hexadecimal numbers, such as 005CFF. In the *red-green-blue (RGB) color model,* the first two digits represent the red (R) intensity in 256 levels ranging from 00 to FF. The middle two digits represent the green (G) intensity, and the last two digits represent the blue (B) intensity.

Logic

Logic refers to methods of reasoning used by people and electronic machines. The term is also sometimes used in reference to the circuits that comprise most digital devices and systems.

Boolean algebra

Boolean algebra is a system of mathematical logic using the numbers 0 and 1 with the operations AND (multiplication), OR (addition), and NOT (negation). Combinations of these operations are NAND (NOT AND) and NOR (NOT OR). Boolean functions are used in the design of digital logic circuits.

In boolean algebra, X AND Y is written XY or $X * Y$. NOT X is written with a line or tilde over the quantity or as a minus sign followed by the quantity. X OR Y is written $X + Y$. Table 4.1 shows the values of these functions, where 0 indicates "falsity" and 1 indicates "truth." The statements on either side of the equal sign are logically equivalent.

Table 4.2 shows several logic equations. These are facts, or *theorems*. Boolean theorems can be used to analyze complicated logic functions.

Trinary logic

Trinary logic allows for a neutral condition, neither true nor false, in addition to the usual true/false (high/low) states. These three values are representable by logic -1 (false), 0 (neutral), and $+1$ (true).

Trinary logic can be easily represented in electronic circuits by positive, zero, and negative currents or voltages.

Fuzzy logic

In *fuzzy logic,* values cover a continuous range from "totally false," through neutral, to "totally true." Fuzzy logic is well

TABLE 4.1   Boolean Operations

X	Y	$-X$	$X * Y$	$X + Y$
0	0	1	0	0
0	1	1	0	1
1	0	0	0	1
1	1	0	1	1

TABLE 4.2 Common Theorems in Boolean Algebra

Equation	Name (if applicable)
$X + 0 = X$	OR identity
$X * 1 = X$	AND identity
$X + 1 = 1$	
$X * 0 = 0$	
$X + X = X$	
$X * X = X$	
$-(-X) = X$	Double negation
$X + (-X) = X$	
$X * (-X) = 0$	Contradiction
$X + Y = Y + X$	Commutativity of OR
$X * Y = Y * X$	Commutativity of AND
$X + (X * Y) = X$	
$X * (-Y) + Y = X + Y$	
$X + Y + Z = (X + Y) + Z = X + (Y - Z)$	Associativity of OR
$X * Y * Z = (X * Y) * Z = X * (Y * Z)$	Associativity of AND
$X * (Y + Z) = (X * Y) + (X * Z)$	Distributivity
$-(X + Y) = (-X) * (-Y)$	DeMorgan's theorem
$-(X * Y) = (-X) + (-Y)$	DeMorgan's theorem

suited for the control of certain processes. Its use will probably become more widespread as the relationship between computers and robots matures. Fuzzy logic can be represented digitally in discrete steps. For a smooth range of values, analog systems are used.

Binary Logic Gates

All binary digital devices and systems employ switches that perform various boolean functions. These switches are called *logic gates*.

Positive and negative logic

Usually, the binary digit 1 stands for true and is represented by about +5 V. The binary digit 0 stands for false and is represented by about 0 V. This is *positive logic*. There are other logic forms, the most common of which is *negative logic* (in which the digit 1 is represented by a more negative voltage than the digit 0). The remainder of this discussion deals with positive logic.

Basic gates

An *inverter,* or *NOT gate,* has one input and one output. It reverses the state of the input. An *OR gate* can have two or more inputs. If both, or all, of the inputs are 0, the output is 0. If any of the inputs are 1, the output is 1. An *AND gate* can have two or more inputs. If both, or all, of the inputs are 1, the output is 1. Otherwise the output is 0.

Other gates

Sometimes an inverter and an OR gate are combined. This produces a *NOR gate.* If an inverter and an AND gate are combined, the result is a *NAND gate.*

An *exclusive OR gate,* also called an *XOR gate,* has two inputs and one output. If the two inputs are the same (either both 1 or both 0), the output is 0. If the two inputs are different, the output is 1.

The functions of logic gates are summarized in Table 4.3. Their schematic symbols are shown in Fig. 4.2.

Digital Circuits

Digital circuits are designed to deal with signals that attain discrete, well-defined levels. Many digital circuits are nothing more than sophisticated electronic switches.

TABLE 4.3   Logic Gates and Their Characteristics

Gate type	Number of inputs	Remarks
NOT	1	Changes state of input.
OR	2 or more	Output high if any inputs are high. Output low if all inputs are low.
AND	2 or more	Output low if any inputs are low. Output high if all inputs are high.
NOR	2 or more	Output low if any inputs are high. Output high if all inputs are low.
NAND	2 or more	Output high if any inputs are low. Output low if all inputs are high.
XOR	2	Output high if inputs differ. Output low if inputs are the same.

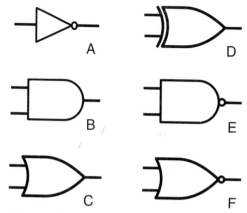

Figure 4.2 (A) An inverter or NOT gate; (B) an AND gate; (C) an OR gate; (D) an XOR gate; (E) a NAND gate; (F) a NOR gate.

Binary data

Binary (two-level) signals are used in many communications systems. Binary data is less susceptible to noise and other interference than analog or multilevel digital data.

Morse code is the oldest binary means of sending and receiving messages. It is a binary code because it has only two possible states: ON (key down) and OFF (key up).

Baudot, also called the *Murray code,* is a five-unit digital code, which is not widely used by today's digital equipment, except in some amateur radio communications.

American National Standard Code for Information Interchange (*ASCII*) is a seven-unit code for the transmission of text and some programs. Letters, numerals, symbols, and control operations are represented. ASCII is designed primarily for computer applications. There are 2^7, or 128, possible representations. Both upper- and lowercase letters can be represented, along with certain symbols.

Flip-flops

A *flip-flop* is a form of *sequential logic gate.* In a sequential gate, the output state depends on both the inputs and the outputs. A flip-flop has two states, called *set* and *reset.* Usually, the set state is logic 1 (high), and the reset state is logic 0 (low).

R-S flip-flop inputs are labeled R (reset) and S (set). The outputs are Q and −Q. (Often, rather than −Q, you will see Q′ or perhaps Q with a line over it.) The outputs are always in logically opposite states. The symbol for an R-S flip-flop, also known as an *asynchronous flip-flop,* and the truth table for an R-S flip-flop are shown in Fig. 4.3A and Table 4.4

Synchronous flip-flop states change when triggered by the signal from a *clock.* In *static triggering,* the outputs change state only when the clock signal is either high or low. This type of circuit is sometimes called a *gated flip-flop.* In *positive-edge triggering,* the outputs change state at the instant the clock pulse is positive-going. The term *edge triggering* derives from the fact that the abrupt rise or fall of a pulse looks like the edge of a cliff (Fig. 4.3B). In *negative-edge triggering,* the outputs change state at the instant the clock pulse is negative-going.

Master/slave (M/S) flip-flop inputs are stored before the outputs are allowed to change state. This device essentially consists of two R-S flip-flops in series. The first flip-flop is called the *master,* and the second is called the *slave.* The master flip-flop functions when the clock output is high, and the

TABLE 4.4   Flip-Flop States

R-S flip-flop			
R	S	Q	−Q
0	0	Q	−Q
0	1	1	0
1	0	0	1
1	1	?	?

J-K flip-flop			
J	K	Q	−Q
0	0	Q	−Q
0	1	1	0
1	0	0	1
1	1	−Q	Q

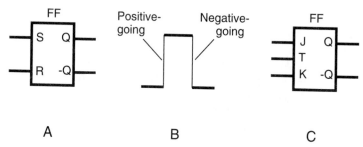

Figure 4.3 (A) Symbol for an R-S flip-flop; (B) pulse edges are either negative-going or positive-going; (C) symbol for a J-K flip-flop.

slave acts during the next low portion of the clock output. This time delay prevents confusion between the input and output.

J-K flip-flop operation is similar to that of an R-S flip-flop, except that the J-K flip-flop has a predictable output when the inputs are both 1. Table 4.4 shows the input and output states for this type of flip-flop. The output changes only when a triggering pulse is received. The symbol for a J-K flip-flop is shown in Fig. 4.3C.

R-S-T flip-flop operation is similar to that of an R-S flip-flop, except that a high pulse at the T input causes the circuit to change state.

T flip-flop operation employs only one input. Each time a high pulse appears at the T input, the output state is reversed.

Clocks

In electronics, the term *clock* refers to a circuit that generates pulses at high speed and at precise intervals. It sets the tempo for the operation of certain kinds of digital devices. In a computer, the clock acts like a metronome for the *microprocessor.* Clock speeds are typically measured in megahertz or gigahertz.

Counters

A *counter* consists of a set of flip-flops or equivalent circuits. Each time a pulse is received, the binary number stored by the counter increases by 1.

Frequency counters measure the frequency of a wave by tallying the cycles in a given interval of time. The circuit consists of a *gate,* which begins and ends each counting cycle at defined intervals. The accuracy is a function of the length of the *gate time*; the longer the time base, the better the accuracy. The readout is in base-10 digital numerals.

Binary Digital Communications

The use of binary data yields optimum communications efficiency. If multilevel signaling is required, all the levels can be represented by groups of binary digits. A group of n binary digits, for example, can represent 2^n levels.

Bits and bytes

A *bit* is an elementary unit of digital data, represented by either logic 0 or logic 1. A group of 8 bits is a *byte.* In communications, a byte is sometimes called an *octet.*

One *kilobit* (Kb) is equal to 1024 bits. A *megabit* (Mb) is 1024 kilobits, or 1,048,576 bits. A *gigabit* (Gb) is 1024 megabits, or 1,073,741,824 bits.

Data quantity is usually specified in *kilobytes* (units of 2^{10} = 1024 bytes) *megabytes* (units of 2^{20} = 1,048,576 bytes), and *gigabytes* (units of 2^{30} = 1,073,741,824 bytes). The abbreviations for these units are KB, MB, and GB, respectively. Alternatively you might see them abbreviated as K, M, and G.

There are larger data units. The *terabyte* (TB) is 2^{40} bytes, or 1024 GB. The *petabyte* (PB) is 2^{50} bytes, or 1024 TB. The *exabyte* (EB) is 2^{60} bytes, or 1024 PB.

Baud versus bits per second

The most common method of measuring data speed is *bits per second* (bps). *Baud* refers to the number of times per second that a signal changes state. These parameters are not equivalent, even though people often speak of them as if they are.

When computers are linked in a *network,* each computer has a *modem* (modulator/demodulator) connecting it to the communications medium. The slowest modem determines the speed at which the machines communicate. Table 4.5 shows common data speeds and the time required to send 1, 10, and 100 pages of double-spaced, typewritten text at each speed.

Data Types and Conversions

Many communications systems "digitize" the analog signals at the *source* and "undigitize" the signals the *destination.* Digital data can be transferred bit by bit (serial) or in bunches (parallel).

Analog to digital

Any analog, or continuously variable, signal can be converted into a string of pulses, whose amplitudes have a finite number of states, usually some power of 2. This is *analog-to-digital (A/D) conversion.*

An A/D converter samples the instantaneous amplitude of an analog signal and outputs pulses having discrete levels, as shown in Fig. 4.4. The number of levels is called the *sampling resolution* and is usually an integral power of 2.

TABLE 4.5   Time Needed to Send Data at Various Speeds

Bits per second	One page	10 pages	100 pages
1,200	9.00 s	1 min 30 s	15 min
2,400	4.50 s	45.0 s	7 min 30 s
4,800	2.25 s	22.5 s	3 min 45 s
9,600	1.13 s	11.3 s	1 min 53 s
14,400	0.75 s	7.5 s	1 min 15 s
19,200	0.56 s	5.6 s	56 s
28,800	0.38 s	3.8 s	38 s
38,400	0.28 s	2.8 s	28 s
57,600	0.19 s	1.9 s	19 s
115,200	0.09 s	0.94 s	9.4 s
230,400	0.05 s	0.47 s	4.7 s
460,800	0.03 s	0.24 s	2.4 s

The number of samples per second is the *sampling rate*. In this example, there are eight levels, represented by three-digit binary numbers from 000 to 111.

In general, the digital sampling rate is approximately twice the highest analog data frequency. For a signal with components as high as 3 kHz, the minimum sampling rate is 6 kHz. The commercial voice standard is 8 kHz. For hi-fi music transmission, the standard sampling rate is 44.1 kHz.

Digital to analog

The scheme for *digital-to-analog (D/A) conversion* depends on whether the signal is binary or multilevel.

With a binary signal, a microprocessor reverses the A/D process done in recording or transmission. The high-pitched incoming tone is converted to one logic state; the low-pitched tone is converted to the other logic state.

Multilevel digital signals can be converted back to analog form by "smoothing out" the pulses. This can be intuitively seen by examining Fig. 4.4. Imagine the train of pulses being smoothed into the continuous curve.

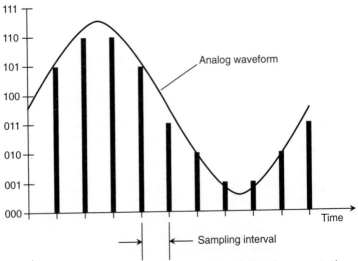

Figure 4.4 An analog waveform and an eight-level digital representation.

Serial versus parallel

Binary data can be sent and received 1 bit at a time along a single line or channel. This is *serial data transmission.* Higher data speeds can be obtained by using multiple lines or a wideband channel, sending independent sequences of bits along each line or subchannel. This is *parallel data transmission.*

Parallel-to-serial (P/S) conversion receives bits from multiple lines or channels and transmits them one at a time along a single line or channel. A *buffer* stores the bits from the parallel lines or channels while they are awaiting transmission along the serial line or channel.

Serial-to-parallel (S/P) conversion receives bits from a serial line or channel and sends them in batches along several lines or channels. The output of an S/P converter cannot go any faster than the input, but the circuit is useful when it is necessary to interface between a serial-data device and a parallel-data device.

Figure 4.5 illustrates a circuit in which a P/S converter is used at the source and an S/P converter is used at the destination. In this example, the words are 8-bit bytes. However, the words could have 16, 32, or even 64 bits.

Data Compression

Data compression is a way of maximizing the amount of digital information that can be stored in a given space or sent in a certain period of time.

Text files can be compressed by replacing often-used words and phrases with symbols such as =, #, &, @, and so on, as long as none of these symbols occurs in the uncompressed file. When the data is received, it is uncompressed by substituting the original words and phrases for the symbols.

Digital images can be compressed in either of two ways. In *lossless image compression,* detail is not sacrificed; only the redundant bits are eliminated. In *lossy image compression,* some detail is lost, although the loss is usually not significant.

Text and programs can generally be reduced in size by about 50 percent. Images can be reduced to a much larger extent.

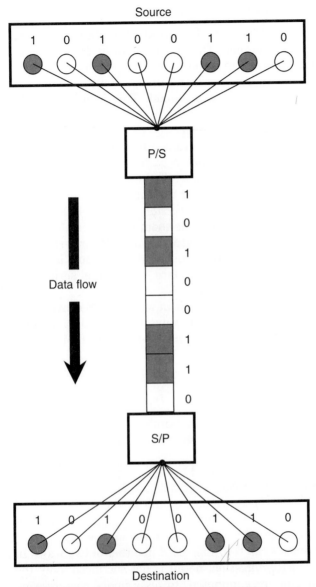

Figure 4.5 A communications circuit employing P/S conversion at the source and S/P conversion at the destination.

Some advanced image-compression schemes can output a file that is only a small percentage of the original file size.

Packet Communications

Packet communications is the most common way in which computers send and receive signals to and from each other.

Packet networks

A *packet* is a block of digital data sent from a source computer to one or more destination computers. The data speed is high, allowing long messages to be sent in short signal bursts. Individual packets might not all follow the same route from the source to the destination, but the destination computer "knows" the correct way to reassemble the packets to obtain the original message.

Packet communications is self-correcting. The destination computer detects discrepancies and instructs the source to retransmit packets that appear corrupted.

Protocol

In a packet network, all computers use a data format, or *protocol*. The universal standard is the *Open Systems Interconnection Reference Model,* abbreviated *OSI-RM,* which has seven levels of activity, called *layers.*

When you send a packet, the computers do all the routing work, once the source computer knows the *local node* of the desired destination station. The operator need only enter destination information to establish the route. The protocol keeps the connection intact via *rerouting,* in case there is disruption along a given route.

Packet wireless

In *packet wireless,* the computer is connected to a radio transceiver using a *terminal node controller* (TNC). An example is shown in Fig. 4.6A. The computer has a telephone modem as well as a TNC, so messages can be sent and received via con-

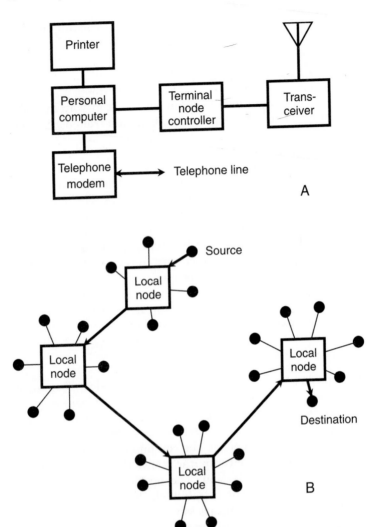

Figure 4.6 (A) A packet-wireless station; (B) passage of a packet through nodes in a wireless communications circuit.

well as a TNC, so messages can be sent and received via conventional on-line services as well as via radio.

Figure 4.6B shows how a packet-wireless message is routed. Black dots represent *subscribers*. Rectangles represent local nodes, each of which serves subscribers via short-range links at very-high frequency (VHF), ultrahigh frequency (UHF), or microwave frequency. The nodes are interconnected by terrestrial VHF, UHF, or microwave links if they are relatively near each other. If the nodes are widely separated, satellite links are used.

RGB Color Model

All visible colors can be obtained by combining red, green, and blue light. The *RGB color model* is a scheme for digital video imaging that takes advantage of this.

Hue, saturation, and brightness

Color is a function of wavelength. When energy is concentrated near a single wavelength, you see an intense *hue*. The vividness of a hue is *saturation*. The *brightness* of a color is a function of how much total energy the light contains. In most video displays, there is a control for adjusting the brightness, also called *brilliance*.

3-D color

A color *palette* is obtained by combining pure red, green, and blue in various ratios. Assign each primary color an axis in three-dimensional space as shown in Fig. 4.7. Call the axes R (for red), G (for green), and B (for blue). Color brightness can range from 0 to 255, or binary 00000000 to 11111111. The result is 16,777,216 (256^3) possible colors. Any point within the cube represents a unique color.

Some RGB systems use only 16 levels for each primary color (binary 0000 through 1111). This results in 4096 possible colors.

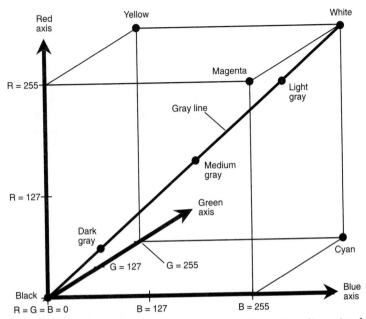

Figure 4.7 The RGB color model, depicted as a cube in three-dimensional space.

Digital Signal Processing

Digital signal processing (DSP) is a scheme for improving the precision of digital data. It can be used to clarify or enhance signals of all kinds.

Analog data cleanup

When DSP is used in analog modes, the signals are changed into digital form by A/D conversion. Then the digital signal is "tidied up" so that the pulse timing and amplitude adhere strictly to protocol. Finally, the digital signal is changed back to analog form via D/A conversion.

Digital signal processing can extend the workable range of a communications circuit, because it allows reception under worse conditions than would be possible without it. Digital signal processing also improves the quality of fair signals, so the receiving equipment or operator makes fewer errors.

Digital data cleanup

In circuits that use only digital modes, A/D and D/A conversions are irrelevant, but DSP can still "tidy up" the signal. This improves the accuracy of the system and also makes it possible to copy data many times (that is, to produce multi-generation copies).

The DSP circuit minimizes confusion between digital states (Fig. 4.8). A hypothetical signal before processing is shown at the top; the signal after processing is shown at

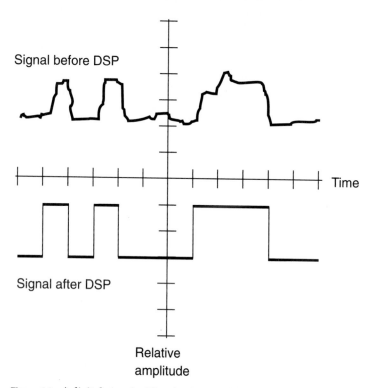

Figure 4.8 A digital signal with noise (top) and the same signal after DSP (bottom).

the bottom. If the input amplitude is above a certain level for an interval of time, the output is high (logic 1). If the input amplitude is below the critical point for a time interval, the output is low (logic 0). A strong burst of noise might fool the circuit into thinking the signal is high when it is really low, but overall, errors are less frequent with DSP than without it.

DSP in computers

A DSP system can be etched onto a single integrated circuit, similar in size to a microprocessor chip. Some DSPs serve multiple functions in a computer system, acting as an assistant to the microprocessor, so that the computer's "mind" can devote itself to doing its primary work without having to bother with extraneous tasks.

A DSP chip can work as a facsimile (fax) machine, a modem, or a TNC. It can compress and decompress data, help a computer recognize and generate speech, translate from one spoken language to another (such as from English to Chinese or vice versa), and even recognize and compare patterns.

Wire and Cable

Wire is the most common vehicle for transferring electrical energy from one place to another. Wire is also used for physical reinforcement in a variety of situations. Some types of wire are used to make lamps, resistors, and heating elements.

A *cable* is a special cord that is designed to carry power or signals. When interconnecting electronic equipment and computer peripherals, it is important that the right types of cables and connectors be used.

Wire Conductors

For a given type of metal and construction, higher wire-gauge numbers correlate with higher *resistivity* and lower *carrying capacity*. The type of metal is also important. Silver is the best known conductor, followed by copper and aluminum.

For dc and low-frequency ac, solid wire can handle more current than stranded wire of the same gauge. At radio frequency (RF), the situation is reversed because of *skin effect*, in which current tends to flow mostly on the outside of a conductor.

The gauge of a wire is correlated with its tensile strength, or ability to withstand mechanical stress. The higher the gauge, the weaker the wire for a given metal and wire construction. Steel is the strongest material available at moderate cost. Copper and aluminum, although better electrical conductors than steel, are weaker mechanically. Stranded wire resists breakage better than solid wire of the same gauge.

American wire gauge

The American wire gauge (AWG) is expressed as a whole number. The larger the number, the smaller the diameter. Table 5.1 shows wire diameters, in millimeters and inches, for AWG No. 1 through AWG No. 40. These diameters do not include any insulation or enamel that might exist on the wire.

British standard wire gauge

In some countries, *British standard wire gauge* (NBS SWG) is used. The higher the number, the thinner the wire. The British standard wire gauge sizes for designators 1 through 40 are shown in Table 5.2.

The British standard wire gauge designator does not take into account any coatings on the wire, such as enamel, rubber, or plastic insulation.

Birmingham wire gauge

The *Birmingham wire gauge* (BWG) designators differ from the American and British standard designators, but the sizes are nearly the same. The higher the designator number, the thinner the wire.

Table 5.3 shows the diameter versus BWG number for designators 1 through 20. The BWG designator does not include any coatings or insulation that might surround the wire.

TABLE 5.1   American Wire Gauge Diameters

AWG	Millimeters	Inches
1	7.35	0.289
2	6.54	0.257
3	5.83	0.230
4	5.19	0.204
5	4.62	0.182
6	4.12	0.163
7	3.67	0.144
8	3.26	0.128
9	2.91	0.115
10	2.59	0.102
11	2.31	0.0909
12	2.05	0.0807
13	1.83	0.0720
14	1.63	0.0642
15	1.45	0.0571
16	1.29	0.0508
17	1.15	0.0453
18	1.02	0.0402
19	0.912	0.0359
20	0.812	0.0320
21	0.723	0.0285
22	0.644	0.0254
23	0.573	0.0226
24	0.511	0.0201
25	0.455	0.0179
26	0.405	0.0159
27	0.361	0.0142
28	0.321	0.0126
29	0.286	0.0113
30	0.255	0.0100
31	0.227	0.00894
32	0.202	0.00795
33	0.180	0.00709
34	0.160	0.00630
35	0.143	0.00563
36	0.127	0.00500
37	0.113	0.00445
38	0.101	0.00398
39	0.090	0.00354
40	0.080	0.00315

TABLE 5.2   British Standard Wire Gauge Diameters

NBS SWG	Millimeters	Inches
1	7.62	0.300
2	7.01	0.276
3	6.40	0.252
4	5.89	0.232
5	5.38	0.212
6	4.88	0.192
7	4.47	0.176
8	4.06	0.160
9	3.66	0.144
10	3.25	0.128
11	2.95	0.116
12	2.64	0.104
13	2.34	0.092
14	2.03	0.080
15	1.83	0.072
16	1.63	0.064
17	1.42	0.056
18	1.22	0.048
19	1.02	0.040
20	0.91	0.036
21	0.81	0.032
22	0.71	0.028
23	0.61	0.024
24	0.56	0.022
25	0.51	0.020
26	0.46	0.018
27	0.42	0.0164
28	0.38	0.0148
29	0.345	0.0136
30	0.315	0.0124
31	0.295	0.0116
32	0.274	0.0108
33	0.254	0.0100
34	0.234	0.0092
35	0.213	0.0084
36	0.193	0.0076
37	0.173	0.0068
38	0.152	0.0060
39	0.132	0.0052
40	0.122	0.0048

TABLE 5.3 **Birmingham Wire Gauge Diameters**

BWG	Millimeters	Inches
1	7.62	0.300
2	7.21	0.284
3	6.58	0.259
4	6.05	0.238
5	5.59	0.220
6	5.16	0.203
7	4.57	0.180
8	4.19	0.165
9	3.76	0.148
10	3.40	0.134
11	3.05	0.120
12	2.77	0.109
13	2.41	0.095
14	2.11	0.083
15	1.83	0.072
16	1.65	0.064
17	1.47	0.058
18	1.25	0.049
19	1.07	0.042
20	0.889	0.035

Resistivity

The ease with which a wire carries current is expressed in terms of its *resistivity,* or resistance per unit length. A common unit is the *microhm per meter* ($\mu\Omega$/m). For a given metal, larger gauges (smaller diameters) of wire have greater resistivity than smaller gauges (larger diameters). In the case of solid copper wire carrying dc at room temperature, approximate resistivity in microhm per meter for even-numbered solid-copper wire sizes from AWG Nos. 2 through 30 are shown in Table 5.4.

Carrying capacity

The ability of a wire to handle electric current safely is called its *carrying capacity.* This specification is usually given in amperes. Table 5.5 shows approximate dc carrying capacity for even-numbered solid-copper wire sizes from AWG Nos. 8 through 20 in open air at room temperature. Wires might

TABLE 5.4   Resistivity of Various
Gauges of Solid Copper Wire

Wire size, AWG	$\mu\Omega/m$
2	523
4	831
6	1,320
8	2,100
10	3,340
12	5,320
14	8,450
16	13,400
18	21,400
20	34,000
22	54,000
24	85,900
26	137,000
28	217,000
30	345,000

TABLE 5.5   Maximum Safe Continuous
DC Carrying Capacity for Various Wire
Sizes in Open Air

Wire size, AWG	Current, A
8	73
10	55
12	41
14	32
16	22
18	16
20	11

intermittently carry somewhat larger currents than those
shown in the table, but the danger of softening or melting,
with consequent breakage, rises rapidly as the current
increases beyond these values.

When wire is run alongside other electronic components,
the figures in the table should be reduced somewhat. The
same is true when wires are bundled into cables and/or when
wires are run near flammable materials or surrounded by
insulation.

Wire Splicing

In wiring of electrical circuits and antennas, it is often necessary to splice two lengths of wire. There are various splicing methods; two are described here.

Twist splice

The simplest way to splice two wires is to bring the exposed ends close together and parallel; then the ends are twisted over each other several times (Fig. 5.1). This scheme, called a *twist splice,* can be used with solid or stranded wire. If the wires are of unequal diameter, the smaller wire is twisted around the larger wire (Fig. 5.2). Electrical tape can be put over the connection if insulation is important. This type of splice has poor mechanical strength. Even if a twist splice is soldered, the connection cannot withstand much strain.

Western Union splice

When a splice must have good tensile strength, the wires are brought together end to end, overlapping about 2 inches (in). The wires are hooked around each other and then twisted several times (Fig. 5.3). This is known as a *Western Union splice.*

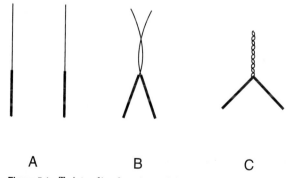

A   B   C

Figure 5.1 Twist splice for wires of the same gauge. Wires are brought parallel (A), looped around each other (B), and then twisted over each other several times (C).

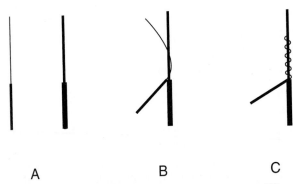

A	B	C

Figure 5.2 Twist splice for wires of differing gauge. Wires are brought parallel (A), and then the smaller wire is looped (B) and twisted (C) around the larger one.

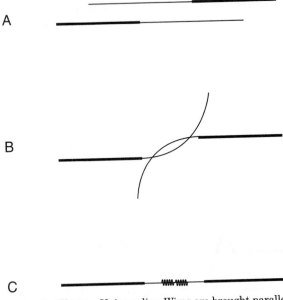

Figure 5.3 Western Union splice. Wires are brought parallel (A), looped around each other (B), and finally twisted over each other (C).

For guy wires, splices should be avoided if possible. If splices are necessary, the Western Union method is recommended. Each end should be twisted around 10 to 12 times. Needle-nosed pliers can be used to secure the extreme ends. Protruding ends are removed using a diagonal cutter.

Splices should be soldered if a good electrical bond is needed, and a layer of electrical tape or other insulation should be applied. For large-diameter wires, the ends can be "tinned" with solder before the splice is made to optimize the electrical bond. For maximum physical strength, both wires should be the same size and the same type (both solid or both stranded).

Electrical Cable

The simplest cable is so-called *lamp cord,* used with common appliances. Two or three wires are embedded in rubber or plastic insulation (Fig. 5.4A). The individual conductors are usually stranded. This makes the conductors resistant to breakage from repeated flexing.

Multiconductor cables

When a cable has several wires, they can be individually insulated, bundled together, and enclosed in an insulating jacket (Fig. 5.4B). If the cable must be flexible, each wire is stranded. Some cables of this type have dozens of conductors.

If there are only a few conductors, they might be run parallel to each other in a flat configuration as shown in Fig. 5.4C.

Sometimes, several conductors are molded into a common insulating jacket as shown in Fig. 5.4D. Such *ribbon cable* is used inside commercially manufactured electronic devices, particularly computers. This cable is physically sturdy, is ideal for saving space, and efficiently radiates heat away.

Shielding

The above-mentioned cable types are unshielded. For dc or low-speed data transmission, unshielded cables are usually all right. But radio signals, video, and other high-speed data

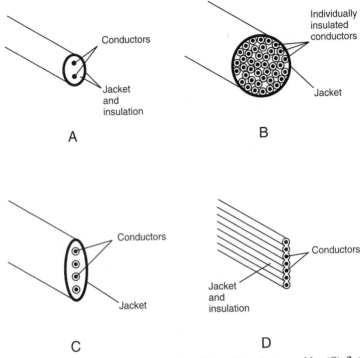

Figure 5.4 (A) Simple two-wire cable; (B) multiconductor cable; (C) flat cable; (D) ribbon cable.

create electromagnetic fields that can be transferred among cable conductors and even between a cable and the surrounding environment. In these situations, *electromagnetic shielding* is necessary.

A wire, or group of wires, is shielded by enclosing it in a conductive cylinder of solid metal, metallic braid (usually copper), or metal foil. The shield is separated from the conductor(s) by dielectric material such as polyethylene.

In some multiconductor cables, a single shield surrounds all the wires. In other cables, each wire has its own shield. The entire cable might be surrounded by a braid in addition to individual shielding of the wires. *Double-shielded cable* is surrounded by two concentric braids separated by dielectric.

Coaxial cable

Coaxial cable, also called *coax* (pronounced *co*-ax), is especially designed for high-frequency signal transmission. It is
used in community-antenna television (CATV) networks. It
is the cable of choice for baseband computer local area networks (LANs). Coaxial cable is employed by amateur and
citizens band radio operators to connect transceivers, transmitters, and receivers to antennas. It is used in high-fidelity
audio systems to interconnect components.

In coaxial cable, a single center conductor is surrounded
by a tubular shield. In some cases, a solid or foamed polyethylene dielectric keeps the center conductor at the central
axis of the cable (Fig. 5.5A). Other cables have a bare center
conductor and a thin layer of polyethylene just inside the
braid (Fig. 5.5B), so most of the dielectric is air. Technically,
this is not true coaxial cable, although it is often called such.

Some coaxial cables have a solid metal pipe surrounding
the center conductor. This is called *hard line.* This type
of cable is available in larger diameters than coaxial
cables and has lower loss per unit length. It is used in
high-power, fixed transmitting installations, especially at
VHF and UHF.

In a coaxial cable, the signal is carried by the center
conductor. The shield is connected to ground, so it keeps signals from "leaking out." The shield also keeps unwanted
signals or noise from getting in.

Serial versus parallel cable

A cable that carries data along one line is a *serial cable.* A
cable that carries data along several lines at once is a *parallel cable.*

Serial cables can generally be longer than parallel cables,
because parallel cables are prone to *crosstalk,* a condition in
which signals in the different conductors interfere or combine with one another. Crosstalk can be prevented by individually shielding each conductor within a parallel cable,
but this increases the physical bulk of the cable and also
increases the cost per unit length.

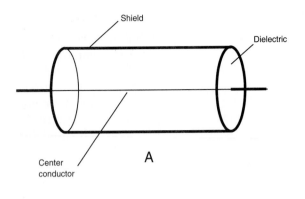

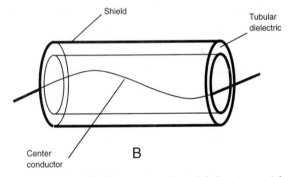

Figure 5.5 (A) Solid-dielectric or foamed-dielectric coaxial cable; (B) a similar cable in which most of the dielectric is air.

An example of a serial cable is the coaxial line used to carry television signals in CATV. These cables can be miles long. The cord connecting the main unit and the printer in a personal computer system is a parallel cable. Such cables are only a few feet in length.

Cable splicing

When splicing cord, multiconductor cable, or ribbon cable, the Western Union splice is preferable. Insulation is important; all splices should be wrapped with electrical tape, and the combination wrapped afterward. For additional insulation, the splices can be made at slightly different points along the cord.

Two-conductor cord or ribbon can be twist-spliced if necessary (Fig. 5.6). After the twists have been soldered, the splices are trimmed to about $\frac{1}{2}$ in, folded back parallel to the cable axes, and insulated. The whole junction is then carefully wrapped with electrical tape or coated with sealant.

Coaxial cables, and other cables in which a constant *characteristic impedance* (Z_0) must be maintained, are generally spliced by using special connectors. A male connector is soldered to each of the two ends to be spliced; a female-to-female adaptor is used between them. The whole joint is covered with insulating tape or sealant.

Fiber-Optic Cable

A *fiber-optic cable* is a bundle of transparent, solid strands designed to carry modulated light or infrared (IR) signals. This type of cable can carry millions of signals at high bandwidth.

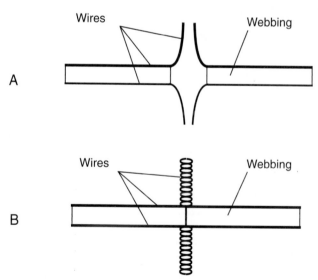

Figure 5.6 Twist splice for two-wire cord or cable. Ends are brought together (A) and the conductors are twisted at right angles (B). The twists are then soldered, trimmed, folded back, and insulated.

Manufacture

Optical fibers are made from glass, to which impurities have been added. The impurities change the *refractive index* of the glass. An optical fiber has a *core* surrounded by a tubular *cladding,* as shown in Fig. 5.7. The cladding has a lower refractive index than the core.

In a *step-index optical fiber* (shown in Fig. 5.7A), the core has a uniform index of refraction and the cladding has a lower index, also uniform. The transition at the boundary is abrupt. In the *graded-index optical fiber* (shown in Fig. 5.7B), the core has a refractive index that is greatest along the central axis and steadily decreases outward from the center. At the boundary, there is an abrupt drop in the refractive index.

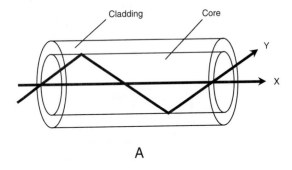

A

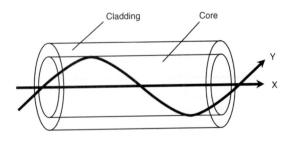

B

Figure 5.7 (A) Step-index optical fiber; (B) graded-index fiber.

Operation

In Fig. 5.7A, showing a step-index fiber, ray X enters the core parallel to the fiber axis and travels without striking the boundary unless there is a bend in the fiber. If there is a bend, ray X veers off center and behaves like Y. Ray Y strikes the boundary repeatedly. Each time ray Y encounters the boundary, *total internal reflection* occurs, so ray Y stays within the core.

In Fig. 5.7B, showing a graded-index fiber, ray X enters the core parallel to the fiber axis and travels without striking the boundary unless there is a bend in the fiber. If there is a bend, ray X veers off center and behaves like ray Y. As ray Y moves farther from the center of the core, the index of refraction decreases, bending the ray back toward the center. If ray Y enters at a sharp enough angle, it might strike the boundary, in which case total internal reflection occurs. Therefore, ray Y stays within the core.

Bundling

Optical fibers can be bundled into cable in the same way that wires are bundled. The individual fibers are protected from damage by plastic jackets. Common coverings are polyethylene and polyurethane. Steel wires or other strong materials are often used to add strength to the cable. The whole bundle is encased in an outer jacket. This outer covering might also be reinforced with wire or tough plastic compounds.

Each fiber in the bundle can carry several rays of visible light and/or IR, each ray having a different wavelength. Each ray can in turn contain a large number of signals. Because the frequencies of visible light and IR are much higher than the frequencies of RF currents, the attainable bandwidth of an optical/IR cable link can be far greater than that of any RF cable link. This allows much greater data speed.

Connectors

A *connector* is a device that provides and maintains a good electrical connection with wire or cable. It is important to use the proper connector for a given application. Often, the choice is dictated by convention.

Single-wire connectors are simple. Multiconductor-cable connectors are more complex. In RF cables, special connectors must be used to maintain *impedance continuity*. This section describes several of the most common types of wire and cable connectors used in electronics.

Clip leads

A *clip lead* is a short length of flexible wire, equipped at one or both ends with a simple, temporary connector. Clip leads are not suitable for permanent installations, especially outdoors, because corrosion occurs easily, and the connector can slip out of position. The current-carrying capacity is limited. Clip leads are used primarily in dc and low-frequency ac applications; they are not generally suitable at RF.

For testing and experimentation when temporary connections are needed, *alligator clips* are often used. They require no modification to the circuit under test. The name derives from the serrated edges of the clip, resembling the mouth of an alligator or crocodile. Alligator clips come in sizes ranging from less than $\frac{1}{2}$ in long to several inches long. They can be clamped to a terminal or a length of exposed wire.

Banana connectors

A *banana connector* is a convenient single-lead connector that slips easily in and out of its receptacle. The banana plug looks something like a banana skin, and this is where it gets its name (Fig. 5.8).

Banana jacks are frequently found in screw terminals of low-voltage dc power supplies. If frequent changing of leads is necessary, the banana jacks allow it with a minimum of trouble. Banana connectors have fairly low dc loss and high current-handling capacity. But they are not generally used at high voltages because of the possibility of shock from exposed conductors.

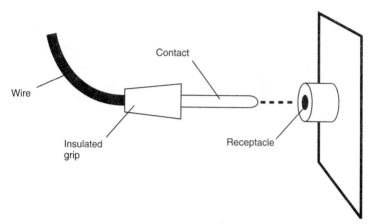

Figure 5.8 Banana plugs are convenient for low-voltage dc use.

Hermaphroditic connectors

A *hermaphroditic connector* is an electrical plug/jack that mates with another plug/jack exactly like itself. Such a connector has an equal number of male and female contacts.

Hermaphroditic connectors are equipped with special pins and holes, so they can be inserted into one another in only one way. This makes them useful in polarized circuits such as dc power supplies. They are rarely used in RF applications.

BNC connectors

A *bayonet Neil-Concelman (BNC) connector* consists of a small jack and plug, used frequently in RF lab applications (Fig. 5.9). The jack and plug are designed to provide a constant impedance for coaxial-cable connections. This minimizes losses that can be caused by impedance "bumps," especially at VHF and UHF.

The BNC connector has a quick-connect, quick-release feature. This type of connector is not intended for high levels of RF power or for permanent installation outdoors. But it can be convenient when frequent wiring changes are necessary.

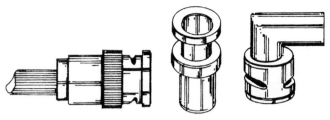

Figure 5.9 BNC connectors are used with test instruments, especially at RF.

UHF connectors

The *UHF connector* is widely used with coaxial cables and is popular among amateur radio operators. The adjective "UHF" is imprecise because this type of connector is designed for low, medium, and high radio frequencies. A UHF connector has a central pin or receptacle and an outer screw-on shell, separated by solid dielectric. The dimensions are tailored to present a Z_0 of 50 Ω. A UHF connector can handle up to several kilowatts of RF power if the SWR on the line is reasonable.

Figure 5.10A shows a male UHF connector on the end of a length of coaxial cable. The male connector is sometimes called a *PL-259,* the catalog number of a plug manufactured by Amphenol. Figure 5.10B shows a female UHF connector. It is sometimes called an *SO-239,* the catalog number of a chassis-mounted jack manufactured by Amphenol. There are other UHF connectors, such as male-to-male, female-to-female, right-angle, and tee adapters.

When a UHF connector is used outdoors, some means must be devised to keep water from entering the cable through the connector. Electrical tape can be used to wrap a connection. Special tapes and sealants are available from commercial sources.

N-type connectors

The *N-type connector* is similar to the UHF connector and is used with coaxial cables in RF applications. N-type connectors are known for low loss and constant Z_0. These features

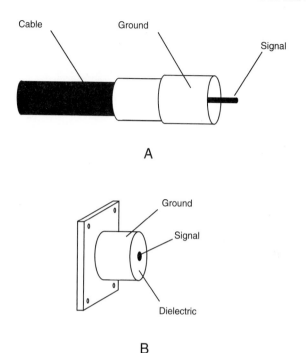

Figure 5.10   UHF connectors. (A) Male; (B) female.

are important at VHF and UHF, and therefore the N-type connector is preferred at these frequencies.

N-type connectors are available in either male or female form and in adapter configurations such as male-to-male and female-to-female. Adaptors are also available for coupling an N-type connector to a cable or chassis that has a different type of connector.

Phone jack and plug

A *phone jack and plug* (Fig. 5.11) is a connector pair originally designed for patching telephone circuits; now it is widely used in electronics and instrumentation from dc through audio frequencies.

In its conventional form, the plug (shown in Fig. 5.11) has a rod-shaped neck that serves as one contact and a ball on the tip of the neck but insulated from it that serves as the

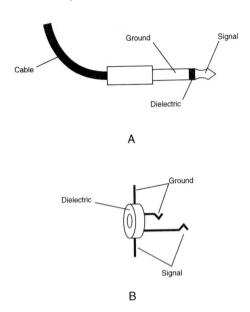

Figure 5.11   (A) Phone plug; (B) phone jack. These are two-conductor units.

other contact. Typical diameters are $\frac{1}{8}$ and $\frac{1}{4}$ in. The jack (Fig. 5.11B) is equipped with contacts that mate securely with the plug when the plug is inserted.

The original phone jack and plug was equipped for two conductors. But in recent decades, three-conductor connectors have become common. They are used in high-fidelity stereo sound systems, audio tape recorders, and in the audio circuits of multimedia computers, radio receivers, transmitters, and transceivers.

Phono jack and plug

A *phono jack and plug* resembles a coaxial connector set, except it is smaller. The plug is designed for ease of connection and disconnection; it is pushed on and pulled off rather than screwed on and off.

Phono connectors are used primarily in audio applications along with low-cost coaxial cable. This so-called audio cable is not suitable for use at RF because the shielding continuity is insufficient at high ac frequencies.

D-shell connectors

Data cables, of the type used in computer systems, have several (sometimes many) wires. If there are more than three or four wires in a cable, a *D-shell connector* is often used at either end. D-shell connectors come in various sizes, depending on the number of wires in the cable.

The connector has a characteristic appearance (Fig. 5.12). This shape forces the user to insert the plug the right way. The female socket has holes into which the pins of the male plug slide. Screws or clips secure the plug once it has been put in place. Some D-shell connectors have metal shells that help keep out dust and moisture and also serve to maintain shielding continuity if needed.

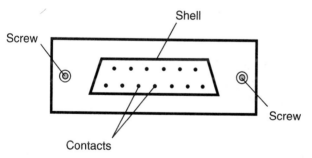

Figure 5.12 A D-shell connector is recognizable by its shape.

6

Power Supplies

Most electronic equipment requires dc electricity as the power source. In the United States, the electricity from the utility company is 60-Hz ac at 117 V rms or 234 V rms. This must be converted to dc, and tailored to the proper voltage to be suitable for electronic equipment.

Parts of a Power Supply

Most power supplies consist of several stages. Figure 6.1 is a block diagram of a typical dc power supply.

First, the ac encounters a *transformer* that steps the voltage either down or up, depending on the needs of the electronic circuits. Then the current passes through a *rectifier* to become pulsating dc. Next, the pulsating dc undergoes *filtering* to become a continuous voltage having either positive or negative polarity with respect to ground. Finally, the dc undergoes *voltage regulation* if the equipment is sensitive to voltage fluctuations. Some power supplies also have *current limiting* to protect against damage in case of a short circuit.

Power supplies that provide more than a few volts must have features that protect the user from receiving an electric shock. All power supplies need *fuses* and/or *circuit breakers* to minimize the fire hazard in case the equipment shorts out.

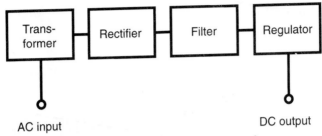

Figure 6.1 Block diagram of a typical dc power supply.

Power Transformers

Power-supply transformers are available in two types: the *step-down transformer,* which converts ac to a lower voltage, and the *step-up transformer,* which converts ac to a higher voltage.

Step-down transformer

Most solid-state electronic devices, such as radios, need only a few volts. The power supplies for such equipment use step-down power transformers (Fig. 6.2A). The physical size of the transformer depends on the current. The input-to-output voltage is directly proportional to the primary-to-secondary turns ratio. That is,

$$\frac{V_{in}}{V_{out}} = \frac{N_{pri}}{N_{sec}}$$

where V_{in} is the rms input voltage, V_{out} is the rms output voltage, N_{pri} is the number of turns in the primary winding, and N_{sec} is the number of turns in the secondary winding.

Some devices need only a low current and a low voltage. The transformer in a radio receiver can be small physically. An amateur-radio transmitter or high-fidelity amplifier needs more current. This means that the secondary winding of the transformer must be of heavy-gauge wire, and the core must be bulky to contain the magnetic flux.

Step-up transformer

Some circuits require high voltage (more than 117 V dc). The picture tube in a TV set needs several hundred volts. Some amateur-radio power amplifiers use vacuum tubes working at kilovolts dc. The transformers in these appliances are step-up types (Fig. 6.2B).

The voltage-transformation formula for a step-up transformer is the same as the formula (above) for step-down transformers.

If a step-up transformer must supply only a small current, it need not be large physically. But most step-up transformers are used with vacuum-tube equipment that consumes considerable power, such as TV sets and radio transmitting amplifiers. For this reason, step-up transformers are generally larger than the step-down units in low-voltage systems such as computers and radio receivers.

Transformer ratings

Transformers are rated according to output voltage and current. For a given unit, the *volt-ampere* (VA) *capacity* is often

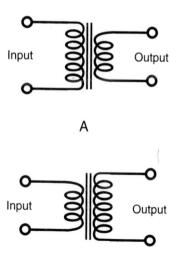

Figure 6.2 (A) Step-down power transformer; (B) step-up power transformer.

specified. This is the product of the voltage and current. A transformer with a 12-V output, capable of delivering 10 A, would have 12 V × 10 A = 120 VA of capacity.

The nature of power-supply filtering, discussed later in this chapter, makes it necessary for the power-transformer VA rating to be greater than the actual wattage consumed by the load.

A rugged power transformer, capable of providing the necessary current and/or voltage on a continuous basis, is crucial in any power supply. The transformer is usually the most expensive component to replace in the event of a power-supply failure.

Rectifiers

A *rectifier* converts ac to pulsating dc. This is usually accomplished by means of one or more semiconductor diodes following a power transformer.

Half-wave rectifier

The simplest type of rectifier circuit, known as the *half-wave rectifier* (Fig. 6.3A), uses one diode (or a series or parallel combination) to "chop off" half of the ac input cycle. The effective, or rms, output voltage is approximately 45 percent of the rms ac input voltage (Fig. 6.4A). But the peak inverse voltage (PIV) across the diode can be as much as 2.8 times the rms ac input voltage. It is recommended to use diodes whose PIV ratings are at least 1.5 times the maximum expected PIV; therefore, with a half-wave power supply, the diodes should be rated for at least 4.2 times the rms ac input voltage.

Half-wave rectification has some shortcomings. First, the output is difficult to filter. Second, the output voltage can diminish considerably when the power supply is connected to a load. Third, half-wave rectification puts a disproportionate strain on the power transformer and the diodes.

Half-wave rectification is useful in power supplies that need not deliver much current or that do not have to be

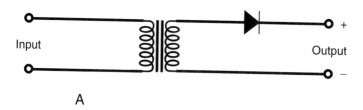

A

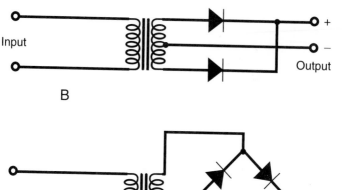

B

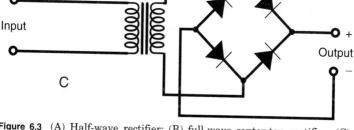

C

Figure 6.3 (A) Half-wave rectifier; (B) full-wave center-tap rectifier; (C) full-wave bridge rectifier.

especially well regulated. The main advantage of using a half-wave circuit in these situations is that it costs less than more sophisticated circuits.

Full-wave center-tap rectifier

A better scheme for changing ac to dc takes advantage of both halves of the ac cycle. A *full-wave center-tap rectifier*

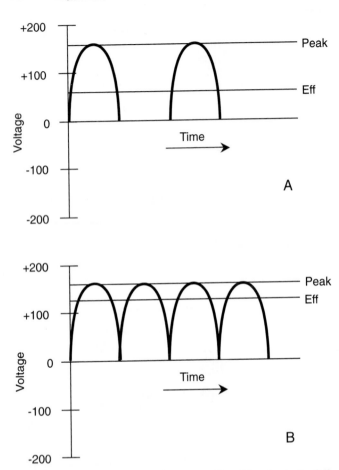

Figure 6.4 (A) Output of a half-wave rectifier; (B) output of a full-wave rectifier. The effective or rms input voltage is 117 (the peak is about 165).

has a transformer with a tapped secondary, as shown in Fig. 6.3B. The center tap is connected to ground. This produces out-of-phase waves at the ends of the winding. These two waves can be individually half-wave rectified, cutting off either half of the cycle.

In this rectifier circuit, the average dc output voltage is about 90 percent of the rms ac input voltage (Fig. 6.4B). The

PIV across the diodes can nevertheless be as much as 2.8 times the rms input voltage. Therefore, the diodes should have a PIV rating of at least 4.2 times the rms ac input.

The output of a full-wave rectifier is easier to filter than that of a half-wave rectifier. The full-wave rectifier is easier on the transformer and diodes than a half-wave circuit. If a load is applied to the output of the full-wave circuit, the voltage will drop less than is the case with a half-wave supply.

Full-wave bridge rectifier

Another way to get full-wave rectification is the *full-wave bridge rectifier,* often called simply a *bridge.* It is diagrammed in Fig. 6.3C. The output waveform is similar to that of the full-wave center-tap circuit (Fig. 6.4B).

The average dc output voltage in the bridge circuit is 90 percent of the rms ac input voltage, as with center-tap rectification. The PIV across the diodes is 1.4 times the rms ac input voltage. Therefore, each diode needs to have a PIV rating of at least 2.1 times the rms ac input voltage.

The bridge does not need a center-tapped transformer secondary. Electrically, the bridge circuit uses the entire secondary on both halves of the wave cycle; the center-tap circuit uses one side of the secondary for one-half of the cycle and use the other side for the other half of the cycle. For this reason, the bridge circuit makes more efficient use of the transformer. The bridge is also easier on the diodes than half-wave or full-wave center-tap circuits.

The main disadvantage of the bridge is that it needs four diodes rather than two. This does not always amount to much in terms of cost, but it can be important when a power supply must deliver high current. Then, the extra diodes—two for each half of the cycle, rather than one—dissipate more overall heat energy. When current is used up as heat, it cannot go to the load.

Voltage doubler

By using diodes and capacitors connected in certain ways, a power supply can be made to deliver a multiple of the peak ac input voltage. Theoretically, large whole-number multiples are possible. But it is rare to see power supplies that make use of multiplication factors larger than 2.

In practice, voltage multipliers are practical only when the load draws low current. Otherwise the regulation is extremely poor. In high-current, high-voltage power applications, the best approach is to use a large step-up transformer in the power supply, not a voltage multiplier.

A voltage-doubler circuit is shown in Fig. 6.5. This circuit works on the entire ac cycle and is called a *full-wave voltage doubler*. Its dc output voltage, when the current drain is low, is about twice the peak ac input voltage, or about 2.8 times the rms ac input voltage. This circuit subjects the diodes to a PIV of 2.8 times the rms ac input voltage. Therefore, they should be rated for PIV of at least 4.2 times the rms ac input voltage.

Note the capacitors. Proper operation of the circuit depends on the ability of these capacitors to hold a charge, even under maximum load. Thus, the capacitors must have large values. The capacitors serve two purposes: to boost the voltage and to filter the output. Also note the resistors. These have low values, similar to those needed when diodes are connected in parallel. When the supply is switched ON, the capacitors draw a large initial *surge current*. Without the resistors in the circuit, this surge could destroy the diodes.

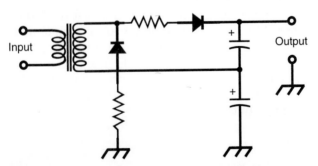

Figure 6.5 An unregulated, full-wave voltage-doubler power supply.

Filters

Most electronic equipment does not work well with the pulsating dc that comes straight from a rectifier. The ripple in the rectifier output is minimized or eliminated by a *filter.*

Capacitors alone

The simplest filter is one or more large-value capacitors, connected in parallel with the rectifier output (Fig. 6.6). Electrolytic capacitors are almost always used. They are polarized; they must be hooked up in the correct direction. Typical values are in the hundreds or thousands of microfarads.

The more current drawn, the more capacitance is needed for effective filtering, because the load resistance decreases as the current increases. The lower the load resistance, the faster the filter capacitors discharge. Larger capacitances hold charge for a longer time with a given load, as compared with smaller capacitances.

Filter capacitors work by "trying" to maintain the dc voltage at its peak level (Fig. 6.7). This is easier to do with the output of a full-wave rectifier (Fig. 6.7A) than with the output of a half-wave rectifier (Fig. 6.7B). With a fill-wave rectifier, the ripple frequency is 120 Hz; with a half-wave rectifier it is 60 Hz. Filter capacitors are recharged twice as often with a full-wave rectifier, as compared with a half-wave rectifier, so the ripple is less severe, for a given capacitance, with full-wave circuits.

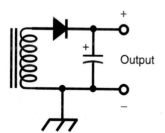

Figure 6.6 A simple capacitor power-supply filter.

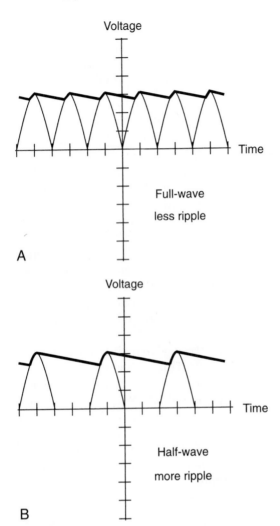

Figure 6.7 Filtering of ripple output from a full-wave rectifier (A) and from a half-wave rectifier (B).

Capacitors and chokes

Another way to smooth out the dc from a rectifier is to use a large inductance in series with the output and a large capacitance in parallel. The inductor, called a *filter choke*, has a value on the order of several henrys.

Sometimes the capacitor is placed on the rectifier side of the choke. This circuit is a *capacitor-input filter* (Fig. 6.8A). If the filter choke is placed on the rectifier side of the capacitor, the circuit is a *choke-input filter* (Fig. 6.8B). Capacitor-input filtering can be used when the load is not expected to be great. The output voltage is higher with a capacitor-input circuit than with a choke-input circuit. If the power supply needs to deliver large or variable amounts of current, a choke-input filter is a better choice, because the output voltage is more stable.

If the output of a dc power supply must have an absolute minimum of ripple, two or three capacitor/choke pairs can be cascaded (Fig. 6.9). Each pair constitutes a *section*. Multisection filters can consist of either capacitor-input or choke-input sections, but the two types are never mixed.

In the example of Fig. 6.9, both capacitor and choke pairs are called *L sections* because of their arrangement in the

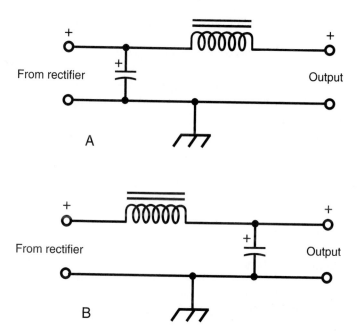

Figure 6.8 (A) Capacitor-input filter; (B) choke-input filter.

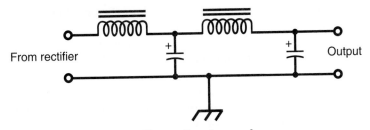

Figure 6.9 Two choke-input filter sections in cascade.

schematic diagram. If the second capacitor is omitted, the filter becomes a *T section*. If the second capacitor is moved to the input and the second choke is omitted, the filter becomes a *pi section*.

Voltage Regulation

If a *zener diode* is connected in parallel with the output of a power supply, the diode limits the output voltage of the supply by "brute force" as long as the diode has a high enough power rating. The limiting voltage depends on the particular zener diode used. There are zener diodes to fit any reasonable power-supply voltage. Figure 6.10 is a diagram of a full-wave, center-tap dc power supply including a zener diode for voltage regulation. The zener diode must be reverse-biased; the cathode should be connected to the positive supply line.

A zener-diode voltage regulator is inefficient when the power supply is used with equipment that draws high current. When a power supply must deliver a high level of current, a *power transistor* is used along with the zener diode to obtain regulation. A circuit diagram of such a scheme is shown in Fig. 6.11.

Voltage regulators are available in integrated-circuit (IC) form. Such an IC, sometimes along with some external components, is installed in the power-supply circuit at the output of the filter. This provides excellent regulation at low and moderate voltages.

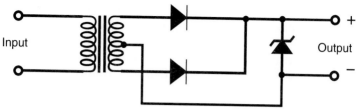

Figure 6.10   A full-wave center-tap supply with a zener-diode regulator.

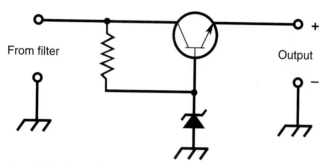

Figure 6.11   A regulator circuit using a zener diode and an *NPN* transistor.

In high-voltage power supplies, *electron-tube voltage regulators* are sometimes used. These are usually gas-filled, cold-cathode devices. They can function under conditions that would destroy semiconductor devices.

Protection of Equipment

The output of a power supply should be free of sudden changes that can damage equipment or components or interfere with their proper performance. The two most common problems are *surge currents* and *transients*.

Surge currents

At the instant a power supply is switched on, a rush (surge) of current occurs, even with no load at the output. This is because the filter capacitors need an initial charge, so they draw a large current for a short time. The surge current is far greater than the operating current. This can destroy the rectifier diodes. The phenomenon is worst in high-voltage supplies and voltage-multiplier circuits. Diode failure can be prevented in at least four different ways:

1. Use diodes with a current rating of many times the operating level. The main disadvantage is cost; high-voltage, high-current diodes can be expensive.

2. Connect several units in parallel wherever a diode is called for in the circuit. Current-equalizing resistors are necessary (Fig. 6.12). The resistors should have a small ohmic value. The diodes should all be identical.

3. Apply the input voltage gradually at power-up. A variable transformer, called a *variac*, is useful for this. Start at zero input and turn a knob to reach full voltage.

4. Use an automatic switching circuit in the transformer primary. This applies a reduced ac voltage for 1 or 2 s and then applies full input voltage.

Transients

The ac on the utility line is a sine wave with a constant rms voltage near 117 or 234 V. But there are often *voltage spikes,* known as *transients,* that can attain peak values of several thousand volts. Transients are caused by sudden changes in the load in a utility circuit. Lightning can also produce them. Unless they are suppressed, they can destroy the diodes in a power supply. Transients can also befuddle the operation of sensitive equipment like personal computers.

The simplest way to get rid of common transients is to place a capacitor of about 0.01 microfarads (μF), rated for 600 V or more, between each side of the transformer primary and electrical ground (Fig. 6.13). Commercially made *transient suppressors* are available. (These are often mistakenly called *surge protectors.*) It is a good idea to use them with all sensitive electronic devices, including computers, high-fidelity stereo systems, TV sets, and VCRs. In the

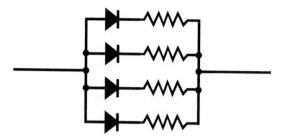

Figure 6.12 Diodes in parallel with current-equalizing resistors.

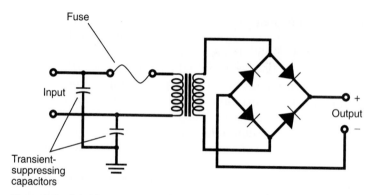

Figure 6.13 A bridge power supply with transient-suppression capacitors and a fuse in the transformer primary circuit.

event of a local thunderstorm, the best way to protect such equipment is to physically unplug it from the wall outlet until the storm has passed.

Fuses

A *fuse* is a piece of soft wire that melts, breaking a circuit if the current exceeds a certain level. Fuses are placed in series with the transformer primary (shown in Fig. 6.13 along with transient-suppressing capacitors). Any component failure, short circuit, or overload will burn the fuse out. Fuses are easy to replace, although it is inconvenient if a fuse blows and replacements are not on hand. If a fuse blows, it must be replaced with another of the same rating.

Fuses are available in two types: the *quick-break fuse* and the *slow-blow fuse*. A quick-break fuse is a straight length of wire or a metal strip; a slow-blow fuse usually has a spring inside along with the wire or strip. Quick-break fuses in slow-blow situations might burn out needlessly; slow-blow units in quick-break environments might not provide adequate protection.

Circuit breakers

A *circuit breaker* performs the same function as a fuse, except that a circuit breaker can be reset by turning off the power supply, waiting a moment, and then pressing a button or flipping a switch. Some circuit breakers reset automatically when the equipment has been shut off for a certain length of time.

If a fuse or circuit breaker keeps blowing out or tripping, or if it blows or trips immediately after it has been replaced or reset, something is probably wrong with the power supply or with the equipment connected to it. Burned-out diodes, a bad transformer, and shorted filter capacitors in the supply can cause this. A short circuit in the equipment connected to the power supply, or the connection of a device in the wrong direction (polarity), can cause repeated fuse blowing or circuit-breaker tripping.

One should *never* replace a fuse or breaker with a larger-capacity unit to overcome the inconvenience of repeated fuse or breaker blowing or tripping. Find the cause of the trouble and repair the equipment as needed. The "penny in the fuse box" scheme is no solution. It can result in more serious equipment damage and might cause an electrical fire.

Electrochemical Power Sources

A *cell* is an electrochemical unit source of dc power. When two or more cells are connected in series, the result is a *battery*. Cells and batteries are extensively used in portable electronic equipment, in communications satellites, and as sources of emergency power.

Electrochemical energy

Figure 6.14 shows an example of a *lead-acid cell*. An electrode of lead and an electrode of lead dioxide, immersed in a sulfuric-acid solution, show a potential difference between them. This voltage can drive a current through a load. The maximum available current depends on the volume and mass of the cell. The voltage does not depend on the volume or mass.

If this cell is connected to a load for a long time, the current will gradually decrease, and the electrodes will become coated. The nature of the acid will change. All the potential energy in the acid will be converted into dc electrical energy and ultimately into heat, visible light, radio waves, sound, mechanical motion, and so on.

Primary and secondary cells

Some cells, once all their chemical energy has been changed to electricity and used up, must be discarded. These are *primary cells*. Other kinds of cells, like the lead-acid unit described above, can get their chemical energy back again by recharging. Such a component is a *secondary cell*.

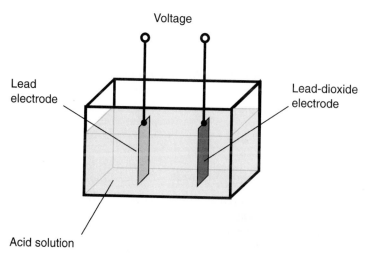

Figure 6.14 A lead-acid cell.

Primary cells contain a dry electrolyte paste along with metal electrodes. They go by names such as *dry cell, zinc-carbon cell,* and *alkaline cell.* They are commonly found in supermarkets and department stores. Some secondary cells can also be found in stores. *Nickel-cadmium (Ni-Cd or NICAD) cells* are one common type. These cost more than ordinary dry cells, and a charging unit also costs a few dollars. But these rechargeable cells can be used hundreds of times and can pay for themselves and the charger several times over.

An *automotive battery* is made from secondary cells connected in series. These cells recharge from the alternator or from an outside charging unit. This battery has cells like the one shown in Fig. 6.14. It is dangerous to short-circuit the terminals of such a battery because the acid can boil out. In fact, it is unwise to short-circuit any cell or battery, because it might explode or cause a fire.

Standard cell

Most cells produce between 1.0 and 1.8 V dc. Some types of cells generate predictable and precise voltages. These are known as *standard cells.* One example is the *Weston cell.* It produces 1.018 V at room temperature. This cell uses a solution of cadmium sulfate. The positive electrode is mercury sulfate, and the negative electrode is mercury and cadmium (Fig. 6.15).

When a Weston cell is properly constructed and used at room temperature, its voltage is always the same. This allows it to be employed as a dc voltage standard.

Storage capacity

Common units of electrical energy are the *watthour* (Wh) and the *kilowatthour* (kWh). Any cell or battery has a certain amount of electrical energy that can be specified in watthours or kilowatthours. Often it is given in terms of the mathematical integral of deliverable current with respect to time, in units of *ampere-hours* (Ah). The energy capacity in watthours is the ampere-hour capacity multiplied by the battery voltage.

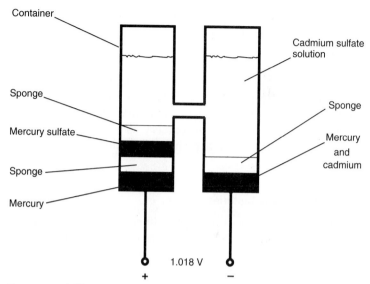

Figure 6.15 A Weston standard cell.

A battery with a rating of 20 Ah can provide 20 A for 1 hour (h), or 1 A for 20 h, or 100 mA for 200 h. The limitations are *shelf life* at one extreme and *maximum deliverable current* at the other. Shelf life is the length of time the battery will remain usable if it is never connected to a load; this might be months or years. The maximum deliverable current is the highest current a battery can drive through a load without the voltage dropping significantly because of the battery's own internal resistance.

Small cells have storage capacity of a few milliampere-hours (mAh) up to 100 or 200 mAh. Medium-sized cells might supply 500 to 1000 mAh (1 Ah). Large automotive lead-acid batteries can provide upward of 100 Ah.

Discharge curves

When an *ideal cell* or *ideal battery* is used, it delivers a constant current for a while and then the current starts to decrease. Some types of cells and batteries approach this ideal behavior, exhibiting a *flat discharge curve* (Fig. 6.16A).

Others have current that decreases gradually from the beginning of use; this is known as a *declining discharge curve* (Fig. 6.16B).

When the current that a battery can provide has decreased to about half of its initial value, the cell or battery is said to be "weak" or "low." At this time, it should be replaced. If it is allowed to run down until the current drops to nearly zero, the cell or battery is said to be "dead" (although, in the case of a rechargeable unit, a better term might be "sleeping").

Common cells and batteries

The cells sold in stores, and used in convenience items like flashlights and transistor radios, are usually of the zinc-carbon or alkaline variety. These provide 1.5 V and are available in sizes AAA (very small), AA (small), C (medium), and D (large). Batteries made from these cells are usually rated at 6 or 9 V.

Zinc-carbon cells have a fairly long shelf life. The zinc forms the outer case and is the negative electrode. A carbon rod serves as the positive electrode. The electrolyte is a paste of manganese dioxide and carbon. Zinc-carbon cells are inexpensive and are usable at moderate temperatures and in applications where the current drain is moderate to high. They do not work well in extremely cold environments.

Alkaline cells have granular zinc for the negative electrode, potassium hydroxide as the electrolyte, and a polarizer as the positive electrode. An alkaline cell can work at lower temperatures than a zinc-carbon cell. It also lasts longer in most electronic devices and is therefore preferred for use in transistor radios, calculators, and portable cassette players. Its shelf life is much longer than that of a zinc-carbon cell.

Transistor batteries are small, 9-V, box-shaped batteries with clip-on connectors on top. They consist of six tiny zinc-carbon or alkaline cells in series. Each of the cells supplies 1.5 V. These batteries are used in low-current electronic devices, such as portable earphone radios, radio garage-door openers, television and stereo remote-control boxes, and electronic calculators.

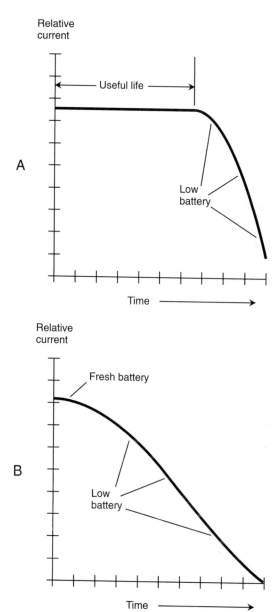

Figure 6.16 (A) Flat discharge curve; (B) declining discharge curve.

Lantern batteries are rather massive and can deliver a fair amount of current. One type has spring contacts on the top. The other type has thumbscrew terminals. Besides keeping an incandescent bulb lit for a while, these batteries, usually rated at 6 V and consisting of four zinc-carbon or alkaline cells, can provide enough energy to operate a low-power communications radio.

Silver-oxide cells are usually made into a button-like shape and can fit inside a wristwatch. They come in various sizes and thicknesses, all with similar appearance. They supply 1.5 V and offer excellent energy storage for the weight. They have a flat discharge curve. Silver-oxide cells can be stacked to make batteries about the size of an AAA cylindrical cell.

Mercury cells, also called *mercuric-oxide cells,* have advantages similar to silver-oxide cells. They are manufactured in the same general form. The main difference, often not of significance, is a somewhat lower voltage per cell: 1.35 V. There has been a decrease in the popularity of mercury cells and batteries in recent years because mercury is toxic and is not easily disposed of.

Lithium cells supply 1.5 to 3.5 V, depending on the chemistry used. These cells, like their silver-oxide cousins, can be stacked to make batteries. Lithium cells and batteries have superior shelf life, and they can last for years in very-low-current applications. They provide excellent energy capacity per unit volume.

Lead-acid cells and batteries have a solution or paste of sulfuric acid, along with a lead electrode (negative) and a lead-dioxide electrode (positive). Paste-type lead-acid batteries can be used in consumer devices that require moderate current, such as laptop computers and portable VCRs. They are also used in uninterruptible power supplies.

Nickel-based cells and batteries

NICAD cells are made in several types. *Cylindrical cells* look like dry cells. *Button cells* are used in cameras, watches, memory backup applications, and other places where miniaturization is important. *Flooded cells* are used in heavy-duty applications and can have a storage capacity of as much as

1000 Ah. *Spacecraft cells* are made in packages that can withstand extraterrestrial temperatures and pressures.

NICAD batteries are available in packs of cells that can be plugged into equipment to form part of the case for a device. An example is the battery pack for a handheld radio transceiver.

NICAD cells and batteries should never be left connected to a load after the current drops to zero. This can cause the polarity of a cell, or of one or more cells in a battery, to reverse. Once this happens, the cell or battery will no longer be usable. When a NICAD is "dead," it should be recharged as soon as possible.

Nickel-metal-hydride (NiMH) cells and batteries can directly replace NICAD units in most applications.

Specialized Power Systems

Various electronic equipment uses specialized power sources, including *power inverters, uninterruptible power supplies,* and *solar-electric energy systems.*

Power inverter

A power inverter, sometimes called a *chopper power supply,* is a circuit that delivers high-voltage ac from a dc source. The input is typically 12 V dc, and the output is usually 117 V rms ac.

A simplified block diagram of a power inverter is shown in Fig. 6.17. The *chopper* consists of a low-frequency oscillator that opens and closes a high-current switching transistor. This interrupts the battery current, producing pulsating dc. The transformer converts the pulsating dc to ac and also steps up the voltage.

The output of a low-cost power inverter might not be a sine wave but instead might resemble a sawtooth or square wave. The frequency might also be considerably higher or lower than the standard 60 Hz. More sophisticated inverters produce fairly good sine-wave output and have a frequency close to 60 Hz. Although such inverters are expensive, they are a good investment if they are to be used with sensitive equipment such as computers.

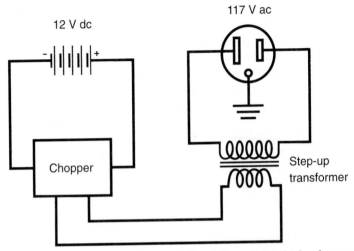

Figure 6.17   A power inverter converts low-voltage dc into high voltage ac.

If the transformer shown in the block diagram (Fig. 6.17) is followed by a rectifier and filter, the device becomes a *dc transformer*, also called a *dc-to-dc converter*. Such a circuit can provide hundreds of volts dc from a 12-V battery source.

Uninterruptible power supplies

When a piece of electronic equipment is operated from utility power, there is a possibility of a system failure resulting from a power blackout, brownout, or dip. If the power to a computer fails, all data in the random-access memory (RAM) will be lost. To prevent this, an *uninterruptible power supply* (UPS) can be used.

Figure 6.18 is a block diagram of a UPS. Under normal conditions, the equipment gets its power via the transformer and regulator. The regulator eliminates transients, surges, and dips in the utility current. A lead-acid battery is kept charged by a small current via the rectifier and filter. If the power goes out, a power-interrupt signal causes the switch to disconnect the equipment from the regulator and connect it to the power inverter, which converts the battery dc output to ac. When utility power returns, the switch disconnects the equipment from the battery and reconnects it to the regulator.

If power to a computer fails and you have a UPS, save all your work immediately on the hard disk and also on an external medium such as a diskette if possible. Then switch the entire system, including the UPS, off until utility power returns.

Solar-electric energy systems

Electric energy can be obtained directly from sunlight by means of *solar cells.* Most solar cells produce about 0.1 W of

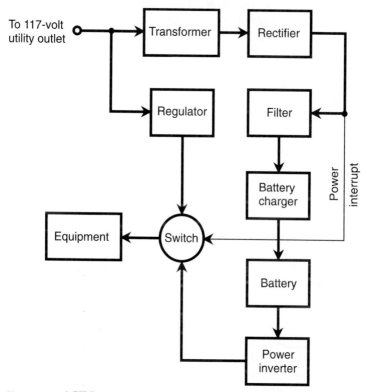

Figure 6.18 A UPS prevents system failure in case of utility power interruptions or irregularities.

power for each square inch of surface area exposed to bright sunlight. Solar cells produce dc, whereas most household appliances require ac at 117 V and 60 Hz. Most solar-electric energy systems intended for general home and business use must therefore employ power inverters.

In some situations, power inverters are not necessary. For example, if you plan to operate a notebook computer from a battery and to keep the battery charged using a solar panel, you will not need a power inverter. But the charging circuit must be carefully designed so that the battery is not damaged during the charging process.

There are two types of solar electric energy systems capable of powering homes and small businesses: the *stand-alone system* and the *interactive system.*

A stand-alone system uses banks of rechargeable batteries, such as the lead-acid type, to store electric energy as it is supplied by photovoltaics during hours of bright sunshine. The energy is released by the batteries at night or in gloomy daytime weather. This system is independent of the electric utility, or *power grid.*

An interactive system is connected into the power grid. This type of system does not normally use storage batteries. Any excess energy is sold to utility companies during times of daylight and minimum usage. Energy is bought from the utility at night, during gloomy daytime weather, or during times of heavy usage.

Personal Safety

Power supplies can be dangerous. This is especially true of high-voltage circuits, but anything over 12 V should be treated as potentially lethal.

Most manufacturers supply safety instructions and precautions with equipment carrying hazardous voltages. But do not assume something is safe merely because dangers are not mentioned in the instructions. *If you have any doubt about your ability to safely work with a power supply, leave it to a professional.*

7

Diodes

Most diodes are manufactured with semiconductor materials. There are two principal categories of semiconductors: *N type,* in which the charge carriers are mainly electrons, and *P type,* in which the charge carriers are primarily holes (atoms having electron shortages).

P-N Junction

When wafers of *N*- and *P*-type semiconductor material are in direct physical contact, the result is a *P-N junction* with certain properties. Figure 7.1 shows the schematic symbol for a semiconductor diode. The *N*-type material is represented by the short, straight line in the symbol and is the *cathode.* The *P*-type material is represented by the arrow and is the *anode.*

In a diode, electrons normally flow in the direction opposite the arrow, not in the direction that the arrow points. If a battery and a resistor are connected in series with the diode, current will flow if the negative terminal of the battery is connected to the cathode and the positive terminal is connected to the anode (Fig. 7.2A). Virtually no current will flow if the battery is reversed (Fig. 7.2B).

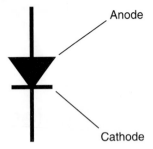

Figure 7.1 Anode and cathode symbology for a diode.

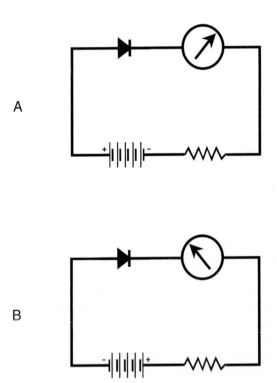

Figure 7.2 (A) Forward bias of a diode results in a flow of current; (B) reverse bias normally produces near zero current.

Forward breakover

It takes a certain minimum voltage for conduction to occur. This is called *forward breakover.* Depending on the type of material, it varies from about 0.3 to 1 V. If the voltage across the junction is not at least as great as the forward breakover value, the diode will not conduct. This effect can be of use in amplitude limiters, waveform clippers, and threshold detectors.

Bias

When the *N*-type material is negative with respect to the *P*-type, electrons flow easily from *N* to *P*. This is *forward bias.* The diode conducts well.

When the polarity is switched so the *N*-type material is positive with respect to the *P* type, it is *reverse bias.* The diode conducts poorly. Electrons in the *N*-type material are pulled toward the positive charge, away from the junction. In the *P*-type material, holes are pulled toward the negative charge, also away from the junction. The electrons (in the *N*-type material) and holes (in the *P* type) become depleted in the vicinity of the junction. This impedes conduction, and the resulting *depletion region* behaves as a dielectric or electrical insulator.

Junction capacitance

Under conditions of reverse bias, a *P-N* junction can act as a capacitor. The *varactor diode* is made with this property specifically in mind. The *junction capacitance* can be varied by changing the reverse-bias voltage, because this voltage affects the width of the depletion region. The greater the reverse voltage, the wider the depletion region gets, and the smaller the capacitance becomes.

Avalanche

If a diode is reverse biased, and the voltage becomes high enough, the *P-N* junction will conduct. This is known as *avalanche effect.* The reverse current, which is near zero at lower voltages, rises dramatically. The *avalanche voltage* varies among different kinds of diodes. Figure 7.3 is a graph of the characteristic current-versus-voltage curve for a typical semiconductor diode, showing the *avalanche point.* The avalanche voltage is considerably greater than, and is of the opposite polarity from, the forward breakover voltage.

Avalanche effect is undesirable in rectifiers because it degrades the efficiency of a power supply. The phenomenon can be prevented by choosing diodes with an avalanche voltage higher than the PIV produced by the power-supply circuit. For extremely high-voltage power supplies, it is necessary to use series combinations of diodes to prevent the avalanche effect.

A *zener diode* makes use of avalanche effect. Zener diodes are specially manufactured to have precise avalanche voltages. They form the basis for voltage regulation in many power supplies.

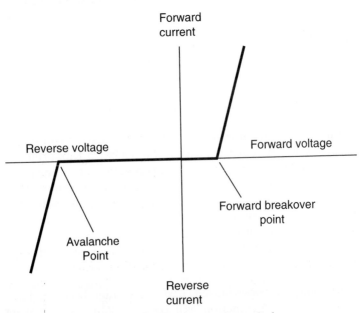

Figure 7.3 Characteristic curve for a semiconductor diode.

Power-Supply Applications

In power supplies, diodes are used for rectification, voltage regulation, switching, and certain control functions.

Rectification

A *rectifier diode* passes current in only one direction under normal operating conditions. This makes it useful for changing ac to dc.

Generally speaking, when the cathode is negative with respect to the anode, current flows; when the cathode is positive relative to the anode, current does not flow. The constraints on this behavior are the forward breakover and avalanche voltages. During half the cycle, the diode conducts, and during the other half, it does not. This cuts off half of every cycle. Depending on which way the diode is connected in the circuit, either the positive half or the negative half of the ac cycle is removed.

Common rectifier circuits are discussed in Chap. 6.

Regulation

Most diodes have avalanche voltages that are much higher than the reverse bias ever gets. The value of the avalanche voltage depends on how a diode is manufactured. *Zener diodes* are made to have well-defined, constant avalanche voltages.

Suppose a certain zener diode has an avalanche voltage, also called the *zener voltage,* of 50 V. If a reverse bias is applied to the *P-N* junction, the diode acts as an open circuit below 50 V. When the voltage reaches 50 V, the diode starts to conduct. The more the reverse bias tries to increase, the more current flows through the *P-N* junction. This effectively prevents the reverse voltage from exceeding 50 V.

There are other ways to get voltage regulation besides the use of zener diodes, but zener diodes often provide the simplest and least-expensive alternative. Zener diodes are available with a wide variety of voltage and power-handling ratings. Power supplies for solid-state equipment commonly employ zener diode regulators. Refer to Chap. 6 for a discussion of power supplies.

Transient protection

A *thyrector* is a semiconductor device consisting essentially of two diodes connected in reverse series. The thyrector is used for protection of equipment against the effects of *transients* in the voltage supply.

Thyrectors are connected between the ac power terminals of a piece of electronic apparatus and electrical ground (Fig. 7.4). The devices do not conduct until the voltage reaches a certain value. If the voltage across a thyrector exceeds the critical level, even for a tiny fraction of a second, the device conducts. This in effect clips the transient, greatly reducing its peak voltage and eliminating its adverse effect on the equipment. The conducting voltage does not depend on the polarity.

Thyrectors are effective against most, but not all, transients. A severe spike, such as might be induced by the *electromagnetic pulse* (EMP) from a nearby lightning strike, might destroy a thyrector and the equipment with which it is used. In thunderstorms, the best way to protect equipment is to physically unplug its power supply from the wall outlets.

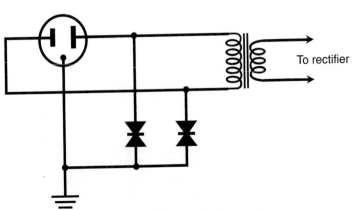

Figure 7.4 Thyrectors can be used for transient suppression.

Signal Applications

Diodes are used in various types of audio-frequency (AF) and RF circuits. The following sections outline some common applications.

Envelope detection

One of the earliest diodes, existing even before vacuum tubes, was a semiconductor. Known as a *cat whisker,* it consisted of a fine piece of wire in contact with a small piece of the mineral substance *galena.* It acted as a rectifier for small RF currents. When the cat whisker was connected in a circuit like that of Fig. 7.5, the result was a *crystal radio receiver* capable of picking up amplitude-modulated (AM) radio signals. Using an RF diode, this circuit can still be used today.

The diode acts to recover the audio from the radio signal. This is *simple envelope detection.* If the detector is to be effective, the diode must be of the right type. It should have low capacitance so that it works as a rectifier at RF, passing current in one direction but not in the other. Some modern RF diodes are actually microscopic versions of the old cat whisker, enclosed in a glass case with axial leads. These are known as *point-contact diodes* and are designed to minimize the junction capacitance.

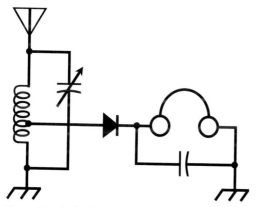

Figure 7.5 A diode as an envelope detector in a crystal radio receiver.

Electronic switching

The ability of diodes to conduct with forward bias, and to insulate with reverse bias, makes them useful for *electronic switching* in some applications. Diodes can switch at extremely high rates, much faster than any mechanical device.

One type of diode, made for use as an RF switch, has a special semiconductor layer sandwiched in between the *P*- and *N*-type material. This layer, called an *intrinsic semiconductor,* reduces the capacitance of the diode so that it can work at higher frequencies than an ordinary diode. The intrinsic material is called *I-type* material. A diode with an *I*-type semiconductor is called a *PIN diode* (Fig. 7.6).

Direct-current bias, applied to one or more PIN diodes, allows RF currents to be effectively channeled without using complicated relays and cables. A PIN diode also makes a good RF detector, especially at frequencies above approximately 30 MHz.

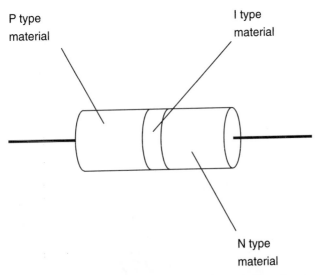

Figure 7.6 A PIN diode has a thin layer of intrinsic (*I*-type) material between the *P* and *N* type semiconductors.

Frequency multiplication

When current passes through a diode, half of the cycle is cut off. The output wave is much different from the input wave. This produces *nonlinearity* that results in an output rich in *harmonics,* and is ideal for frequency multiplication.

A simple frequency-multiplier circuit is shown in Fig. 7.7. The output inductance-capacitance (LC) circuit is tuned to the desired nth harmonic frequency, nf_0, rather than to the input or fundamental frequency, f_0.

For a diode to work as a frequency multiplier, it must be of a type that would also work well as a detector at the same frequencies. This means that the component should behave as a rectifier and not as a capacitor. Point-contact and PIN diodes are well designed for this purpose. Power-supply rectifier diodes, in general, are not suitable for use as RF multipliers because the junction capacitance is too high.

Step-recovery diodes

A *step-recovery diode,* also called a *charge-storage* or *snap diode,* is a special form of semiconductor device that is used chiefly as a harmonic generator. When a signal is passed through a step-recovery diode, dozens of harmonics occur at the output. When this type of device is connected in the output circuit of a radio transmitter, it can be used in conjunction with a resonator to obtain signals at frequencies up to

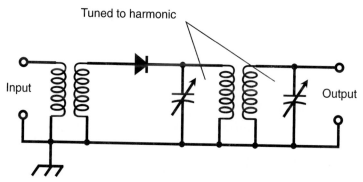

Figure 7.7 A simple diode frequency-multiplier circuit.

several gigahertz. Step-recovery diodes are also used to square off the rise and decay characteristics of digital pulses.

A step-recovery diode is manufactured in a manner similar to the PIN diode. Charge storage is accomplished by means of a very thin layer of intrinsic semiconductor material between the *P*- and *N*-type wafers. The charge carriers accumulate very close to the junction.

When the step-recovery diode is forward-biased, current flows in the same manner as it would in any forward-biased *P-N* junction. However, when reverse bias is applied, the conduction does not immediately stop. The step-recovery diode stores a large number of charge carriers while it is forward-biased, and it takes a short time for these charge carriers to be drained off. Current flows in the reverse direction until the charge carriers have been removed from the *P-N* junction, and then the current diminishes extremely rapidly. The transition time between maximum and zero current can be less than 0.1 nanosecond (ns). This rapid change in current is responsible for the harmonic output.

Hot-carrier diodes

The *hot-carrier diode* (HCD) is used in various circuits, especially mixers and detectors at VHF and UHF. The HCD will work at much higher frequencies than most other semiconductor diodes.

The HCD generates relatively little noise and exhibits a high breakdown or avalanche voltage. The reverse current is minimal. The internal capacitance is low because the diode consists of a tiny point contact (Fig. 7.8). The wire is gold-plated to minimize corrosion. An *N*-type silicon wafer is used as the semiconductor.

Single balanced mixer

A *single balanced mixer* is a mixer circuit that is easily built for a minimum amount of expense. The input and output ports are not completely isolated; some of either input signal leaks through to the output. If isolation is needed, a *double balanced mixer* is preferable.

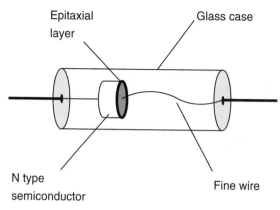

Figure 7.8 Construction of a hot-carrier diode.

A typical single-balanced-mixer circuit is shown in Fig. 7.9. The diodes are generally of the hot-carrier type. This circuit will work at frequencies up to several gigahertz. The circuit shown is a passive circuit, and therefore some loss will occur. However, the loss can be overcome by an amplifier following the mixer.

Double balanced mixer

The input energy in a double balanced mixer (Fig. 7.10) does not leak through to the output. All of the output and input ports are completely isolated, so the coupling between the input oscillators is negligible, and the coupling between the inputs and the output is also negligible.

As in the single balanced mixer, hot-carrier diodes are used in this circuit. They can handle large signal amplitudes without distortion, and they generate very little noise. This circuit is passive rather than active, so it has about a 6-decibel (dB) conversion loss.

Amplitude limiting

The forward breakover voltage of a germanium diode is about 0.3 V; for a silicon diode it is about 0.6 V. A diode will not conduct until the forward bias voltage is at least as great

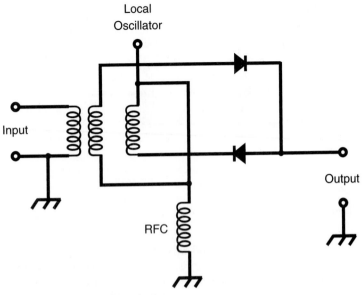

Figure 7.9 A single-balanced mixer.

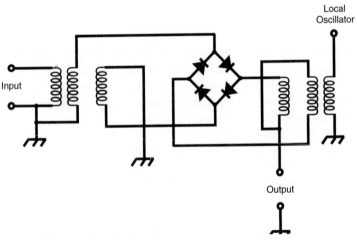

Figure 7.10 A double-balanced mixer.

as the forward breakover voltage. The corollary is that the diode will always conduct when the forward bias exceeds the breakover value, and the voltage across the diode will be constant: 0.3 V for germanium and 0.6 V for silicon.

This property can be used to advantage when it is necessary to limit the amplitude of a signal, as shown in Fig. 7.11. By connecting two identical diodes in reverse parallel with the signal path (Fig. 7.11A), the maximum peak amplitude is limited, or clipped, to the forward breakover voltage of the diodes. The input and output waveforms of a clipped signal are illustrated in Fig. 7.11B. This scheme is sometimes used in the audio stages of communications receivers to prevent "blasting" when a strong signal comes in.

The chief disadvantage of the diode limiter circuit is that it introduces distortion when limiting is taking place. This might not be a problem with signals that are two-tone analog renditions of digital data (for example, radioteletype) or with analog voice or video signals that rarely reach the limiting voltage. But for analog voice or video signals with amplitude peaks that rise well past the limiting voltage, it can seriously degrade reception.

Noise limiting

In a radio receiver, a *noise limiter* can consist of a pair of clipping diodes with variable bias for control of the clipping level (Fig. 7.12). The bias is adjusted until clipping occurs at the signal amplitude. Noise pulses then cannot exceed the signal amplitude. This makes it possible to receive a signal that would otherwise be drowned out by the noise.

A noise limiter is typically installed between two intermediate-frequency (IF) stages of a superheterodyne receiver. In a direct-conversion receiver, the best place for the noise limiter is immediately prior to the detector stage.

Frequency control

When a diode is reverse-biased, there is a *depletion region* at the P-N junction with dielectric properties. The width of

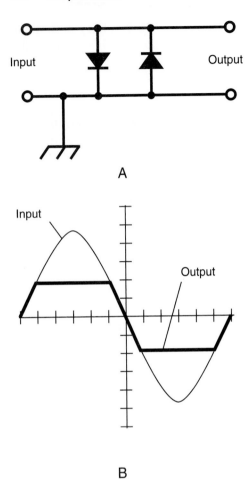

Figure 7.11   (A) A simple diode limiter; (B) clipping of a signal passing through the limiter.

this zone depends on several factors, including the reverse-bias voltage.

As long as the reverse bias is less than the avalanche voltage, the width of the depletion region can be changed by varying the bias. This causes a change in the capacitance of the junction. The capacitance, which is on the order of a few picofarads, varies inversely with the square root of the reverse bias.

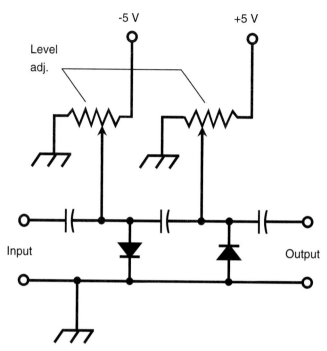

Figure 7.12 A variable-threshold noise-limiter circuit.

Some diodes are manufactured especially for use as variable capacitors. These are *varactor diodes,* also called *varactors* or *varicaps.* They are made from silicon or gallium arsenide.

A common use for a varactor diode is in a *voltage-controlled oscillator* (VCO). A parallel voltage-tuned *LC* circuit using a varactor is shown in Fig. 7.13. The fixed capacitors, whose values are large compared with that of the varactor, prevent the coil from short-circuiting the control voltage across the varactor. Notice that the symbol for the varactor has two lines, rather than one, on the cathode side.

Oscillation and Amplification

Under some conditions, diodes can generate or amplify microwave radio signals. Certain diodes are specifically manufactured for this purpose.

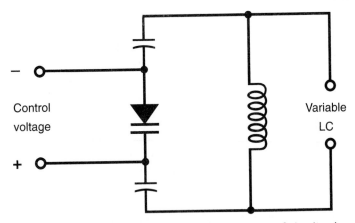

Figure 7.13 A variable inductance-capacitance (*LC*) tuned circuit using a varactor.

Gunn diodes

A *Gunn diode* can produce up to 1 W of RF power output, but more commonly it works at levels of about 0.1 W. Gunn diodes are usually made from gallium arsenide (GaAs). The device oscillates because of the so-called Gunn effect, named after J. Gunn of International Business Machines (IBM), who observed it in the 1960s. A Gunn diode does not function like a rectifier, detector, modulator, mixer, or clipper. Instead, oscillation takes place as a result of a property called *negative resistance*. In a certain portion of the characteristic curve, the current decreases as the voltage increases.

Gunn-diode oscillators are often tuned using varactor diodes. A Gunn-diode oscillator, connected directly to a microwave horn antenna, is known as a *gunnplexer*. These devices are popular with amateur-radio experimenters.

IMPATT diodes

The acronym *IMPATT* (pronounced IM-pat) comes from the words *impact avalanche transit time*. This effect is similar to negative resistance. An IMPATT diode is a microwave oscillating device like a Gunn diode, except that it uses silicon rather than gallium arsenide.

An IMPATT diode can be used as an amplifier for a microwave transmitter that employs a Gunn-diode oscillator. As an oscillator, an IMPATT diode produces about the same amount of output power, at comparable frequencies, as a Gunn diode.

Tunnel diodes

Another type of diode that will oscillate at microwave frequencies is the *tunnel diode,* also known as the *Esaki diode.* It produces only a very small amount of power, but it can be used as a local oscillator in a microwave radio receiver.

Tunnel diodes work well as amplifiers in microwave receivers because they generate very little noise. This is especially true of gallium-arsenide devices.

Photoemission

Some semiconductor diodes emit radiant energy when a current passes through the *P-N* junction in a forward direction. This phenomenon, called *photoemission,* occurs as electrons fall from higher to lower energy states within atoms.

LEDs and IREDs

Depending on the exact mixture of semiconductors used in manufacture, visible light of almost any color can be produced. Infrared-emitting devices also exist. The most common color for a *light-emitting diode* (LED) is bright red. An *infrared-emitting diode* (IRED) produces wavelengths too long to see.

The intensity of energy emission from an LED or IRED depends to some extent on the forward current. As the current rises, the brightness increases up to a certain point. If the current continues to rise, no further increase in brilliance takes place. The LED or IRED is then in *saturation.*

Injection lasers

The *injection laser,* also called a *laser diode,* is a form of LED or IRED with a relatively large and flat *P-N* junction. The injection laser emits coherent electromagnetic waves, provided the applied current is sufficient. If the current is below a certain level, the injection laser behaves like an ordinary LED or IRED, but when the threshold current is reached, the charge carriers recombine in such a manner that *lasing* occurs.

Most injection-laser devices are fabricated from gallium arsenide (GaAs) and have a primary emission wavelength of about 905 nanometers (nm) in the IR range. Other types of injection lasers are available for different wavelengths, some in the visible-light range. The maximum peak power of pulses can be as great as 100 W, but the pulse duration is very short. Injection lasers produce emission at exceedingly narrow bandwidth. This is characteristic of laser devices. The construction of a typical injection-laser diode is shown in Fig. 7.14.

Injection lasers are used mainly in optical communications systems. The coherent light output from the injection laser suffers less attenuation than incoherent light as it passes through the atmosphere.

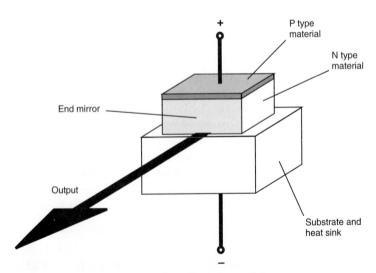

Figure 7.14 An injection laser, also called a laser diode.

Communications

Both LEDs and IREDs are useful in communications because their intensity can be modulated to carry information. When the current through the device is sufficient to produce output, but not enough to cause saturation, the LED or IRED output will follow along with rapid current changes. Voices, music, video, and digital signals can be conveyed over light beams in this way. Some modern telephone systems make use of modulated light, transmitted through clear fibers. This is known as *fiber-optic technology*.

Photosensitive Diodes

Virtually all *P-N* junctions exhibit characteristics that change when electromagnetic rays strike them. Conventional diodes are usually not affected by these rays because the *P-N* junctions are enclosed in opaque packages.

Some photosensitive diodes have variable resistance that depends on the illumination intensity. Others actually generate dc voltages in the presence of electromagnetic radiation.

Silicon photodiodes

A silicon diode, housed in a transparent case and constructed in such a way that visible light can strike the barrier between the *P-* and *N*-type materials, forms a *photodiode*. A reverse bias is applied to the device. When light falls on the junction, current flows. The current is proportional to the intensity of the light, within certain limits. Silicon photodiodes are more sensitive at some wavelengths than at others. The greatest sensitivity is in the near IR part of the spectrum, at wavelengths a little longer than visible red light.

When energy of varying intensity strikes the *P-N* junction of a reverse-biased silicon photodiode, the output current follows the fluctuations. This makes silicon photodiodes useful for receiving modulated-light signals, of the kind used in fiber-optic systems.

Optoisolators

An LED or IRED and a photodiode can be combined in a single package to form an *optoisolator*, also known as an *optical coupler*. With an electrical-signal input, the LED or IRED generates a modulated-light beam and sends it over a small, clear gap to the photodiode, which converts the visible light or IR energy back into an electrical signal.

A traditional source of trouble for engineers has been the fact that when a signal is electrically coupled from one circuit to another, the impedances of the two stages interact. This can lead to nonlinearity, unwanted oscillation, loss of efficiency, or other problems. Optoisolators overcome this effect because the coupling is not done electrically. If the electrical input impedance of the second circuit changes, the impedance that the first circuit "sees" is not affected.

Photovoltaic cells

A silicon diode, with no bias voltage applied, will generate dc all by itself if sufficient electromagnetic energy strikes its *P-N* junction. This is known as the *photovoltaic effect* and is the principle by which solar cells work.

Photovoltaic (PV) cells have a large *P-N* junction surface area (Fig. 7.15). This maximizes the amount of visible and IR energy that falls on the junction. A single silicon PV cell produces approximately 0.6 V dc in direct sunlight with no load. The amount of current that it can deliver depends on the surface area of the junction. For every square inch of junction area, the cell can produce 150 to 200 mA in bright sunlight.

Silicon PV cells are connected in series-parallel combinations to provide solar power for solid-state electronic devices such as portable radios. A large assembly of such cells constitutes a *solar panel*. The dc voltages of the cells add when they are connected in series. A typical *solar battery* supplies 6, 9, or 12 V dc. The PV cells can be used to keep storage batteries charged. Power inverters can be used if household-level ac (117 V rms) is required. More information about solar-electric energy systems can be found in Chap. 6.

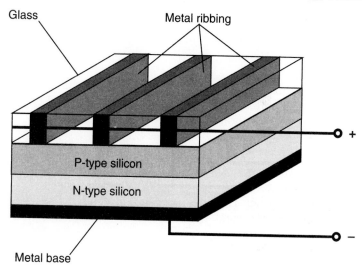

Figure 7.15 Construction of a PV cell.

8

Transistors and Integrated Circuits

The most common applications of discrete *transistors* are to generate signals, change weak or low-power signals into strong or high-power signals, mix signals, and act as electronic switches. *Integrated circuits* can perform myriad different functions. An IC contains many transistors, along with other components, on a single "chip" of semiconductor material.

The Bipolar Transistor

Bipolar transistors have two *P-N* junctions connected together. This is done in either of two ways: a *P*-type layer between two *N*-type layers or an *N*-type layer between two *P*-type layers.

NPN and PNP

A simplified drawing of an *NPN transistor* and its schematic symbol is shown in Fig. 8.1A and B. The *P*-type, or center, layer is the *base*. The thinner of the *N*-type semiconductors is the *emitter*, and the thicker one is the *collector*. Sometimes these are labeled B, E, and C in schematic diagrams, although the transistor symbol indicates which is which (the arrow is at the emitter). A *PNP transistor* (Fig. 8.1C and D) has two *P*-type layers, one on either side of a thin, *N*-type layer. In the *NPN* symbol, the arrow points outward. In the *PNP* symbol, the arrow points inward.

Generally, *PNP* and *NPN* transistors can perform identical tasks. The only difference is the polarities of the voltages and the directions of the currents. In most applications, an *NPN* device can be replaced with a *PNP* device or vice-versa, and with the power-supply polarity reversed, the circuit will still work if the new device has the appropriate specifications.

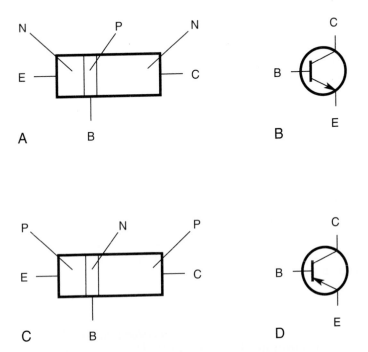

Figure 8.1 (A) Pictorial diagram of *NPN* transistor; (B) schematic symbol for *NPN* transistor; (C) pictorial diagram of *PNP* transistor; (D) schematic symbol for *PNP* transistor.

There are various kinds of bipolar transistors. Some are used for RF amplifiers and oscillators; others are intended for AF. Some can handle high power, and others are made for weak-signal work. Some are manufactured for switching, and others are intended for signal processing.

NPN biasing

The normal method of biasing an *NPN* transistor is to have the emitter negative and the collector positive. This is shown by the connection of the battery in Fig. 8.2. Typical voltages range from 3 to approximately 50 V.

The base is labeled "control" because the flow of current through the transistor depends on the base bias voltage, denoted E_B or V_B, relative to the emitter-collector bias voltage, denoted E_C or V_C.

Zero bias

When the base is not connected to anything, or when it is at the same potential as the emitter, a bipolar transistor is at *zero bias*. Under this condition, no appreciable current

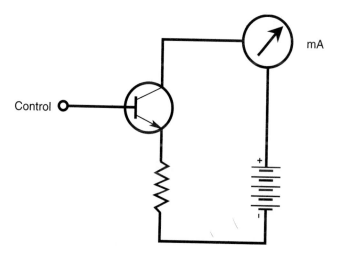

Figure 8.2   Typical biasing of an *NPN* transistor.

can flow through a *P-N* junction unless the forward bias is at least equal to the *forward breakover* voltage. For silicon, this is 0.6 V; for germanium it is 0.3 V.

With zero bias, the emitter-base current I_B is zero, and the emitter-base (*E-B*) junction does not conduct. This prevents current from flowing in the collector circuit unless a signal is injected at the base to change the situation. This signal must have a positive polarity, and its peaks must be sufficient to overcome the forward breakover of the *E-B* junction for at least a portion of the input signal cycle.

Reverse bias

Suppose another battery is connected to the base of the *NPN* transistor at the point marked "control," so the base is negative with respect to the emitter. The addition of this new battery will cause the *E-B* junction to be in a condition of *reverse bias*. It is assumed that this new battery is not of such a high voltage that *avalanche breakdown* takes place at the *E-B* junction.

A signal might be injected to overcome the reverse-bias battery and the forward-breakover voltage of the *E-B* junction, but such a signal must have positive voltage peaks high enough to cause conduction of the *E-B* junction for part of the input signal cycle. Otherwise the device will remain cut off for the entire cycle.

Forward bias

Suppose the bias at the base of an *NPN* transistor is positive relative to the emitter, starting at small levels and gradually increasing. This is *forward bias*. If this bias is less than forward breakover, no current flows. But as the situation reaches breakover, the *E-B* junction begins to conduct.

Despite reverse bias at the base-collector (*B-C*) junction, the emitter-collector current, more often called *collector current* and denoted I_C, flows when the *E-B* junction conducts. A small rise in the positive-polarity signal at the base, attended by a small rise in the base current I_B, will cause a large increase in I_C. This is the principle via which a bipolar transistor amplifies signals.

Saturation

If I_B continues to rise, a point will eventually be reached where I_C increases less rapidly. Ultimately, the I_C versus I_B function, or *characteristic curve,* will level off. The graph in Fig. 8.3 shows a family of characteristic curves for a hypothetical bipolar transistor. The actual current values depend on the particular type of device; they are larger for power transistors and smaller for weak-signal transistors. Where the curves level off, the transistor is in a state of *saturation*. Under these conditions it is not as effective as a weak-signal amplifier.

Bipolar transistors are not normally operated in the saturated state in analog circuits, although they sometimes are biased to that extent in digital switching systems.

PNP biasing

For a *PNP* transistor, the bias situation is a mirror image of the case for an *NPN* device, as shown in Fig. 8.4. The power-supply polarities are reversed. To overcome forward breakover at the emitter-base junction, an applied signal must have sufficient negative polarity.

Either the *PNP* or the *NPN* device can serve as a "current valve." Small changes in the base current I_B induce large fluctuations in the collector current I_C when the device is operated in that region of the characteristic curve where the slope is steep. Although the internal atomic activity is different in the *PNP* device as compared with the *NPN,* the performance of the external circuitry is, in most situations, identical for practical purposes.

Current Amplification

Because a small change in I_B results in a large I_C variation when the bias is correct, a transistor can operate as a *current amplifier.*

Relative
collector
current
I_C

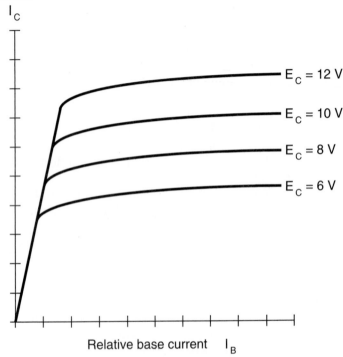

Relative base current I_B

Figure 8.3 A family of characteristic curves for a hypothetical bipolar transistor.

Static current amplification

The maximum obtainable current amplification factor of a bipolar transistor is known as *beta* and can range from a factor of just a few times up to hundreds of times. One method of expressing the beta of a transistor is as the *static forward current transfer ratio,* abbreviated H_{FE}. Mathematically,

$$H_{FE} = \frac{I_C}{I_B}$$

Thus, if a base current I_B of 1 mA results in a collector cur-

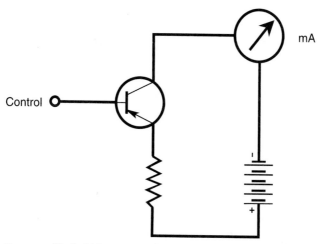

Figure 8.4 Typical biasing of a *PNP* transistor.

rent I_C of 35 mA, H_{FE} = 35/1 = 35. If I_B = 0.5 mA and I_C = 35 mA, H_{FE} = 35/0.5 = 70.

Dynamic current amplification

Another way of specifying current amplification is as the ratio of a difference in I_C to a small, incremental difference in I_B that produces it. This is the *dynamic current amplification,* also known as *current gain.* Abbreviate the words *the difference in* by the Greek letter Δ (delta). Then, according to this second definition,

$$\text{Current gain} = \frac{\Delta I_C}{\Delta I_B}$$

The ratio $\Delta I_C/\Delta I_B$ is greatest where the slope of the characteristic curve is steepest. Geometrically, $\Delta I_C/\Delta I_B$ at a given point is the slope of a line tangent to the curve at that point.

When the operating point is on the steep part of the characteristic curve, a transistor provides the largest possible $\Delta I_C/\Delta I_B$, as long as the input signal is small. This value is close to H_{FE}. Because the characteristic curve is a straight line in this region, the transistor can serve as a *linear amplifier* if the input signal is not too strong.

As the operating point is shifted into the nonlinear part of the characteristic curve, the current gain decreases and the amplifier becomes nonlinear. The same thing can happen if the input signal is strong enough to drive the transistor into the nonlinear part of the curve during any portion of the signal cycle.

Gain versus frequency

In a bipolar transistor, the gain decreases as the signal frequency increases. There are two expressions for gain-versus-frequency behavior.

The *gain bandwidth product,* abbreviated f_T, is the frequency at which the current gain becomes equal to unity (1) with the emitter connected to ground. The *alpha cutoff* is the frequency at which the gain becomes 0.707 times its value at 1 kHz. A transistor might have gain at frequencies above its alpha cutoff, but it cannot produce gain at frequencies higher than f_T.

Basic Bipolar Transistor Circuits

A transistor can be connected in a circuit in three general ways. The emitter, base, or collector can be grounded for signal.

Common emitter

Probably the most often-used arrangement is the *common-emitter circuit.* The basic configuration is shown in Fig. 8.5.

Capacitor C_1 presents a short circuit to the ac signal, so the emitter is at signal ground. But R_1 causes the emitter to have a certain positive dc voltage with respect to ground (or a negative voltage, if a *PNP* transistor is used). The exact dc voltage at the emitter depends on the value of R_1 and on the bias. The bias is set by the ratio of resistances R_2 and R_3. It can be anything from zero, or ground potential, to the power-supply voltage. Normally it is a couple of volts.

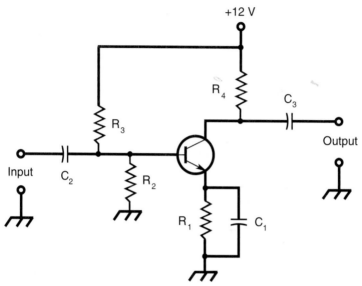

Figure 8.5 Common-emitter circuit configuration.

Capacitors C_2 and C_3 block dc to or from the input and output circuitry while letting the ac signal pass. Resistor R_4 keeps the output signal from being shorted out through the power supply.

A signal voltage enters the common-emitter circuit through C_2, where it causes base current I_B to vary. The small fluctuations in I_B cause large changes in collector current I_C. This current passes through R_4, causing a fluctuating dc voltage to appear across this resistor. The ac part of this passes through C_3 to the output.

The common-emitter configuration produces the largest gain of any arrangement. The output is 180° out of phase with the input.

Common base

As its name implies, the *common-base circuit* (Fig. 8.6) places the base at signal ground. The dc bias on the transistor is the same for this circuit as for the common-emitter circuit.

The input signal is applied at the emitter. This causes fluctuations in the voltage across R_1, causing variations in I_B. The result of these small current fluctuations is a large change in the current through R_4. Therefore, amplification occurs. In the common-base arrangement, the output signal is in phase with the input.

The signal enters through C_1. Resistor R_1 keeps the input signal from being shorted to ground. Bias is provided by R_2 and R_3. Capacitor C_2 keeps the base at signal ground. Resistor R_4 keeps the signal from being shorted out through the power supply. The output is through C_3.

A common-base circuit provides less gain than a common-emitter circuit but is less prone to oscillate.

Common collector

A *common-collector circuit* (Fig. 8.7) operates with the collector at signal ground. The input is applied at the base.

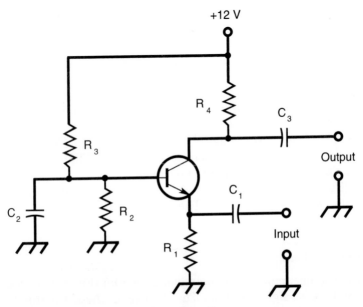

Figure 8.6 Common-base circuit configuration.

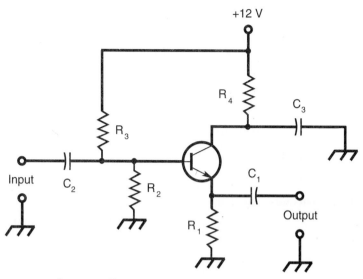

Figure 8.7 Common-collector circuit configuration.

The signal passes through C_2 onto the base of the transistor. Resistors R_2 and R_3 provide the bias. Resistor R_4 limits the current through the transistor. Capacitor C_3 keeps the collector at signal ground. A fluctuating current flows through R_1, and a fluctuating voltage therefore appears across it. The ac component passes through C_1 to the output. Because the output follows the emitter current, this circuit is sometimes called an *emitter follower.*

The output of this circuit is in phase with the input. The input impedance is high, and the output impedance is low. When well designed, an emitter follower works over a wide range of frequencies and is a low-cost alternative to a broadband impedance-matching transformer.

The Field-Effect Transistor

The other major category of transistor, besides the bipolar device, is the *field-effect transistor,* or FET. There are two main types of FET: the *junction FET* (JFET) and the *metal-oxide-semiconductor FET* (MOSFET).

Principle of the JFET

In a JFET, the current varies because of the effects of an electric field within the device. Electrons or holes move along a *channel* from the *source* (S) to the *drain* (D). This results in a drain current I_D that is normally the same as the source current I_S. The current depends on the voltage at the *gate* (G). Fluctuations in gate voltage E_G, cause changes in the current through the channel. Small fluctuations in E_G can cause large variations in the flow of charge carriers through the JFET. This translates into voltage amplification in ac circuits.

N-channel and *P*-channel JFET

A simplified drawing of an *N-channel JFET,* and its schematic symbol, are shown in Fig. 8.8A and B. The *N*-type material forms the path for the current. The majority of carriers are electrons. The drain is positive with respect to the source. The gate consists of *P*-type material. Another, larger section of *P*-type material, called the *substrate,* forms a boundary on the side of the channel opposite the gate. The voltage on the gate produces an electric field that interferes with the flow of charge carriers through the channel. The more negative E_G becomes, the more the electric field chokes off the current though the channel, and the smaller I_D becomes.

A *P-channel JFET* (Fig. 8.8C and D) has a channel of *P*-type semiconductor. Most charge carriers are holes. The drain is negative with respect to the source. The gate and substrate are of *N*-type material. The more positive E_G gets, the more the electric field chokes off the current through the channel, and the smaller I_D becomes.

The *N*-channel device can be recognized by the arrow pointing inward at the gate, and the *P*-channel JFET, by the arrow pointing outward. The power-supply polarity also shows which type of device is used. A positive drain indicates an *N*-channel JFET, and a negative drain indicates a *P*-channel type.

An *N*-channel JFET can almost always be replaced with a *P*-channel JFET, and with the power-supply polarity

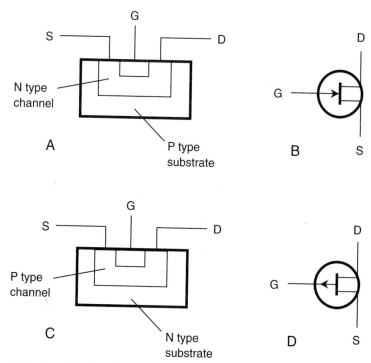

Figure 8.8 (A) Pictorial diagram of *N*-channel JFET; (B) schematic symbol for *N*-channel JFET; (C) pictorial diagram of *P*-channel JFET; (D) schematic symbol for *P*-channel JFET.

reversed, the circuit will still work if the new device has the right specifications.

Depletion and pinch off

A JFET works because the voltage at the gate causes an electric field that interferes, more or less, with the flow of charge carriers along the channel.

As the drain voltage E_D increases, so does the drain current I_D, up to a certain level-off value. This is true as long as the gate voltage E_G is constant and is not too large negatively. As E_G increases (negatively in an *N* channel or positively in a P channel), a *depletion region* begins to form in the channel. Charge carriers cannot flow in this region; they must pass through a narrowed channel. The larger E_G becomes, the wider the depletion region gets. If E_G becomes high enough, the depletion completely obstructs the flow of carriers. This is *pinch off*.

JFET biasing

Two biasing arrangements for an N-channel JFET are shown in Fig. 8.9. In Fig. 8.9A, the gate is grounded through resistor R_2. The source resistor R_1 limits the current through the JFET. Resistors R_1 and R_2 together determine the gate bias. The drain current I_D flows through R_3, producing a voltage across this resistor. The ac output signal passes through C_2.

In Fig. 8.9B, the gate is connected through potentiometer R_2 to a negative power-supply voltage. Adjusting this potentiometer results in a variable negative E_G between R_2 and R_3. Resistor R_1 limits the current through the JFET. The drain current I_D flows through R_4, producing a voltage across it; the ac output signal passes through C_2.

In both of these circuits, the drain is positive relative to ground. For a P-channel JFET, reverse the polarities.

The biasing arrangement in Fig. 8.9A is commonly used for weak-signal amplifiers, low-level amplifiers, and oscillators. The scheme in Fig. 8.9B is employed in class-C RF power amplifiers that have a substantial input signal. Typical JFET power-supply voltages are comparable to those with bipolar transistors. The voltage between the source and drain, abbreviated E_D, can range from approximately 3 to 50 V; most often it is 6 to 12 V.

Voltage Amplification

The graph in Fig. 8.10 shows the drain (channel) current I_D as a function of the gate bias voltage E_G for a hypothetical N-channel JFET. The drain voltage E_D is assumed to be constant.

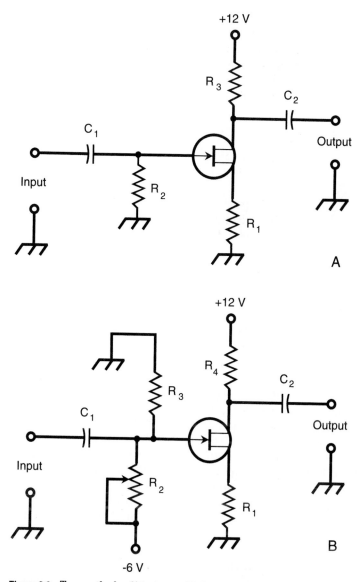

Figure 8.9 Two methods of biasing an *N*-channel JFET. (A) Fixed gate bias; (B) variable gate bias.

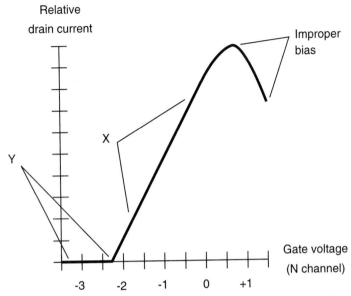

Figure 8.10 Relative drain current as a function of gate voltage, for a hypothetical N-channel JFET.

When E_G is fairly large and negative, the JFET is pinched off, and no current flows through the channel. As E_G gets less negative, the channel opens up, and current begins flowing. As E_G gets still less negative, the channel gets wider and the current I_D increases. As E_G approaches the point where the source-gate (S-G) junction is at forward breakover, the channel conducts as well as it can. If E_G becomes positive enough so that the S-G junction conducts, the JFET will no longer work properly. Some of the current in the channel will be shunted off through the gate.

The best amplification for weak signals is obtained when E_G is such that the slope of the curve in Fig. 8.10 is steepest. This is shown roughly by the range marked X in the figure. For power amplification, results are often best when the JFET is biased at or beyond pinch off, in the range marked Y.

The current I_D passes through the drain resistor, as shown in Fig. 8.9. Small fluctuations in E_G cause large changes in I_D, and these variations in turn produce wide swings in the dc voltage across R_3 (Fig. 8.9A) or R_4 (Fig. 8.9B). The ac part of this voltage goes through capacitor C_2 and appears at the output as a signal of much greater ac voltage than that of the input signal at the gate. This produces voltage amplification.

Drain current versus drain voltage

Drain current I_D can be plotted as a function of drain voltage E_D for various values of gate voltage E_G. The resulting set of curves is called a *family of characteristic curves* for the device. Figure 8.11 shows a family of characteristic curves for a hypothetical N-channel JFET. Also of importance is the curve of I_D versus E_G, one example of which is shown in Fig. 8.10.

Transconductance

Recall the discussion of dynamic current amplification for bipolar transistors earlier in this chapter. The JFET analog of this is called *dynamic mutual conductance,* or *transconductance.*

Refer to Fig. 8.10. Suppose that E_G is a certain value, with a corresponding I_D resulting. If the gate voltage changes by a small amount ΔE_G, the drain current will also change by a certain increment ΔI_D. The transconductance is the ratio $\Delta I_D/\Delta E_G$. Geometrically, this translates to the slope of a line tangent to the curve of Fig. 8.10.

The value of $\Delta I_D/\Delta E_G$ is not the same everywhere along the curve. When the JFET is biased beyond pinch off, in the region marked Y in the figure, the slope of the curve is zero. There is no drain current, even if the gate voltage changes. Only when the channel is conducting will there be a change in I_D when there is a change in E_G. The region where the transconductance is the greatest is the region marked X, where the slope of the curve is steepest. This is where the most gain can be obtained.

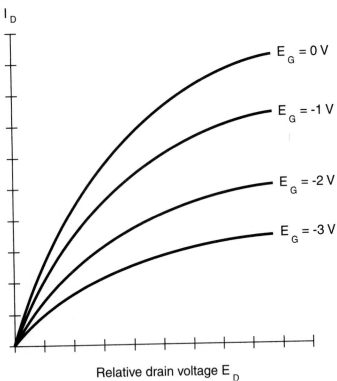

Figure 8.11 A family of characteristic curves for a hypothetical *N*-channel JFET.

The MOSFET

The acronym MOSFET is pronounced moss-fet. A simplified cross-sectional drawing of an *N*-channel MOSFET, along with the schematic symbol, is shown in Fig. 8.12A and B. The *P*-channel device is shown in Fig. 8.12C and D.

Super-high input impedance

When the MOSFET was first developed, it was called an *insulated-gate FET,* or IGFET. This is perhaps more descriptive of the device than the currently accepted name. The gate electrode is actually insulated, by a thin layer of dielectric, from the channel. As a result, the input impedance is even higher than that of a JFET; the gate-to-source

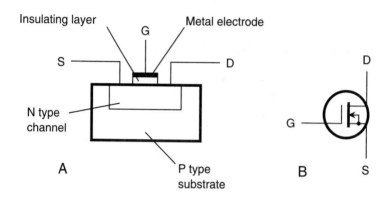

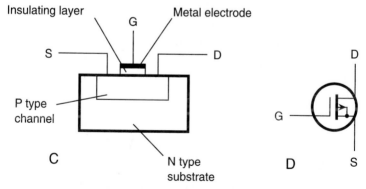

Figure 8.12   (A) Pictorial diagram of *N*-channel MOSFET; (B) schematic symbol for *N*-channel MOSFET; (C) pictorial diagram of *P*-channel MOSFET; (D) schematic symbol for *P*-channel MOSFET.

resistance of a typical MOSFET is comparable to that of a capacitor.

The main problem

One trouble with MOSFETs is that they can be easily damaged by electrostatic discharge. When building or servicing circuits containing MOS devices, technicians must use special equipment to ensure that their hands do not carry electrostatic charges that might ruin the components. If a static discharge occurs through the dielectric of a MOS device, the component will be destroyed permanently. Humid climates do not offer protection against this hazard.

Flexibility

In practical circuits, an N-channel JFET can sometimes be replaced with an N-channel MOSFET; P-channel devices can be similarly interchanged. But the characteristic curves for MOSFETs are not the same as those for JFETs. The S-G junction in a MOSFET is not a P-N junction, so forward breakover does not occur. A family of characteristic curves for a hypothetical N-channel MOSFET is shown in Fig. 8.13.

Depletion and enhancement mode

In a JFET, the channel conducts with zero bias. As the depletion region grows, charge carriers pass through a narrowed channel. This is known as *depletion mode*. A MOSFET can also be made to work in the depletion mode. The drawings and schematic symbols in Fig. 8.12 show depletion-mode MOSFETs.

Metal-oxide-semiconductor technology allows a second mode of operation. An *enhancement-mode MOSFET* has a pinched-off channel at zero bias. It is necessary to apply a gate bias voltage E_G to create a channel. If $E_G = 0$, the drain current I_D is zero when there is no signal input.

The schematic symbols for N- and P-channel enhancement-mode devices are shown in Fig. 8.14.

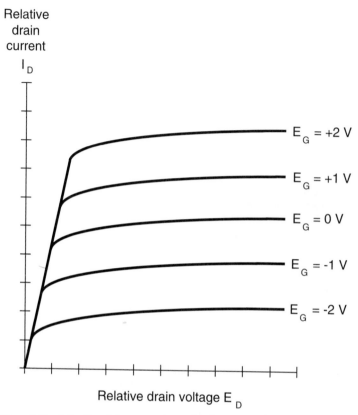

Figure 8.13 A family of characteristic curves for a hypothetical *N*-channel MOSFET.

Basic FET Circuits

There are three general circuit configurations for FETs. These three arrangements have the source, gate, or drain at signal ground.

Common source

In a *common-source circuit,* the signal is applied to the base (Fig. 8.15). An *N*-channel JFET is shown here, but the

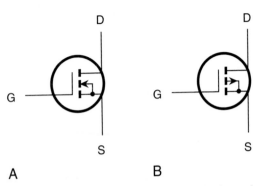

Figure 8.14 (A) The symbol for an N-channel enhancement-mode MOSFET; (B) symbol for a P-channel enhancement-mode MOSFET.

device could be an N-channel, depletion-mode MOSFET and the circuit diagram would be the same. For an N- channel, enhancement-mode device, an extra resistor is necessary, running from the gate to the positive power-supply terminal. For P-channel devices, the power supply must provide a negative, rather than a positive, voltage.

Capacitor C_1 and resistor R_1 place the source at signal ground, while elevating the source above ground for dc. The ac signal enters through C_2; resistor R_2 adjusts the input impedance and provides bias for the gate. The ac signal passes out of the circuit through C_3. Resistor R_3 keeps the output signal from being shorted out through the power supply.

The circuit in Fig. 8.15 is the basis for amplifiers and oscillators, especially at RF. The common-source arrangement provides the greatest gain of the three FET circuit configurations. The output is $180°$ out of phase with the input.

Common gate

The *common-gate circuit* (Fig. 8.16) has the gate at signal ground. The input is applied to the source. The illustration shows an N-channel JFET. For other types of FETs, the

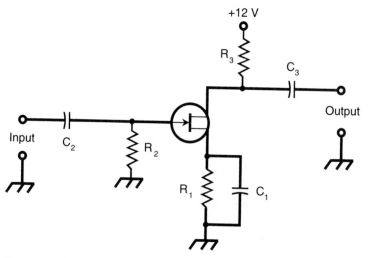

Figure 8.15 Common-source circuit configuration.

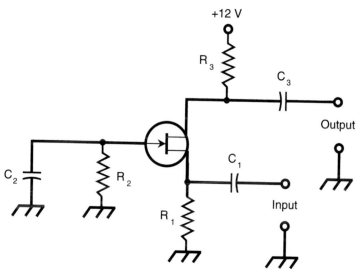

Figure 8.16 Common-gate circuit configuration.

same considerations apply as described above for the common-source circuit. Enhancement-mode devices require a resistor between the gate and the positive power-supply terminal (or the negative terminal if the MOSFET is *P*-channel).

The dc bias for the common-gate circuit is basically the same as that for the common-source arrangement, but the signal follows a different path. The ac input signal enters through C_1. Resistor R_1 keeps the input from being shorted to ground. Gate bias is provided by R_1 and R_2; capacitor C_2 places the gate at signal ground. In some common-gate circuits, the gate electrode is directly grounded, and components R_2 and C_2 are not used. The output leaves the circuit through C_3. Resistor R_3 keeps the output signal from being shorted through the power supply.

The common-gate arrangement produces less gain than its common-source counterpart. But a common-gate amplifier is not likely to break into unwanted oscillation. The output is in phase with the input.

Common drain

A *common-drain circuit* is shown in Fig. 8.17. This circuit has the collector at signal ground. It is sometimes called a *source follower.*

The FET is biased in the same way as for the common-source and common-gate circuits. In the illustration, an *N*-channel JFET is shown, but any other kind of FET can be used, reversing the polarity for *P*-channel devices. Enhancement-mode MOSFETs require a resistor between the gate and the positive power-supply terminal (or the negative terminal if the MOSFET is *P*-channel).

The input signal passes through C_2 to the gate. Resistors R_1 and R_2 provide gate bias. Resistor R_3 limits the current. Capacitor C_3 keeps the drain at signal ground. Fluctuating dc (the channel current) flows through R_1 as a result of the input signal; this causes a fluctuating dc voltage to appear across the resistor. The output is taken from the source, and its ac component passes through C_1.

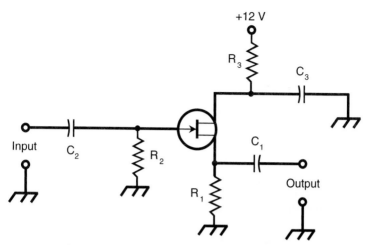

Figure 8.17 Common-drain circuit configuration.

The output of the common-drain circuit is in phase with the input. This circuit is often used for broadband impedance matching.

Integrated Circuits

Most ICs look like plastic boxes with protruding metal pins. Common configurations are the *single inline package* (SIP), the *dual inline package* (DIP), and the *flatpack.* Another package looks like a transistor with too many leads. This is a *metal-can package,* sometimes also called a *TO package.* The schematic symbol for an IC is a triangle or rectangle with the component designator written inside.

Compactness

Integrated-circuit devices and systems are far more compact than equivalent circuits made from discrete components. More complex circuits can be built, and kept down to a reasonable size, using ICs rather than discrete components. Thus, for example, there are notebook computers with capabilities more advanced than early computers, which took up entire rooms.

High speed

In an IC, the interconnections among components are physically tiny, making high switching speeds possible. Electric currents travel fast, but not instantaneously. The faster the charge carriers move from one component to another, the more operations can be performed per unit time and the less time is required for complex operations.

Low power requirement

Integrated circuits generally use less power than equivalent discrete-component circuits. This is important if batteries are used. Because ICs draw so little current, they produce less heat than their discrete-component equivalents. This results in better efficiency and minimizes problems that plague equipment that gets hot with use, such as frequency drift and generation of internal noise.

Reliability

Systems using ICs fail less often, per component-hour of use, than systems that make use of discrete components. This is mainly because all interconnections are sealed within an IC case, preventing corrosion or the intrusion of dust. The reduced failure rate translates into less downtime.

Ease of maintenance

Integrated-circuit technology lowers service costs, because repair procedures are simple when failures occur. Many systems use sockets for ICs, and replacement is simply a matter of finding the faulty IC, unplugging it, and plugging in a new one. Special desoldering equipment is used for servicing circuit boards that have ICs soldered directly to the foil.

Modular construction

Modern IC appliances employ *modular construction*. Individual ICs perform defined functions within a circuit board; the circuit board or card, in turn, fits into a socket and has a specific purpose. Computers, programmed with customized software, are used by technicians to locate the faulty card in a system. The card can be pulled and replaced, getting the system back to the user in the shortest possible time.

Inductors impractical

Devices using ICs must generally be designed to work without inductors, because inductances cannot easily be fabricated onto silicon chips. *Resistance-capacitance* (*RC*) *circuits* are capable of doing most things that *inductance-capacitance* (*LC*) *circuits* can do. Therefore, some inductances can be replaced with resistances, which are easily fabricated onto silicon chips.

Mega-power impossible

High-power amplifiers cannot, in general, be built onto semiconductor chips. High power necessitates a certain minimum physical bulk and mass to allow the conduction and radiation of excess heat energy. Power transistors and, in some systems, vacuum tubes are generally employed for high-power amplification.

Linear ICs

A *linear IC* is used to process analog signals such as voices, music, and most radio transmissions. The term *linear* arises from the fact that the instantaneous output is a linear function of the instantaneous input.

Operational amplifier

An *operational amplifier* (op amp) consists of several transistors, resistors, diodes, and capacitors, interconnected to produce high gain over a wide range of frequencies. An op amp has two inputs and one output.

When a signal is applied to the *noninverting input,* the output is in phase with it; when a signal is applied to the *inverting input,* the output is 180° out of phase with it. An op amp has two power-supply connections, one for the emitters of the transistors (V_{ee}) and one for the collectors (V_{cc}). The symbol for an op amp is a triangle. The inputs, output, and power-supply connections are drawn as lines emerging from the triangle.

The gain characteristics of an op amp are determined by external resistors. Normally, a resistor is connected between the output and the inverting input. This is the *closed-loop configuration*. The feedback is negative, causing the gain to be less than it would be if there were no feedback (*open-loop configuration*). A closed-loop amplifier using an op amp is shown in Fig. 8.18.

When an RC combination is used in the feedback loop of an op amp, the amplification factor varies with the frequency. It is possible to get a low-pass response, a high-pass response, a resonant peak, or a resonant notch using an op amp and various RC feedback arrangements.

Voltage regulator

A *voltage regulator* IC acts to control the output voltage of a power supply. This is important with precision electronic

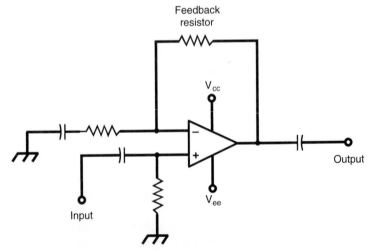

Figure 8.18 Closed-loop op amp circuit.

equipment. These ICs are available in various different voltage and current ratings. Typical voltage regulator ICs have three terminals. They look like power transistors.

Timer

A *timer* IC is a form of oscillator. It produces a delayed output, with the delay being variable to suit the needs of a particular device. The delay is generated by counting the number of oscillator pulses. The length of the delay is adjusted by means of external resistors and capacitors.

Multiplexer/demultiplexer

A *multiplexer* IC allows several different signals to be combined in a single channel via time-division multiplexing (TDM), in a manner similar to that used with pulse modulation. An analog multiplexer can also be used in reverse; then it works as a *demultiplexer.*

Comparator

Like an op amp, a *comparator* IC has two inputs. The device compares the voltages at the two inputs (called A and B). If the input at A is significantly greater than the input at B, the output is about +5 V. This is logic 1, or high. If the input at A is not greater than the input at B, the output voltage is about +2 V. This is designated as logic 0, or low.

Comparators are employed to actuate, or trigger, other devices such as relays and electronic switching circuits. Some can switch between low and high states at high speed; others are slow. Some have low input impedance, and others have high impedance. Some are intended for audio or low-frequency use; others are fabricated for video or high-frequency applications.

Digital ICs

Digital ICs consist of gates that perform logical operations at high speeds. There are several different technologies, each with unique characteristics. Digital-logic technology can employ bipolar and/or MOS devices.

TTL

In *transistor-transistor logic* (TTL), arrays of bipolar transistors, some with multiple emitters, operate on dc pulses. A TTL gate is illustrated in Fig. 8.19. The transistors are either cut off or saturated; there is no "in between." Because of this, TTL circuitry is comparatively immune to extraneous noise.

ECL

Another bipolar-transistor logic form is known as *emitter-coupled logic* (ECL). In ECL, the transistors are not operated at saturation, as they are with TTL. This increases the speed of operation of ECL compared with TTL. But noise pulses have a greater effect in ECL, because unsaturated transistors amplify as well as switch signals. The schematic in Fig. 8.20 shows a simple ECL gate.

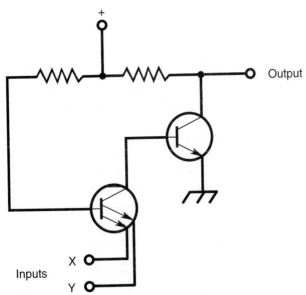

Figure 8.19   Transistor-transistor logic gate.

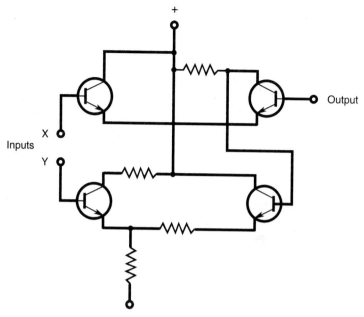

Figure 8.20   Emitter-coupled logic gate.

MOS logic

N-channel MOS logic (NMOS) offers simplicity of design, along with high operating speed. *P-channel MOS logic* (PMOS) is similar to NMOS, but the speed is slower. An NMOS or PMOS digital IC is like a circuit that uses only *N-* channel FETs or only *P-*channel FETs.

    Complementary-metal-oxide-semiconductor logic (CMOS), pronounced seamoss, employs both *N-* and *P-*type silicon on a single chip. This is analogous to using *N-* and *P-*channel FETs in a circuit. The main advantages of CMOS are extremely low current drain, high operating speed, and immunity to noise.

Component Density

The number of elements per chip in an IC is called the *component density*. There is a practical limit on component density imposed by the atomic structure of semiconductor material.

MSI

Medium-scale integration (MSI) is not a high level of component density, relatively speaking. An advantage of MSI (in a few applications) is that fairly large currents can be carried by the gates. Both bipolar and MOS technologies can be adapted to MSI.

LSI

Large-scale integration (LSI) is an order of magnitude more component dense than MSI. Electronic wristwatches, single-chip calculators, and small microcomputers are examples of devices using LSI ICs.

VLSI

Very-large-scale integration (VLSI) devices are an order of magnitude more component dense than LSI. Complex microcomputers and peripheral circuits such as memory storage ICs are made using VLSI.

Higher densities

Some ICs contain far more components per chip than any of the above types. The principal use for these technologies lies in ever-larger memory and ever-more-powerful processors for personal and business computing hardware.

IC Memory

Binary digital data, in the form of high and low levels (logic 1s and 0s), can be stored in ICs in the form of *memory*. In ICs, memory can take various forms.

RAM

A RAM stores binary data in arrays. The data can be addressed from anywhere in the matrix. Data is easily changed and stored back in RAM. There are two kinds of RAM: *dynamic RAM* (DRAM) and *static RAM* (SRAM).

DRAM employs IC transistors and capacitors; data is stored as charges on the capacitors. The charge must be replenished frequently, or it will be lost via discharge. Replenishing is done automatically several hundred times per second. SRAM uses flip-flops to store data. This eliminates the need for constant replenishing of charge, but SRAM ICs require more elements to store a given amount of data.

With any RAM, data vanishes on powering down unless some provision is made for memory backup. A memory whose content disappears when power is removed is *volatile*. If data is retained when power is removed, the memory is *nonvolatile*.

ROM

Read-only memory (ROM) can be accessed in whole or in part, but not written over, in the course of normal operation. A standard ROM chip is programmed at the factory. This permanent programming is known as *firmware*. But there are also ROMs that you can program and reprogram yourself.

EPROM

An *erasable programmable ROM* (EPROM) is an IC whose memory is of the read-only type, but it can be reprogrammed by exposure to ultraviolet. The IC must be taken from the circuit in which it is used, exposed to the ultraviolet for several minutes, and then reprogrammed via a special process.

There are EPROMs that can be erased by electrical means. Such an IC is called an EEPROM, for *electrically erasable programmable ROM*. These do not have to be removed from the circuit for reprogramming.

9

Transducers and Sensors

A *transducer* is a device that converts one form of energy into another. In electronics, transducers convert ac or dc into sound, ultrasound, visible light, IR, RF waves, mechanical motion, or other forms. Transducers also convert non-electrical energy into ac or dc.

Acoustic Transducers

An *acoustic transducer* converts sound or ultrasound waves into an ac electrical signal, or vice-versa. The waveforms of the acoustic energy and the ac signal energy are identical, or nearly so. Two common acoustic transducers are the microphone and the speaker. Other such devices include earphones, headsets, contact microphones, underwater speakers and microphones, and ultrasonic emitters and pickups.

Dynamic transducer

A *dynamic transducer* is a coil-and-magnet device that converts mechanical vibration into electrical currents, or vice-versa. The most common examples are the *dynamic microphone* and the *dynamic speaker.*

Figure 9.1 shows a functional diagram of a dynamic transducer. A diaphragm is attached to a permanent magnet that is mounted so that it can move back and forth rapidly along its axis. The magnet is surrounded by a coil of wire. Sound vibrations cause the diaphragm to move; this moves the magnet, which causes fluctuations in the magnetic field within the coil. The result is ac output from the coil, having the same waveform as the sound waves that strike the diaphragm.

If an audio signal is applied to the coil of wire, it creates a magnetic field that produces forces on the permanent magnet. This causes the magnet to move, pushing the diaphragm back and forth and creating acoustic waves in the surrounding medium.

Electrostatic transducer

An *electrostatic transducer* takes advantage of electrostatic forces and is shown in the functional diagram in Fig. 9.2.

In an *electrostatic pickup,* incoming sound waves cause vibration of the flexible plate. This produces rapid (although small) changes in the spacing, and therefore the

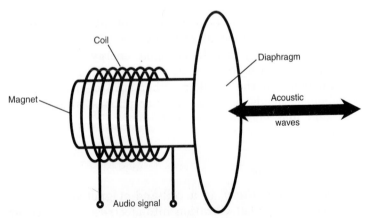

Figure 9.1 Pictorial diagram of a dynamic transducer.

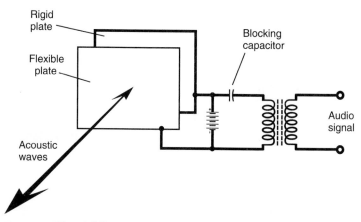

Figure 9.2 Pictorial diagram of an electrostatic transducer.

capacitance, between the two plates. A dc voltage is applied to the plates. As the capacitance changes between the plates, the electric field between them fluctuates. This produces variations in the current through the primary winding of the transformer. Audio signals appear across the secondary winding.

An *electrostatic emitter* works in just the opposite way. Currents in the transformer produce changes in the voltage between the plates. This change results in electrostatic force fluctuations, pulling and pushing the flexible plate in and out. The motion of the flexible plate produces sound waves.

Electrostatic transducers can be used in most applications where dynamic transducers are employed. Advantages of electrostatic transducers include light weight and good sensitivity. They also work with extremely low electric current.

Piezoelectric transducer

Figure 9.3 shows a *piezoelectric transducer.* This device consists of a crystal, such as quartz or ceramic material, sandwiched between two metal plates.

When a sound wave strikes one or both of the plates, the metal vibrates. This vibration is transferred to the crystal.

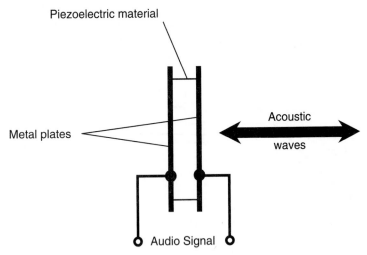

Figure 9.3 Pictorial diagram of a piezoelectric transducer.

The piezoelectric crystal generates weak electric currents when it is subjected to mechanical stress. Therefore, an ac voltage develops between the two metal plates, with a waveform similar to that of the sound waves.

If an ac signal is applied to the plates, it causes the crystal to vibrate in synchronization with the current. The result is that the metal plates vibrate also, producing an acoustic disturbance.

Piezoelectric transducers are common in ultrasonic applications, such as intrusion detectors and alarms.

Electromagnetic Transducers

An *electromagnetic (EM) transducer* converts EM fields into electrical energy, or vice-versa. There are numerous diverse forms.

RF antennas

Radio-frequency antennas can be categorized as receiving or transmitting types. A *receiving antenna* converts an EM field to ac. A *transmitting antenna* converts ac to an EM field. Most transmitting antennas will function effectively for reception. Some receiving antennas function as efficient transmitting antennas, and others do not. Antennas are discussed in Chap. 18.

Visible, IR, and UV transducers

Most transducers for visible light, IR, and ultraviolet (UV) are semiconductor diodes whose *P-N* junctions emit, and/or are sensitive to, EM fields at these wavelengths. Some such devices respond only to changes in the intensity of the EM energy, whereas others generate dc when irradiated. For details, see Chap. 7.

Magnetic media

Electromagnetic transducers are used in *magnetic media* to store and retrieve analog and digital data. As an example, consider the recording, playback, and erase head(s) in a tape recorder. All tape recorders act as transducers between acoustic, digital, or video signals and variable magnetic fields. A simplified rendition of the recording/playback apparatus in a typical audiotape recorder is shown in Fig. 9.4.

In *record mode,* the tape moves past the erase head before anything is recorded. If the tape is not blank (that is, if

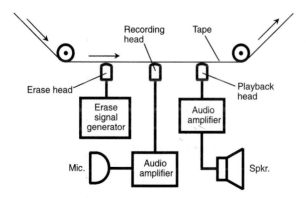

Figure 9.4 Block diagram of a tape recorder.

magnetic impulses already exist on it), the *erase head* removes these before anything else is recorded. This prevents *doubling,* or the simultaneous presence of two programs on the tape. The *recording head* is an electromagnet that generates a fluctuating magnetic field, whose instantaneous flux density is proportional to the instantaneous level of the audio input signal. This magnetizes the tape in a pattern that duplicates the waveform of the signal.

In *playback mode,* the erase and recording heads are not activated. The *playback head* acts as a magnetic-field detector. As the tape moves past, the playback head is exposed to a fluctuating magnetic field whose waveform is identical to that produced by the recording head when the audio was originally recorded on the tape. This magnetic field induces ac in the playback head. The ac is amplified and delivered to a speaker, headset, or other output device.

Electromechanical Transducers

An *electromechanical transducer* converts movement (mechanical energy) into electrical energy, or vice-versa.

Displacement transducer

A *displacement transducer* measures a distance or angle traversed or the distance or angle separating two points. Conversely, it might convert a signal into movement over a certain distance or angle. A device that measures or produces movement in a straight line is a *linear displacement transducer.* If it measures or produces movement through an angle, it is an *angular displacement transducer.*

Suppose a robot arm is to rotate 28° in the horizontal plane. A command is given to the robot controller, such as BR 28 (base rotation = 28°). The robot controller (computer) sends a signal to the robotic end effector, which rotates clockwise. An angular distance transducer keeps track of the angle of rotation, sending a signal back to the controller. This signal increases in linear proportion to the angle that the end effector has turned.

When the robot controller receives the command BR 28, it causes the end effector to begin rotating. The number 28 programs a threshold level for the return signal into the robot controller. As the signal from the displacement transducer increases, it reaches the threshold at 28° of rotation. The controller detects this and issues a stop signal to the end effector.

Pointing and control devices

A *joystick* is a control device capable of producing movement, or controlling variable quantities, in two or three dimensions. Figure 9.5 shows a joystick with two fundamental dimensions of movement, labeled $+x/-x$ and $+y/-y$. The device consists of a movable lever or handle and a ball bearing within a control box. The stick is moved by hand. Joysticks are used in computer games, for entering coordinates into a computer, and for the remote-control of robots. Some joysticks can be rotated clockwise and counterclockwise, in addition to the usual two coordinates, allowing control in a third dimension, labeled $+z/-z$.

A *mouse* is a peripheral commonly used with personal computers. By sliding the mouse around on a flat surface, a cursor or arrow is positioned on the display. Push-button switches on the top of the unit actuate the computer to perform whatever function the cursor or arrow shows. These actions are called "clicks" and "double clicks."

A *trackball* resembles an inverted mouse or a two-dimensional joystick without the stick. Instead of the device being pushed around on a surface, the computer user moves a ball bearing, causing the display cursor to move vertically and horizontally. Push-button switches on the keyboard, in which the trackball is installed, actuate the computer to perform functions.

An *eraser-head pointer* is a rubber button approximately 5 millimeters (5 mm) in diameter, usually placed in the center of a computer keyboard, for example, in between the G, H, and B keys. The computer user moves the cursor on the display by pushing against the button. Pushing vertically

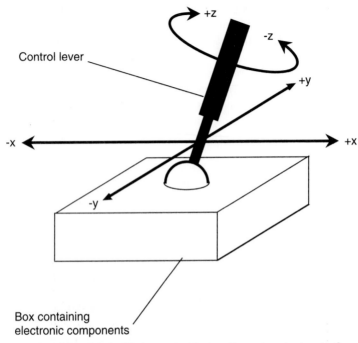

Figure 9.5   A joystick facilitates control in two dimensions (x,y) or in three dimensions (x,y,z).

(toward the back end of the keyboard) causes the cursor to move upward; "pulling" toward the front end of the keyboard causes the cursor to move downward. Pushing in other directions results in intuitive cursor motion. Clicking and double clicking are done via button switches on the keyboard.

A *touch pad* is a sensitive plate measuring approximately 4 by 6 cm. The user places an index finger on the plate and moves the finger around. This results in intuitive movement of the display cursor. Clicking and double clicking are done in the same way as with the trackball and eraser-head pointer.

Electric motor

An *electric motor* converts electrical energy into mechanical energy. Motors can operate from ac or dc, and range in size from tiny devices in wristwatches to huge machines that pull passenger trains.

All motors operate via magnetic effects. Electric current flows through a set of coils, producing magnetic fields. The magnetic forces generate torsion. The greater the current in the coils, the greater the rotational force. The *armature coil* rotates with the motor shaft; the *field coil* is fixed. In a dc motor, a *commutator* reverses the current with each half-rotation of the shaft, so the torsion occurs in the same direction at all times.

When a motor is connected to a load, the torque required to turn the shaft increases. The greater the required torque becomes, the more power is drawn from the source.

Stepper motor

A *stepper motor* turns in small increments, rather than continuously. The *step angle,* or extent of each turn, varies depending on the particular motor. It can range from less than 1° of arc to a quarter of a circle (90°). A stepper motor will turn through its step angle and then stop, even if the current is maintained. In fact, when a stepper motor is stopped with a current going through its coils, the shaft resists applied torque.

Conventional motors run at hundreds, or even thousands, of revolutions per minute (rpm). A stepper motor usually runs at less than 180 rpm, and often much less. In a conventional motor, the torque increases as the motor runs faster. But with a stepper motor, the torque decreases as the motor runs faster. A stepper motor has the most turning power when it is running at slow speed.

The most common stepper motors are of two types: two- and four-phase. A *two-phase stepper motor* has two coils, called phases, controlled by four wires. A *four-phase stepper motor* has four phases and eight wires. The motors are stepped by applying current sequentially to the phases. Figure 9.6 shows schematic diagrams of two-phase (Fig. 9.6A) and four-phase (Fig. 9.6B) stepper motors. Table 9.1 shows control-current sequences for two-phase motors, and Table 9.2 shows them for four-phase motors.

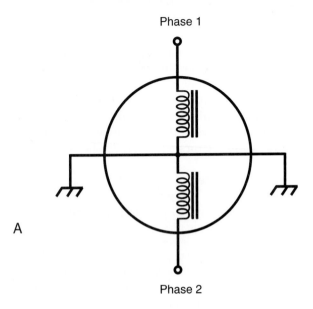

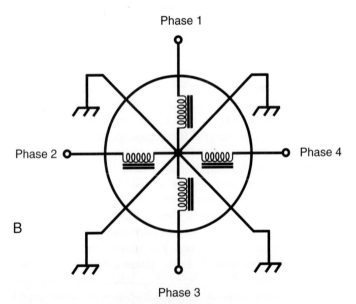

Figure 9.6 Two- (A) and four-phase (B) stepper motors.

TABLE 9.1 Two-Phase Stepper Control

Read down for clockwise rotation; read Up for counterclockwise rotation.

Step	Phase 1	Phase 2
1	Off	Off
2	On	Off
3	On	On
4	Off	On

TABLE 9.2 Four-Phase Stepper Control

Read down for clockwise rotation; read up for counterclockwise rotation.

Step	Phase 1	Phase 2	Phase 3	Phase 4
1	On	Off	On	Off
2	Off	On	On	Off
3	Off	On	Off	On
4	On	Off	Off	On

When a pulsed current is supplied to a stepper motor, with the current rotating through the phases as shown in the tables, the motor will rotate in increments, one step for each pulse. In this way, a precise speed can be maintained. Because of the braking effect, this speed will be constant for a wide range of mechanical turning resistances.

Stepper motors can be controlled using microcomputers. Several stepper motors, all under the control of a single microcomputer, are typical in robot arms of all geometries. Stepper motors are especially well suited for point-to-point motion. Complicated, intricate tasks are done by computer-controlled robotic arms and end effectors using stepper motors.

Selsyn

A *selsyn* is an indicating device that shows the direction in which an object is oriented. It consists of a transmitting unit and an indicator unit. As the shaft of the transmitting unit rotates, the shaft of the indicator unit follows exactly. A common application is as a direction indicator for a rotatable antenna (Fig. 9.7). When the antenna is turned, the indicator unit shaft moves through the same number of angular degrees as the transmitting unit shaft. A selsyn for azimuth (compass) bearings has a range of 0 to 360°; a selsyn for elevation bearings has a range of 0 to 90°.

A *synchro* is a selsyn used for control of mechanical devices. Synchros are well suited for the remote control of robots (teleoperation). Some synchros are programmable. The operator inputs a number into the generator, and the receiver changes position accordingly. Computers allow complex sequences of movements to be executed.

Electric generator

An *electric generator* is constructed in basically the same way as a conventional motor, although it functions in the opposite sense. Some generators can actually be used as motors; they are called *motor/generators*. A typical generator produces ac by rotating a coil in a strong magnetic field. Alternatively, a permanent magnet can be rotated within a coil of wire. The rotating shaft is driven by a gasoline-powered motor, a turbine, or some other source of mechanical energy. A commutator can be used with a generator to produce pulsating dc output, which can be filtered to obtain pure dc for use with electronic equipment.

Small portable gasoline-powered generators, capable of delivering a few kilowatts, can be purchased in department stores. Larger generators allow homes or buildings to keep their electrical power in the event of an interruption in the utility. The largest generators are found in power plants and can produce in excess of a megawatt of continuous power.

Small generators are commonly used in synchro systems. These specialized generators allow remote control of robotic devices. A generator can be used to measure the speed at which a vehicle or rolling robot is moving. The shaft of the generator is connected to one of the wheels, and the generator output voltage and frequency vary directly with the angular speed of the wheel.

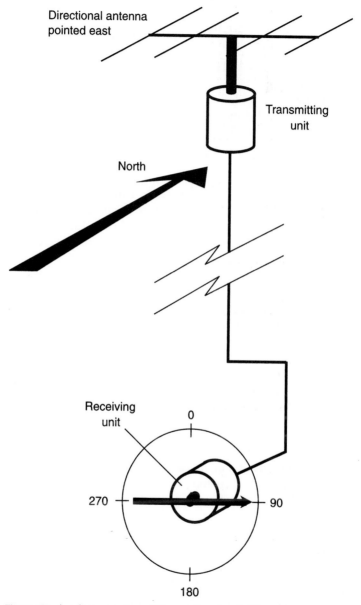

Figure 9.7 A selsyn can be used as a direction indicator for a rotatable antenna.

Optical encoder

In digital radios, frequency adjustment is done in discrete steps. In high-frequency (HF) communications gear, the usual increment is 10 Hz. At VHF and UHF, 10- or 100-Hz increments are used for radiotelegraphy, radioteletype, and single sideband; 5-kHz increments are common for frequency modulation (FM). Most high-fidelity receivers are digitally tuned. The increment is normally 10 kHz in the standard AM broadcast band, and 200 kHz in the FM broadcast band.

For tuning, an alternative to mechanical switches (which corrode or wear out with time) is the *optical encoder*. This consists of an LED, a photodetector, and a device called a *chopping wheel*. The LED shines on the photodetector through the wheel. The wheel has radial bands, alternately transparent and opaque (Fig. 9.8). The wheel is attached to the tuning knob. As the knob is turned, the light beam is interrupted. Each interruption causes the frequency to change by a specified increment. The difference between "frequency up" and "frequency down" (clockwise and counterclockwise shaft rotation, respectively) is determined by using two LEDs and two photodetectors side by side. The sense of rotation can be determined according to which photodetector senses the beam interruptions first.

Sensors

A *sensor* employs one or more transducers to detect and/or measure phenomena such as temperature, texture, proximity, and the presence of certain substances.

Capacitive pressure sensor

A *capacitive pressure sensor* is shown in Fig. 9.9. Two metal plates are separated by a layer of nonconductive foam, forming a capacitor. This component is connected in parallel with an inductor. The resulting LC circuit determines

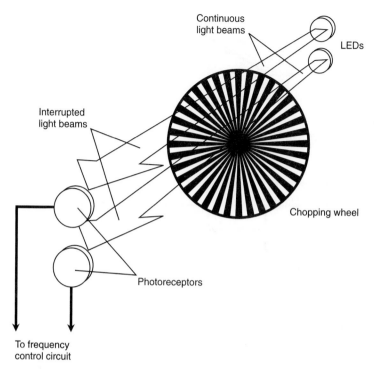

Figure 9.8 An optical encoder uses LEDs and photodetectors to sense the direction and extent of rotation.

the frequency of an oscillator. If an object strikes the sensor, the plate spacing momentarily decreases. This increases the capacitance, causing a drop in the oscillator frequency. When the object moves away from the transducer, the foam springs back, the plates return to their original spacing, and the oscillator frequency returns to normal.

The output of a capacitive pressure sensor can be converted to digital data using an *analog-to-digital converter* (ADC). This signal can be sent to a microcomputer such as a robot controller. Pressure sensors can be mounted in various places on a mobile robot, such as the front, back, and sides. Then, for example, physical pressure on the sensor in

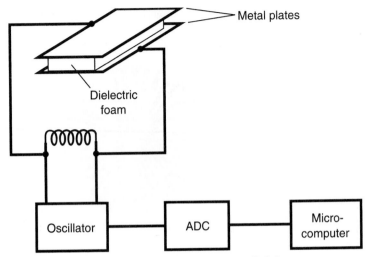

Figure 9.9 A capacitive pressure sensor detects applied force.

the front of the robot might send a signal to the controller, which tells the machine to move backward.

A capacitive pressure sensor can be fooled by massive conducting or semiconducting objects in its vicinity. If such a mass comes near the transducer, the capacitance changes, even if direct contact is not made. This phenomenon is known as *body capacitance*. When the effect must be avoided, an *elastomer* device can be used for pressure sensing.

Elastomer

An *elastomer* is a flexible substance resembling rubber or plastic that can be used to detect the presence or absence of mechanical pressure. Figure 9.10 illustrates how an elastomer can be used to detect, and locate, a pressure point. The elastomer conducts electricity fairly well, but not perfectly. It has a foamlike consistency, so it can be compressed. Conductive plates are attached to the pad.

When pressure appears at some point in the elastomer pad, the material compresses, and this lowers its electrical

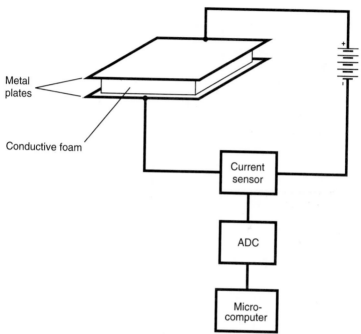

Figure 9.10 An elastomer-type pressure sensor detects applied force without unwanted capacitive effects.

resistance. This is detected as an increase in the current between the plates. The greater is the pressure, the more the elastomer is compressed, and the greater is the increase in the current. The current-change data can be sent to a microcomputer such as a robot controller.

Back-pressure sensor

Any motor produces a measurable pressure that depends on the torque being applied. A *back-pressure sensor* detects and measures the torque that the motor is applying at any given time. The sensor produces a signal, usually a variable voltage, that increases as the torque increases (Fig. 9.11).

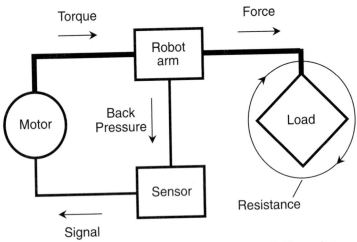

Figure 9.11   A back-pressure sensor governs the force applied by a robot arm or other mechanical device.

Back-pressure sensors are used to limit the force applied by a robot gripper, arm, drill, hammer, or other device. The back voltage, or signal produced by the sensor, reduces the torque applied by the motor. This can prevent damage to objects being handled by the robot. It also helps to ensure the safety of people working around the robot.

Capacitive proximity sensor

A *capacitive proximity sensor* uses an RF oscillator, a frequency detector, and a metal plate connected to the oscillator circuit (Fig. 9.12). The oscillator is designed so that a change in the capacitance of the plate, with respect to the environment, will cause the oscillator frequency to change. This change is sensed by the frequency detector, which sends a signal to a microcomputer.

Objects that conduct electricity to some extent, such as house wiring, people, cars, or refrigerators, are sensed more easily by capacitive transducers than are things that do not conduct, like wooden or plastic furniture. Therefore,

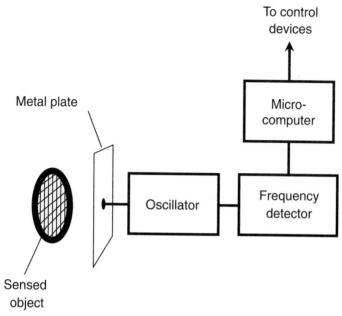

Figure 9.12 A capacitive proximity sensor can detect nearby conducting or semiconducting objects.

other kinds of proximity sensors are often needed for a robot to navigate well in a complex environment such as a home or office.

Photoelectric proximity sensor

Reflected light can provide a way for a robot to tell if it is approaching something. A *photoelectric proximity sensor* uses a light-beam generator, a photodetector, a frequency-sensitive amplifier, and a microcomputer (Fig. 9.13).

The light beam reflects from the object and is picked up by the photodetector. The light beam is modulated at a certain frequency, say 1000 Hz. The amplifier responds only to light modulated at that frequency. This prevents false imaging that might otherwise be caused by lamps or sunlight. If the robot is approaching an object, its controller

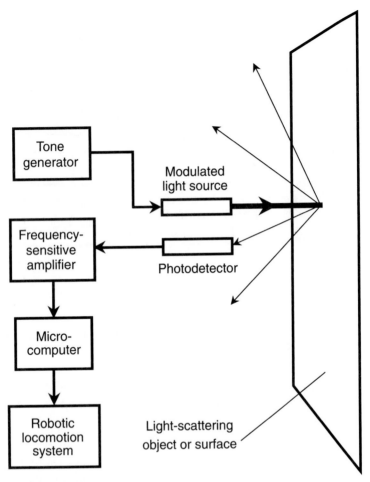

Figure 9.13 A photoelectric proximity sensor detects scattered light.

(microcomputer) senses that the reflected beam is getting stronger. The robot can then steer clear of the object.

This method of proximity sensing does not function for objects that do not reflect light or for windows or mirrors approached at a sharp angle. In these scenarios, the light beam is not reflected toward the photodetector.

Fluxgate magnetometer

When other position sensors do not function in a particular environment, a *fluxgate magnetometer* can be used. This system employs sensitive magnetic receptors and a micro-computer to sense the presence of, and detect changes in, an artificially generated magnetic field.

Navigation within a defined area can be done by check-ing the orientation of magnetic lines of flux generated by electromagnets in the walls, floor, and/or ceiling of the room. For each point in the room, the magnetic flux lines have a unique direction and intensity. There exists an *iso-morphism,* or one-to-one correspondence, between the magnetic flux intensity and direction and the points with-in the robot's operating environment. The robot controller is programmed to "know" this relation. This makes it pos-sible for the machine to pinpoint its position with extreme accuracy.

Texture sensor

Texture sensing is the ability of a machine to determine whether a surface is "shiny" or "rough." Basic texture

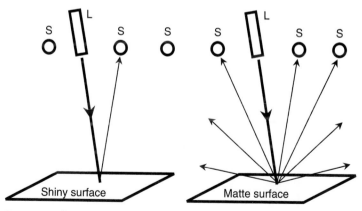

Figure 9.14 In texture sensing, lasers (L) and sensors (S) analyze a shiny surface (A) and a matte surface (B).

sensing involves the use of a laser and several light-sensitive sensors.

Figure 9.14 shows how a laser (L) and sensors (S) can be used to tell the difference between a shiny surface (Fig. 9.14A) and a rough or matte surface (Fig. 9.14B). The shiny surface, such as the polished hood of a car, reflects light at the incidence angle. But the matte surface, such as a sheet of paper, scatters light. The shiny surface reflects the beam entirely to the sensor in the path of the beam whose reflection angle equals its incidence angle. The matte surface reflects the beam to all the sensors.

Smoke detector

"Normal" air has a characteristic *dielectric constant,* which is a measure of how well the air can hold an electric

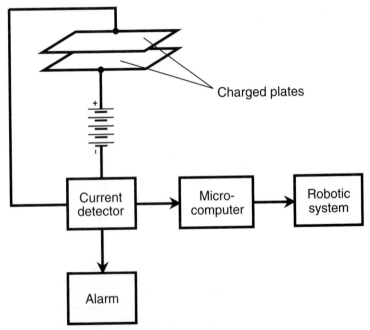

Figure 9.15 A simple smoke detector senses changes in the characteristics of the air.

charge. Air also has an *ionization potential*; this is the energy needed to strip electrons from the atoms. Many factors can affect these properties. Common influences are the relative humidity, the barometric pressure, the presence of smoke, and changes in the relative concentrations of gases.

A *smoke detector* can work by sensing a change in the dielectric constant and/or the ionization potential of the air. Two electrically charged plates are spaced a fixed distance apart (Fig. 9.15). If the properties of the air change, the plates charge or discharge. This causes a momentary electric current that can be used, for example, to actuate an alarm, notify the fire department, and operate a set of water sprinklers.

10

Electron Tubes

Electron tubes, also called *tubes* or *valves* (in England), are used in some electronic equipment. In a tube, the *charge carriers* are free electrons that travel through space between electrodes inside the device. This makes tubes fundamentally different from semiconductor devices, in which charge carriers move among atoms in a solid medium.

Tube Forms

There are two basic types of electron tube: the *vacuum tube* and the *gas-filled tube.* As their names imply, vacuum tubes have virtually all the gases removed from their envelopes. Gas-filled tubes contain elemental vapor at low pressure.

Vacuum tube

Vacuum tubes accelerate electrons to high speeds, resulting in large electric currents. This current can be focused into a beam and guided in a particular direction. The intensity and/or beam direction can be changed with extreme rapidity, producing various useful effects such as rectification, detection, oscillation, amplification, signal mixing, waveform displays, spectral displays, and video imaging. Vacuum tubes are the predecessors of bipolar and field-effect transistors.

Gas-filled tube

Gas-filled tubes have a constant voltage drop, no matter what the current. This makes them useful as voltage regulators for high-voltage, high-current power supplies. Gas-filled tubes can withstand conditions that would destroy semiconductor regulating devices. Gas-filled tubes emit IR, visible light, and/or UV. This property can be put to use for decorative lighting. A small *neon lamp* can be employed to construct an audio-frequency *relaxation oscillator* (Fig. 10.1).

Diode tube

Before the start of the twentieth century, scientists knew that electrons could carry a current through a vacuum. They

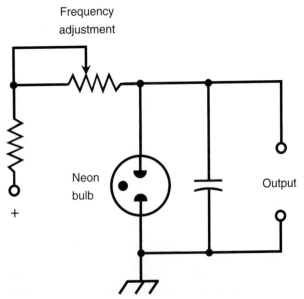

Figure 10.1 A neon-tube oscillator, also known as a relaxation oscillator.

also knew that hot electrodes would emit electrons more easily than cool ones. These phenomena were put to use in the first electron tubes, known as *diode tubes,* for the purpose of rectification.

Diode tubes are rarely used nowadays, although they can still be found in some power supplies that are required to deliver several thousand volts for long periods at a 100 percent duty cycle.

Electrodes

In any tube, the electron-emitting electrode is the *cathode.* The cathode is usually heated by means of a wire *filament,* similar to the glowing element in an incandescent bulb. The heat drives electrons from the cathode. The electron-collecting electrode is the *anode,* also called the *plate.* Intervening *grids* control the flow of electrons from the cathode to the plate. In a cathode ray tube, (CRT), the electron beam is manipulated magnetically using *deflecting coils* or electrostatically using *deflection plates.*

Directly heated cathode

In some tubes, the filament also serves as the cathode. This type of electrode is called a *directly heated cathode.* The negative power-supply voltage is applied directly to the filament. The filament voltage for most tubes is 6 or 12 V dc. The schematic symbol for a diode tube with a directly heated cathode is shown in Fig. 10.2A.

Indirectly heated cathode

The filament is commonly enclosed within a cylindrical cathode, and the cathode gets hot from IR radiation. This is an *indirectly heated cathode.* The cathode itself is usually grounded. The filament normally receives 6 or 12 V ac. The schematic symbol for a diode tube with an indirectly heated cathode is shown in Fig. 10.2B.

In either the directly or indirectly heated cathode, electrons are driven off the element by the heat of the filament.

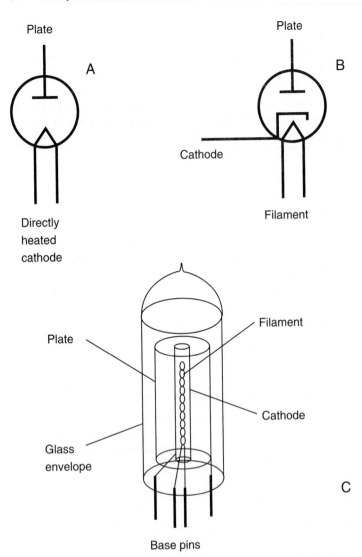

Plate

A

Directly
heated
cathode

Plate

B

Cathode

Filament

Plate

Filament

Cathode

Glass
envelope

C

Base pins

Figure 10.2 (A) Schematic symbol for diode tube with directly heated cathode; (B) symbol for diode tube with indirectly heated cathode; (C) simplified rendition of the construction of a diode tube.

The cathode of a tube is analogous to the source of an FET or to the emitter of a bipolar transistor.

Because the electron emission in a tube depends on the filament or "heater," tubes need a certain amount of time to "warm up." This time can vary from a few seconds (for a small tube with a directly heated cathode) to a couple of minutes (for massive power-amplifier tubes with indirectly heated cathodes).

Cold cathode

In a gas-filled tube, the cathode does not have a filament to heat it. Such an electrode is called a *cold cathode*. Various chemical elements are used in gas-filled tubes. In fluorescent devices, neon, argon, and xenon are common. In gas-filled voltage-regulator (VR) tubes, mercury vapor is used. In a mercury-vapor VR tube, the warm-up period is the time needed for the elemental mercury, which is a liquid at room temperature, to vaporize (approximately 2 minutes, or 2 min).

Plate

The plate, or anode, of a tube is a cylinder concentric with the cathode and filament (Fig. 10.2C). The plate is connected to the positive dc power-supply voltage. Tubes operate at plate voltages ranging from 50 V to more than 3 kV. These voltages are potentially lethal. Technicians unfamiliar with vacuum tubes should not attempt to service equipment that contains them.

The output of a tube-type amplifier circuit is almost always taken from the plate circuit. The plate exhibits high impedance for signal output, similar to that of a JFET.

The Triode

In a diode tube, the flow of electrons from cathode to plate depends on the dc power-supply voltage. The greater this voltage, the greater the current through the device, when all other parameters are held constant.

The flow of current can also be controlled via an electrode between the cathode and the plate. This electrode, the *control grid* (or simply the *grid*), is a wire mesh or screen that lets electrons pass through. The grid impedes the flow of electrons if it is provided with a negative voltage relative to the cathode. The greater the negative *grid bias*, the more the grid obstructs the flow of electrons through the tube.

A tube with one grid is a *triode,* whose schematic symbol is shown in Fig. 10.3A. In this case the cathode is indirectly heated; the filament is not shown. This omission is standard in schematics showing tubes with indirectly heated cathodes. When the cathode is directly heated, the filament symbol serves as the cathode symbol.

Multigrid Tubes

Some tubes have more than one grid. These extra elements allow for improved gain and stability in tube-type amplifiers.

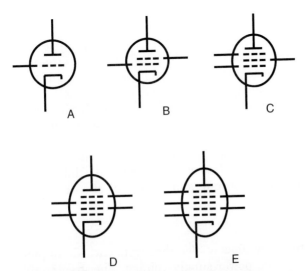

Figure 10.3 Schematic symbols for vacuum tubes with grids: (A) triode, (B) tetrode, (C) pentode, (D) hexode, (E) heptode.

Tetrode

A second grid can be added between the control grid and the plate. This is a spiral of wire or a coarse screen and is called the *screen grid* or *screen*. This grid normally carries a positive dc voltage at 25 to 35 percent of the plate voltage.

The screen grid reduces the capacitance between the control grid and plate, minimizing the tendency of a tube amplifier to oscillate. The screen grid can also serve as a second control grid, allowing two signals to be injected into a tube. This tube has four elements and is known as a *tetrode*. Its schematic symbol is shown in Fig. 10.3B.

Pentode

The electrons in a tetrode can bombard the plate with such force that some of them bounce back or knock other electrons from the plate. Some of these *secondary electrons* leave the tube through the screen circuit rather than the plate circuit. The result is diminished plate current and increased screen current. This so-called secondary emission can hinder tube performance and, at high power levels, cause screen current so high that the electrode is destroyed.

Excessive screen current can be controlled by placing another grid, called the *suppressor grid,* or *suppressor,* between the screen and the plate. The suppressor repels secondary electrons emanating from the plate, preventing most of them from reaching the screen. The suppressor also reduces the capacitance between the control grid and the plate more than a screen grid does by itself.

Greater gain and stability are possible with a *pentode,* or tube with five elements, than with a tetrode or triode. The schematic symbol for a pentode is shown in Fig. 10.3C. The suppressor is generally connected to the cathode. It carries a negative charge with respect to the screen grid and the plate.

Hexode and heptode

In some older radio and TV receivers, tubes with four or five grids were sometimes used. These tubes had six and seven elements, respectively, and were called *hexode* and *heptode*. The usual function of such tubes was signal mixing. The schematic symbol for a hexode is shown in Fig. 10.3D; the symbol for a heptode is in Fig. 10.3E.

You will not encounter hexodes and heptodes in modern electronics because solid-state components are used for signal mixing. But if you like to work with "antique" radios, you should be familiar with them. Hexodes and heptodes are relics. It is difficult to find a replacement should such a component go bad.

Interelectrode capacitance

In a vacuum tube, the cathode, grid(s), and plate have mutual capacitance. This *interelectrode capacitance* is the primary limiting factor on the frequency range in which the device can produce gain. The interelectrode capacitance in a typical tube is a few picofarads. This is negligible at low frequencies, but at VHF and UHF it becomes a significant consideration. Vacuum tubes intended for use at VHF and UHF are designed to minimize this capacitance.

Circuit Configurations

The most common application of vacuum tubes is in amplifiers, especially in VHF and UHF radio and television transmitters, and/or at power levels of more than 1 kW. Some high-fidelity audio systems also employ vacuum tubes. In recent years they have regained favor with some popular music bands. There are two basic vacuum-tube circuit arrangements: *grounded cathode* and *grounded grid*.

Grounded cathode

A simplified schematic diagram of a grounded-cathode, triode circuit is shown in Fig. 10.4. This circuit is the basis for many tube-type RF power amplifiers and audio amplifiers. The input impedance is moderate and the output impedance is high. Impedance matching between the amplifier and the

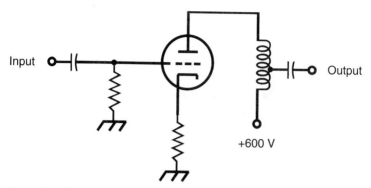

Figure 10.4   Basic grounded-cathode circuit.

load can be obtained by tapping a coil in the output circuit or by using a transformer.

Grounded grid

The grounded-grid configuration requires more driving power than the grounded-cathode scheme. A grounded-cathode amplifier might produce 1 kW of RF output for 10 W input, but a grounded-grid amplifier needs about 50 to 100 W of drive to produce 1 kW of RF output.

A basic grounded-grid circuit is shown in Fig. 10.5. The cathode input impedance is low, and the plate output impedance is high. The output impedance is matched by the same means as with the grounded-cathode arrangement.

The plate voltages (+600 V dc) in the circuits of Figs. 10.4 and 10.5 are given as examples. The amplifiers shown could produce 75- to 150-W signal output provided they receive sufficient drive and are properly biased. An amplifier rated at 1-kW output would require a plate voltage of 2 to 5 kV, depending on the tube characteristics and the class of amplifier operation.

Beam-power tube

A *beam-power tube* is a vacuum tube with an electrode that confines the electrons to certain directions. The

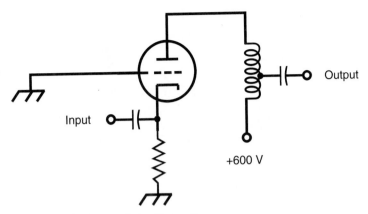

Figure 10.5 Basic grounded-grid circuit.

result is a concentrated beam, and this is where the tube gets its name. The beam-power tube is more efficient than an ordinary vacuum tube, especially when large amounts of power output are required, such as in RF transmitting equipment. The grids of a beam-power tube are aligned, which further aids in efficient operation. A beam-power tube can be a triode, tetrode, or pentode. The pentode is preferred, because a suppressor grid is needed to prevent excessive screen current that can result from the high electron speeds.

Cooling

In power amplifiers, vacuum tubes must be provided with some means for cooling, or damage will result. *Air cooling* is a common method of achieving this. A fan, also called a *blower,* forces air over the tubes or through an array of *cooling fins* (Fig. 10.6).

Conduction cooling is sometimes used in place of an air blower. A *heat sink,* which consists of a mass of metal, is placed in direct contact with the tube envelope. The anode of the tube is in thermal contact with the envelope; the heat sink conducts excess heat away from the anode.

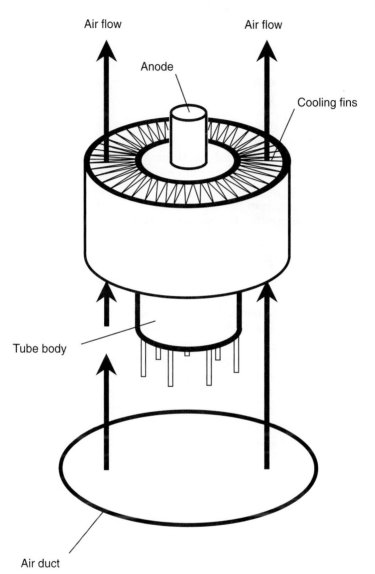

Figure 10.6 Air cooling works by convection.

Cathode-Ray Tubes

Everyone encounters TV screens and desktop computer monitors. Most people have seen an oscilloscope or spectrum-analyzer display. These devices generally use CRTs.

Electron beam

In a CRT, an *electron gun* emits an electron beam that is focused and accelerated as it passes through positively charged *anodes.* The beam then strikes a glass screen whose inner surface is coated with *phosphor.* The phosphor glows visibly as seen from the face of the CRT. The beam scanning pattern is controlled by magnetic or electrostatic fields.

Electromagnetic CRT

A simplified cross-sectional drawing of an *electromagnetic CRT* is shown in Fig. 10.7. There are two sets of *deflecting coils,* one for the horizontal plane and the other for the vertical plane. (In the figure, only the horizontal deflecting coils are shown for clarity.) The greater the current in the coils, the greater the intensity of the magnetic field, and the more the electron beam is deflected. The electron beam is bent at right angles to the magnetic lines of flux.

In an *oscilloscope,* the horizontal deflecting coils receive a sawtooth waveform. This causes the beam to scan or *sweep* at a precise, adjustable speed across the screen from left to right as viewed from in front. After each timed left-to-right sweep, the beam jumps back to the left side of the screen for the next sweep. The vertical deflecting coils receive the waveform to be analyzed. This waveform makes the electron beam undulate up and down. The combination of vertical and horizontal beam motion produces a display of the input waveform as a function of time.

Electrostatic CRT

In an *electrostatic CRT,* charged metal plates are used for deflection of the electron beam. When voltages appear on

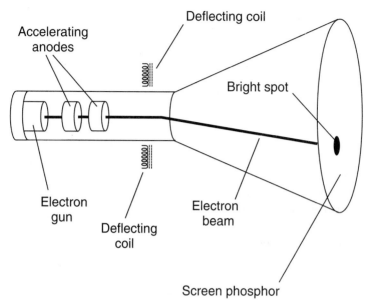

Figure 10.7   Simplified cross-sectional rendition of an electromagnetic CRT.

these *deflecting plates,* the beam is bent in the direction of the electric lines of flux. The greater the voltage applied to a deflecting plate, the stronger the electric field, and the greater the extent to which the beam is deflected. The principal advantage of an electrostatic CRT is that it generates much less magnetic-field energy than an electromagnetic CRT.

Camera Tubes

Video cameras use a form of electron tube that converts visible light into varying electric currents. The two most common types of camera tube are the *vidicon* and the *image orthicon.*

Vidicon

Virtually every *videocassette recorder* (VCR) makes use of a vidicon camera tube. Closed-circuit TV systems also employ vidicons. The main advantage of the vidicon is its small physical size and mass.

In the vidicon, a lens focuses the incoming image onto a photoconductive screen. An electron gun generates a beam that sweeps across the screen via deflecting coils, in a manner similar to the operation of an electromagnetic CRT. The sweep in the vidicon is synchronized with any CRT that displays the image.

As the electron beam scans the photoconductive surface, the screen becomes charged. The rate of discharge in a certain region on the screen depends on the intensity of the visible light striking that region. A simplified cutaway view of a vidicon tube is shown in Fig. 10.8.

A vidicon is sensitive, but its response can be sluggish when the level of illumination is low. This causes images to persist for a short while, resulting in poor portrayal of motion scenes.

Image orthicon

Another type of camera tube, also quite sensitive but having a quicker response to image changes, is the image orthicon. It is constructed much like the vidicon, except that there is

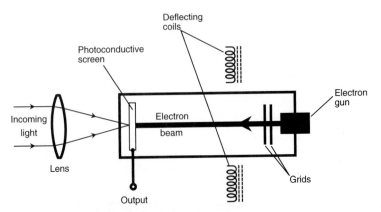

Figure 10.8 Functional diagram of a vidicon.

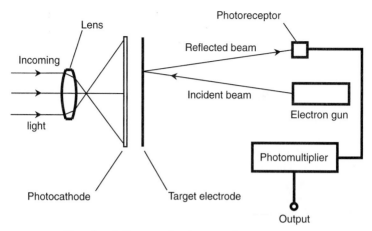

Figure 10.9 Functional diagram of an image orthicon.

a target electrode behind the *photocathode* (Fig. 10.9). When an electron from the photocathode strikes this target electrode, several secondary electrons are emitted as a result. The image orthicon thus acts as a video signal amplifier as well as a camera.

A fine beam of electrons, emitted from the electron gun, scans the target electrode. The secondary electrons cause some of this beam to be reflected back toward the electron gun. Areas of the target electrode with the most secondary-electron emission produce the greatest return beam intensity; regions with the least emission produce the lowest return beam intensity. The greatest return beam intensity corresponds to the brightest parts of the video image. The return beam is modulated as it scans the target electrode and is picked up by a receptor electrode.

The main disadvantage of the image orthicon is that it produces significant noise in addition to the signal output. But when a fast response is needed and the illumination ranges from dim to very bright, the image orthicon is the camera tube of choice. It is common in commercial broadcasting.

Photomultiplier

A *photomultiplier* is a vacuum-tube device that generates a variable current depending on the intensity of the light that strikes it. It gets its name from the fact that it literally multiplies its own output, thereby obtaining extremely high sensitivity. Photomultipliers are used for measurement of light intensity at low levels.

The photomultiplier consists of a photocathode, which emits electrons in proportion to the intensity of the light striking it. These electrons are focused into a beam, and this beam strikes an electrode called a *dynode*. The dynode emits several secondary electrons for each electron that strikes it. The resulting beam is collected by the anode.

A photomultiplier can have several dynodes, resulting in high gain. The extent to which the sensitivity can be improved by cascading dynodes is limited by the amount of background electron emission, or *dark noise,* from the photocathode.

Dissector tube

A *dissector tube,* also known as an *image dissector,* is a form of photomultiplier in which the light is focused by a lens onto a translucent photocathode. This surface emits electrons in proportion to the light intensity. The electrons from the photocathode are directed to a barrier containing a small aperture. The vertical and horizontal deflection plates, supplied with synchronized scanning voltages, move the beam from the photocathode across the aperture. The electron stream passing through the aperture is modulated depending on the light and dark nature of the image.

The *image resolution* of the dissector tube depends on the size of the aperture. The smaller the aperture, the sharper the image, down to a certain limiting point. There is a limit, however, to how small the aperture can be while still allowing enough electrons to pass and avoiding the generation of *interference patterns.* The image dissector tube produces very little dark noise. This results in an excellent signal-to-noise (S/N) ratio.

Traveling-Wave Tubes

A *traveling-wave tube* is used primarily at UHF and microwave frequencies. There are several variations including the *magnetron* and the *klystron*.

Magnetron

A magnetron contains a cathode and a surrounding anode, as shown in Fig. 10.10. The anode is divided into sections, or *cavities*, by radial barriers. (Here, only two cavities are shown for clarity of illustration; often there are several.) The output is taken from an opening in the anode and passes into a *waveguide* that serves as a transmission line for the RF energy.

The cathode is connected to the negative terminal of a high-voltage source, and the anode is connected to the positive terminal. Therefore, electrons flow radially outward. A magnetic field is applied lengthwise through the cavities. As a result, the electron paths are bent into spirals. The electric field produced by the high voltage, interacting with the

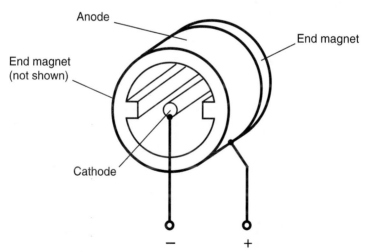

Figure 10.10 Cross-sectional rendition of a magnetron. Only two cavities are shown for clarity; generally there are several cavities.

longitudinal magnetic field and the effects of the cavities, causes the electrons to bunch up into "clouds." The swirling movement of the *electron clouds* causes a fluctuating current in the anode. The frequency depends on the shapes and sizes of the cavities. Small cavities result in the highest oscillation frequencies; larger cavities produce oscillation at relatively lower frequencies.

A magnetron can generate more than 1 kW of RF power at a frequency of 1 GHz. As the frequency increases, the realizable power output decreases. At 10 GHz, a typical magnetron generates about 20 W of RF power output.

Klystron

A klystron is a linear-beam electron tube. It has an electron gun, one or more cavities, and a device that modulates the electron beam. There are several different types. The most common are the *multicavity klystron* and the *reflex klystron*.

A multicavity klystron is shown in Fig. 10.11. In the first cavity, the electron beam is velocity modulated. This causes the density of electrons in the beam to change as the beam moves through subsequent cavities. The intermediate cavi-

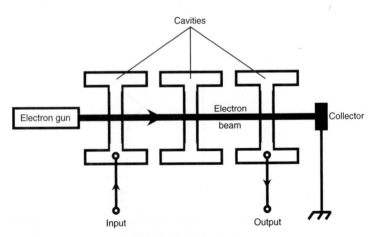

Figure 10.11 Functional diagram of a klystron.

ties increase the magnitude of the electron-beam modulation, resulting in amplification. Output is taken from the last cavity. Peak power levels in some multicavity klystrons can exceed 1 MW (10^6 W), although the average power is much less.

A reflex klystron has a single cavity. A *retarding field* causes the electron beam to periodically reverse direction. This produces a phase reversal that allows large amounts of energy to be drawn from the electrons. A typical reflex klystron produces UHF and microwave signals on the order of a few watts.

11

Oscillators

Some oscillators work at AF, and others are intended to produce RF signals. Most generate sine waves, although some are built to emit square waves, sawtooth waves, or other waveshapes.

In radio communications, oscillators generate signals that are ultimately sent over the air. For data to be sent, the signal from an oscillator must be modulated. Radio-frequency oscillators are used in radio and TV receivers for frequency control, detection, and mixing. Audio oscillators find applications in music synthesizers, fax modems, doorbells, beepers, sirens, alarms, and electronic toys.

RF Oscillators

For a circuit to oscillate, the gain must be high, the feedback must be positive, and the coupling from output to input must be good. Common-emitter and common-source circuits are favored for oscillator design. Common-base and common-gate circuits can be made to oscillate, but these configurations are more often employed as RF power amplifiers.

Feedback

The frequency of an RF oscillator is controlled by means of tuned, or resonant, circuits. These are usually inductance-capacitance (LC) or resistance-capacitance (RC) combinations. The LC scheme is common at RF; the RC method is often used in audio oscillators.

The tuned circuit exhibits low loss at a single frequency but high loss at other frequencies (Fig. 11.1). The result is that the oscillation takes place at a predictable and stable frequency, determined by the inductance and capacitance or by the resistance and capacitance.

Armstrong circuit

A common-emitter or common-source amplifier can be made to oscillate by coupling the output back to the input through a transformer that reverses the phase of the fed-back signal. The phase at a transformer output can be inverted by reversing the secondary terminals. The schematic diagram in Fig. 11.2A shows a common-emitter NPN bipolar-transistor amplifier whose collector circuit is coupled to the emitter circuit via a transformer. In Fig. 11.2B, the equivalent N-channel JFET circuit is shown. These circuits are *Armstrong oscillators*.

The oscillation frequency is controlled by a capacitor across either the primary or the secondary winding of the transformer. The inductance of the winding, along with the capacitance connected in parallel with it, forms a resonant

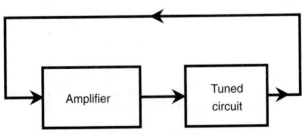

Figure 11.1   Basic concept of an oscillator.

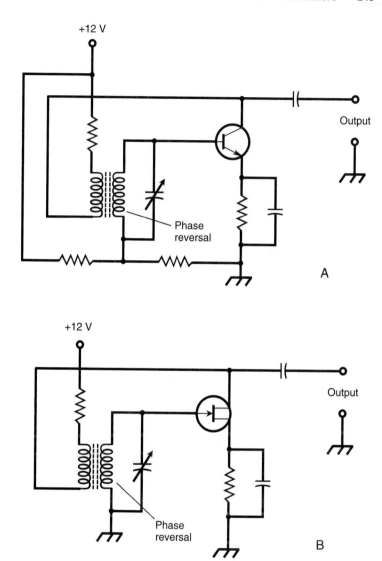

Figure 11.2 Armstrong oscillator. (A) *NPN* bipolar transistor; (B) *N*-channel JFET.

circuit. The fundamental frequency of oscillation is determined according to the formula

$$f = \frac{1}{2\pi \, (LC)^{1/2}}$$

where f is the frequency in megahertz, L is the inductance
of the transformer winding in microhenrys, and C is the value of the parallel capacitor in microfarads. Alternatively, for
low-frequency applications, f can be expressed in hertz, L in
henrys, and C in farads.

Hartley circuit

A method of obtaining controlled feedback at RF is shown in
Fig. 11.3. In Fig. 11.3A, an *NPN* bipolar transistor is used;
in Fig. 11.3B, an *N*-channel JFET is employed. (The *PNP*
and *P*-channel circuits are identical to them, but the power
supply is negative instead of positive.) The circuit uses a
single coil with a tap on the windings to provide the feedback. A variable capacitor in parallel with the coil determines the oscillating frequency and allows for frequency
adjustment. This is a *Hartley oscillator.*

The Hartley circuit employs about 25 percent of its amplified power to produce feedback. The remainder of the power
is available as output. Oscillators do not, in general, produce more than a fraction of a watt of output power. If more
power is needed, the signal can be boosted by one or more
stages of amplification. It is important to use only the minimum amount of feedback necessary to obtain oscillation.
The amount of feedback is controlled by the position of the
coil tap.

Colpitts circuit

Another way to provide RF feedback is to tap the capacitance instead of the inductance in the tuned circuit. In Fig.
11.4, *NPN* bipolar (Fig. 11.4A) and *N*-channel JFET (Fig.
11.4B) Colpitts oscillator circuits are diagrammed.

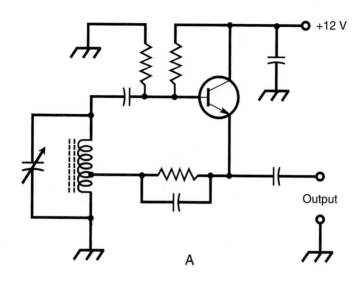

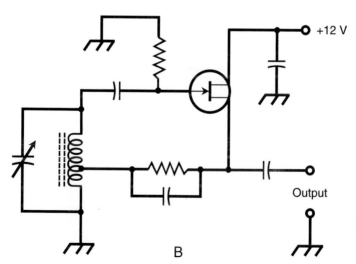

Figure 11.3   Hartley oscillator. (A) *NPN* bipolar transistor; (B) *N*-channel JFET.

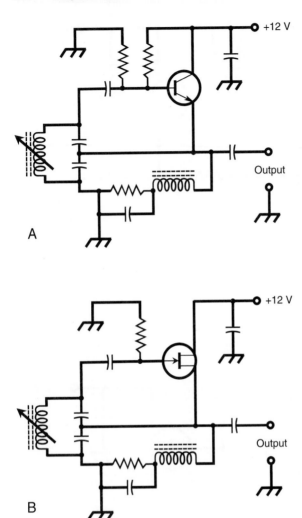

Figure 11.4 Colpitts oscillator. (A) *NPN* bipolar transistor; (B) *N*-channel JFET.

The amount of feedback is controlled by the ratio of capacitances. The oscillation frequency depends on the net capacitance C and also on the value of the inductor L. The general formula for resonant frequency applies (see above). If the series capacitance values are C_1 and C_2, the net capacitance is

$$C = \frac{C_1 C_2}{C_1 + C_2}$$

In the Colpitts circuit, the inductance rather than the capacitance is variable. This is a matter of convenience. It is difficult to find a dual variable capacitor with the proper capacitance ratio between sections for a particular application, but it is easy to adjust the ratio using a pair of fixed capacitors.

The use of variable inductances in RF oscillators is a tricky business. A permeability-tuned coil can be used, but ferromagnetic cores can impair the stability of oscillators at high frequencies. A roller inductor might be employed, but it will probably be bulky and expensive. An inductor with several switch-selectable taps can be used, but this does not allow for continuous frequency adjustment. Despite these complexities, the Colpitts circuit offers exceptional stability and reliability when properly designed. As with the Hartley circuit, the feedback should be kept to the minimum necessary to sustain oscillation.

Clapp circuit

A variation of the Colpitts oscillator employs a series-resonant tuned circuit. A schematic diagram of an *NPN* bipolar-transistor *Clapp oscillator* is shown in Fig. 11.5A; the equivalent *N*-channel JFET circuit is shown in Fig. 11.5B.

The Clapp oscillator offers excellent stability at RF. Its frequency does not fluctuate appreciably when high-quality components are used. The Clapp is a reliable circuit, and it is easy to make it oscillate. Another advantage is that it allows the use of a variable capacitor for frequency control,

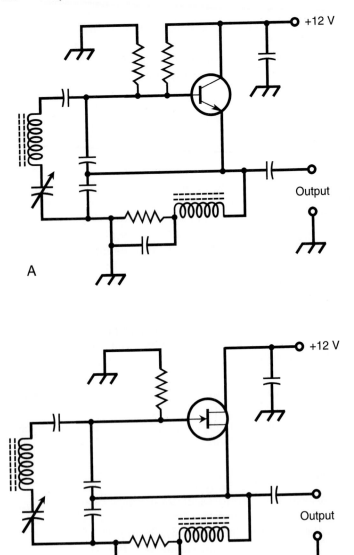

Figure 11.5 Clapp oscillator. (A) *NPN* bipolar transistor; (B) *N*-channel JFET.

while accomplishing feedback through a capacitive voltage divider.

In the Hartley, Colpitts, and Clapp configurations, the output is usually taken from the emitter or source to optimize stability. To prevent the output signal from being short-circuited to ground, an RF choke is connected in series with the emitter or source lead in the Colpitts and Clapp circuits. Typical values for RF chokes are approximately 100 μH at high frequencies such as 15 MHz and 10 mH at low frequencies such as 150 kHz.

Diode oscillator

At ultrahigh and microwave frequencies, certain diodes can be used as oscillators. These components are discussed in Chap. 7.

Oscillator Stability

With respect to oscillators, the term *stability* has two meanings: (1) constancy of frequency and (2) reliability of performance.

Frequency stability

The above-depicted oscillator types allow for frequency adjustment using variable capacitors or variable inductors. The component values are affected by temperature and sometimes by humidity. In a *variable-frequency oscillator* (VFO), it is crucial that the components maintain constant values, as much as possible, under all anticipated conditions.

Some types of capacitors maintain their values better than others as the temperature rises or falls. *Polystyrene capacitors* are excellent in this respect. *Silver-mica capacitors* work well when polystyrene units cannot be found.

Inductors are most temperature-stable when they have air cores. Coils should be wound, when possible, from stiff wire with strips of plastic to keep the windings in place. Some air-core coils are wound on hollow cylindrical forms made of ceramic or phenolic material. Ferromagnetic cores are less desirable for use in RF oscillators because the permeability is affected by temperature. This changes the inductance, in turn affecting the oscillator frequency.

Reliability

An oscillator should begin functioning as soon as the power supply is switched ON. It should keep oscillating under all normal conditions, not quitting if the load changes or the temperature suddenly changes. The failure of a single oscillator can cause an entire communications station to go down.

The oscillator circuits generalized in this chapter are those that engineers have found, through trial and error over the years, to work well. Yet, when an oscillator is built and put to use, *debugging* is often necessary. If two oscillators are built from the same diagram, with the same component types and values in the same geometric arrangement, one circuit might be stable and the other one unstable. The usual reason is a difference in the quality or tolerance of one or more components.

Most oscillators are designed to operate with high load impedances. If the load impedance is low, the load will "try" to draw power from the oscillator. Under these conditions, even a well-designed oscillator can become unstable. Signal power should be delivered via amplifiers following an oscillator, not by the oscillator itself.

Crystal-Controlled Oscillators

Quartz crystals, also called *piezoelectric crystals,* can be used in place of tuned *LC* circuits in RF oscillators, if it is not necessary to change the frequency often. *Crystal-controlled oscillators* offer frequency stability superior to that of *LC*-tuned VFOs. There are several ways that crystals can be connected in bipolar or FET circuits to obtain oscillation.

Pierce circuit

One common crystal-controlled circuit is the *Pierce oscillator.* An active amplifying device and a quartz crystal are connected as shown in Fig. 11.6. In Fig. 11.6A, an *NPN* bipolar transistor is used; in Fig. 11.6B, an *N*-channel JFET is employed.

The frequency of a Pierce oscillator can be varied by about ±0.1 percent via an inductor in series with the crystal or by a capacitor in parallel with the crystal. But the frequency is determined mainly by the thickness of the crystal and by the angle at which it is cut.

Crystals change in frequency as the temperature changes. But they are more stable than *LC* circuits, most of the time. Some crystal oscillators are housed in temperature-controlled chambers called *crystal ovens.* They maintain their frequency so well that they are often used as standards against which other oscillators are calibrated. The accuracy can be within a few hertz at working frequencies of several megahertz.

Reinartz circuit

A *Reinartz crystal oscillator* is characterized by high efficiency and minimal output at harmonic frequencies. The Reinartz configuration can be used with FETs or bipolar transistors.

Figure 11.7 is a schematic diagram of a Reinartz oscillator that employs an *N*-channel JFET. An *LC* circuit is inserted in the source line. This resonant circuit is tuned to approximately half the crystal frequency. The result is enhanced positive feedback, which allows the circuit to oscillate at a lower level of crystal current than would otherwise be possible. The *LC* circuit in the drain line is tuned to a frequency slightly above that of the crystal.

Variable-frequency crystal circuit

The frequency of a *variable-frequency crystal oscillator* (VXO) can be trimmed with the addition of a reactance in series or parallel with the crystal. VXOs are sometimes used in transmitters or transceivers to obtain operation over a small part of a band.

Examples of bipolar-transistor VXO circuits are shown in Fig. 11.8. In Fig. 11.8A, a variable inductor is connected in series with the crystal. In Fig. 11.8B, a variable capaci-

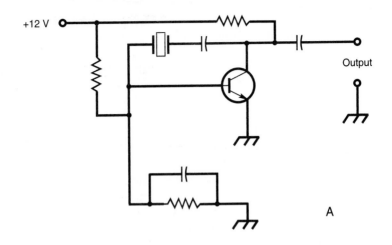

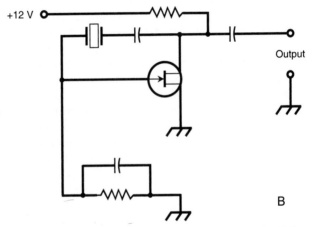

Figure 11.6 Pierce oscillator. (A) *NPN* bipolar transistor; (B) *N*-channel JFET.

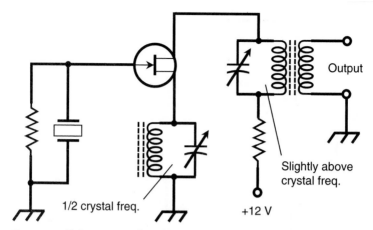

Figure 11.7 Reinartz crystal oscillator using an *N*-channel JFET.

in series with the crystal. In Fig. 11.8B, a variable capacitor is connected in parallel with the crystal.

The main advantage of the VXO over an ordinary VFO is excellent frequency stability. The main disadvantage of the VXO is limited frequency coverage. In most practical cases, the frequency cannot be varied by more than ±0.1 percent of the operating frequency without loss of stability.

Voltage-controlled oscillator

The frequency of a VFO can be adjusted via a *varactor diode* in the tuned *LC* circuit. Hartley and Clapp oscillator circuits are easily adapted to varactor-diode frequency control. The varactor is isolated for dc by blocking capacitors. The frequency is adjusted by applying a variable dc voltage to the varactor.

The schematic diagram of Fig. 11.9 shows a JFET Hartley VCO in which a varactor is used in place of the variable capacitor.

PLL frequency synthesizer

An oscillator that combines the flexibility of a VFO with the stability of a crystal oscillator is known as a *phase-locked-*

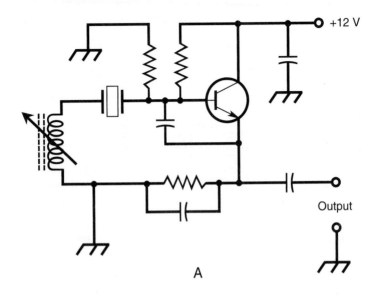

A

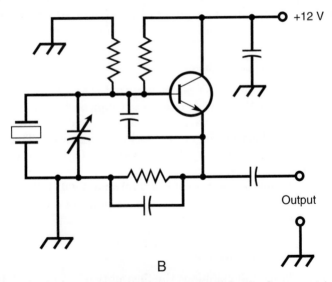

B

Figure 11.8 Bipolar-transistor VXO circuits. (A) Inductive frequency trimming; (B) capacitive frequency trimming.

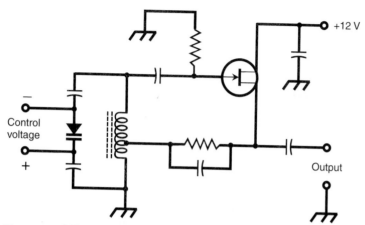

Figure 11.9 A Hartley VCO using an *N*-channel JFET.

loop (PLL) frequency synthesizer. This scheme is extensively used in radio transmitters and receivers.

In a frequency synthesizer, the output of a VCO is passed through a *programmable divider,* a digital circuit that can divide the VCO frequency by any of thousands of numerical values chosen by the operator. The output frequency of the programmable divider is locked, by means of a *phase comparator,* to the signal from a crystal-controlled *reference oscillator.*

As long as the output from the programmable divider is exactly on the reference-oscillator frequency, the two signals are in phase, and the output of the comparator is zero volts dc. If the VCO frequency begins to drift, the output frequency of the programmable divider also drifts (although at a different rate). But even the tiniest frequency change causes the phase comparator to produce a dc *error voltage.* This voltage is either positive or negative, depending on whether the VCO has drifted higher or lower in frequency. The error voltage is applied to a varactor in the VCO, causing the VCO frequency to change in a direction opposite to that of the drift. This forms a dc feedback circuit that maintains the VCO frequency at a precise multiple of the reference-oscillator frequency, that multiple having been chosen by the programmable divider.

The PLL frequency synthesizer is stable because the reference oscillator is crystal-controlled. A block diagram of such a synthesizer is shown in Fig. 11.10. The frequency stability can be enhanced by using an amplified signal from the National Bureau of Standards, transmitted on shortwave by station WWV at 5, 10, or 15 MHz, directly as the reference oscillator.

AF Oscillators

Audio-frequency oscillators are used in doorbells, ambulance sirens, electronic games, personal computers, and toys that play musical tunes. All audio oscillators consist of amplifiers with positive feedback. At AF, oscillators can use *RC* or *LC* combinations to determine frequency.

Waveforms

Radio-frequency oscillators are usually designed to produce a nearly perfect sine-wave output, representing energy at one well-defined frequency. But audio oscillators do not necessarily concentrate all their energy at a single frequency.

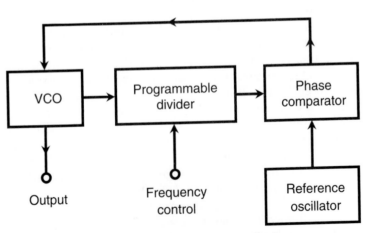

Figure 11.10 Block diagram of a PLL frequency synthesizer.

Various musical instruments sound different, even when they play the same note, because each instrument has its own unique waveform. An instrument's unique sound qualities can be reproduced using an AF oscillator whose waveform output matches that of the instrument. A multimedia computer might have a *musical-instrument digital interface* (MIDI) player that employs audio oscillators capable of duplicating the sounds of a large band or orchestra.

Twin-T oscillator

One form of AF oscillator that is popular for general-purpose use is the *twin-T oscillator* (Fig. 11.11). The frequency is determined by the values of the resistors R and capacitors C. The circuit in this example uses NPN bipolar transistors. Two JFETs could also be used.

Multivibrator

A *multivibrator* uses two amplifier circuits, interconnected so the signal loops between them. N-channel JFETs can be connected as shown in Fig. 11.12. Each JFET amplifies the signal in class A and reverses the phase by $180°$. The frequency is set by means of an LC combination. A *toroidal core,* or *pot core,* is generally used for the inductor. The value of L can range from about 10 μH to 1 H. The capacitance C is chosen to obtain an audio tone at the frequency desired.

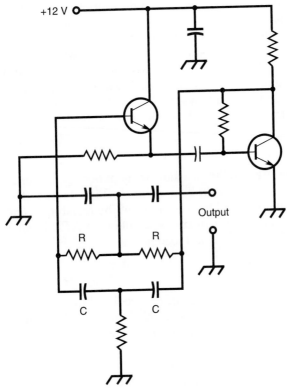

Figure 11.11 A twin-T audio oscillator.

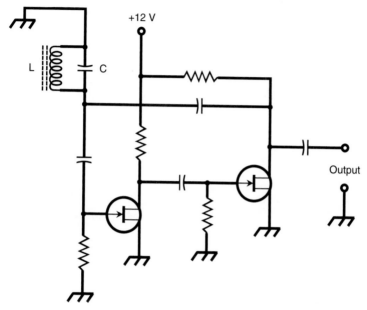

Figure 11.12 A multivibrator-type audio oscillator.

12

Amplifiers

An *amplifier* is a circuit or system that increases the current, voltage, or power level of a signal.

Amplification Factor

The extent to which a circuit amplifies is the *amplification factor,* which is specified in *decibels.* In a given circuit, the amplification factor might differ considerably as measured for voltage, current, and power.

The decibel

Humans perceive most variable quantities in a logarithmic manner. Thus, scientists have devised the *decibel system,* in which amplitude changes are expressed according to the logarithm of the actual value to define relative signal strength. *Gain* is assigned positive decibel values; *loss* is assigned negative values.

An amplitude change of 1 dB is roughly equal to the smallest change a listener or observer can detect if the change is expected. If the change is not expected, the smallest difference a listener or observer can notice is about 3 dB.

For voltage

Suppose there is a circuit with an rms ac input voltage of E_{in} and an rms ac output voltage of E_{out}, both specified in the same units. Then the *voltage gain* of the circuit, in decibels, is given by the formula

$$\text{Gain (dB)} = 20 \log_{10} \frac{E_{out}}{E_{in}}$$

For current

The *current gain* of a circuit is calculated the same way as is voltage gain. If I_{in} is the rms ac input current and I_{out} is the rms ac output current specified in the same units,

$$\text{Gain (dB)} = 20 \log_{10} \frac{I_{out}}{I_{in}}$$

Often, a circuit that produces voltage gain will produce current loss, and vice-versa. An example is an ac transformer.

Some circuits have gain for both the voltage and the current, although not the same decibel figures. The reason for the difference is that the output impedance is higher or lower than the input impedance, altering the ratio of voltage to current.

For power

The *power gain* of a circuit, in decibels, is calculated according to the formula

$$\text{Gain (dB)} = 10 \log_{10} \frac{P_{out}}{P_{in}}$$

where P_{out} is the output signal power and P_{in} is the input signal power, specified in the same units.

Basic Amplifier Circuits

In general, amplifiers must use active components, such as transistors or integrated circuits. Although a transformer can increase the deliverable current or voltage, it cannot produce an output signal that has more power than the input signal.

Generic bipolar amplifier

A generic *NPN* bipolar-transistor amplifier is shown in Fig. 12.1. The input signal passes through a capacitor to the base. Resistors provide bias. In this amplifier, the capacitors must have values large enough to allow the ac signal to pass with ease. But they should not be much larger than the minimum necessary for this purpose. The ideal capacitance values depend on the design frequency of the amplifier and also on the impedances at the input and output. In general, as the frequency and/or circuit impedance increase, less capacitance is needed. Resistor values depend on the input and output impedances.

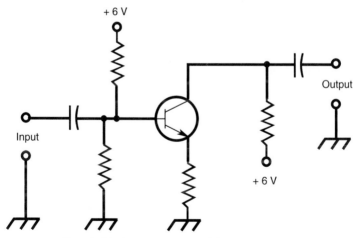

Figure 12.1   Generic bipolar-transistor amplifier circuit.

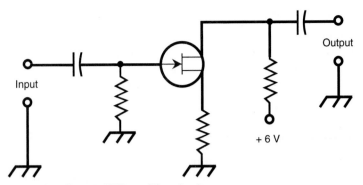

Figure 12.2   Generic FET amplifier circuit.

Generic FET amplifier

A generic *N*-channel JFET amplifier is shown in Fig. 12.2. The same considerations concerning the values of the capacitors apply for this amplifier as for the bipolar circuit. A JFET has a high input impedance, and therefore the value of the input capacitor is small. If the device is a MOS-FET, the input impedance is extremely high, and the input capacitance will be smaller yet, sometimes less than 1 pico-farad (pF). Resistor values depend on the input and output impedances.

Generic IC amplifier

A simple IC amplifier is shown in Fig. 12.3. This circuit uses an op amp. Op amps can perform various electronic functions; they can oscillate, amplify, and filter. The amplifier shown here is an *inverting amplifier,* meaning that the output is 180° out of phase with the input. The gain and frequency response depend on the values of the resistors and capacitors. Op amps are employed at AF but are rarely used at RF.

Amplifier Classes

Amplifier circuits can be categorized as class A, AB, B, or C. Each class has unique characteristics and applications.

Class-A amplifier

Weak-signal amplifiers, such as the kind used in the first stage of a sensitive radio receiver, are always *class-A amplifiers*. This type of amplifier is linear. For class-A operation with a bipolar transistor, the bias is such that, with no signal input, the device is near the middle of the straight-line portion of the collector current (I_C) versus base current (I_B) curve. This is shown for an *NPN* transistor in Fig. 12.4. For *PNP,* reverse the polarity signs.

With a JFET or MOSFET, the bias must be such that, with no signal input, the device is near the middle of the straight-line part of the drain current (I_D) versus gate voltage (E_G) curve. This is shown in Fig. 12.5 for an *N*-channel device. For *P*-channel, reverse the polarity signs.

Class-AB amplifier

When a bipolar transistor is biased close to *cutoff* under no-signal conditions (Fig. 12.4) or when an FET is near *pinch off* (Fig. 12.5), the input signal will drive the device into the nonlinear part of the operating curve. This is *class-AB operation.* The input signal might cause the device to go into cutoff or pinch off for a small part of the cycle. Whether or not this happens depends on the actual bias point and also on

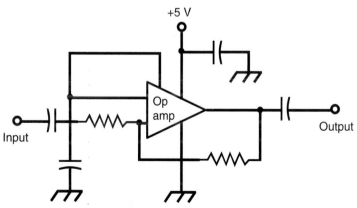

Figure 12.3   Generic operational-amplifier circuit.

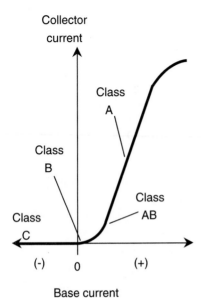

Figure 12.4 Classes of amplification for a bipolar transistor.

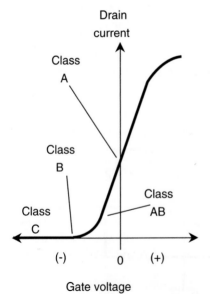

Figure 12.5 Classes of amplification for an *N*-channel JFET.

the strength of the input signal. If the bipolar transistor or FET is never driven into cutoff or pinch off during any part of the signal cycle, the circuit is a *class-AB₁ amplifier*. If the device goes into cutoff or pinch off for any part of the cycle, the circuit is a *class-AB₂ amplifier*.

In any class-AB amplifier, the output waveform differs from the input waveform. But if the signal is modulated, such as in a voice radio transmitter, the modulation envelope will emerge undistorted. Therefore, class-AB operation is useful in RF power amplifier (PA) systems.

Class-B amplifier

When a bipolar transistor is biased exactly at cutoff, or an FET at pinch off, under zero-input-signal conditions, an amplifier is working in class B. These operating points are labeled on the curves in Figs. 12.4 and 12.5. In a *class-B amplifier*, there is no collector or drain current when there is no signal. This saves energy compared with class-A and class-AB circuits. When there is an input signal, current flows in the device during exactly half of the cycle.

Sometimes two bipolar transistors or FETs are used in a class-B circuit, one for the positive half of the cycle and the other for the negative half. In this way, distortion is eliminated. This is a *class-B push-pull amplifier* (Fig. 12.6) and is commonly used in audio applications.

The class-B scheme can be used for RF power amplification. Although the output waveshape is distorted, resulting in harmonic energy, this problem can be overcome by a resonant *LC* circuit in the output. If the signal is modulated, the modulation envelope will not be distorted.

Class-C amplifier

A bipolar transistor or FET can be biased past cutoff or pinch off, and it will work as a PA if the drive is sufficient to overcome the bias during part of the cycle. The operating points for class C are labeled in Figs. 12.4 and 12.5.

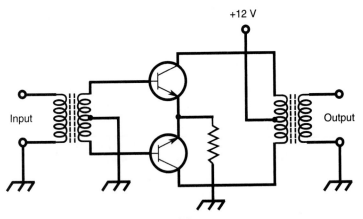

Figure 12.6   A class-B push-pull amplifier.

A class-C RF PA is nonlinear for amplitude-modulation envelopes. Therefore, a class-C circuit is useful only for signals whose amplitudes are either constant or have only two states (ON and OFF, or mark and space). Continuous-wave (CW) *radiotelegraphy, radioteletype* (RTTY), and FM are examples of such signals. A class-C RF PA needs substantial driving power.

Efficiency and Drive

In a PA, *efficiency* is the ratio of the useful power output to the total power input. High efficiency translates into minimal cost, optimal equipment size and weight, and maximum component life.

DC power input

The *dc power input* (P_{in}), in watts, to a bipolar-transistor amplifier circuit is the product of the collector current (I_C) in amperes and the collector voltage (E_C) in volts. For an FET, the dc power input is the product of the drain current (I_D) and the drain voltage (E_D). Mathematically,

$$P_{in} = E_C I_C$$

for bipolar-transistor power amplifiers and

$$P_{in} = E_D I_D$$

for FET power amplifiers.

In some circuits, the dc power input is considerable even when there is no signal applied. In class A, the average dc power input is constant whether there is an input signal or not. In class AB_1 or AB_2, there is a small dc power input with no signal input and significant dc power input with the application of a signal. In classes B and C, there is zero dc power input when there is no input signal and significant dc power input with the application of a signal.

Signal power output

When there is no signal input to an amplifier, there is no signal at the output, and therefore the *signal power output* (P_{out}) is zero. This is true no matter what the class of amplification. The greater the signal input, in general, the greater the power output of a power amplifier, up to a certain point.

The signal power output from an amplifier cannot be directly measured using dc instruments. Specialized ac wattmeters are necessary.

Definition of efficiency

The *efficiency* (Eff) of a PA is the ratio of the signal power output to the dc power input:

$$Eff = \frac{P_{out}}{P_{in}}$$

This ratio is always between 0 and 1. Efficiency is often expressed as a percentage between zero and 100 ($Eff_\%$), so the formula becomes

$$Eff_\% = \frac{100 P_{out}}{P_{in}}$$

Efficiency versus class

Class-A amplifiers are inefficient in general. Their efficiency ratings vary from 25 to 40 percent, depending on the nature of the input signal and the type of bipolar or field-effect transistor used. If the input signal is weak, such as is the case at the antenna input of a radio receiver, the efficiency of a class-A circuit is near zero. But in that application, efficiency is not of primary importance. A high gain and low *noise figure* (internally generated circuit noise) are vastly more important.

A class-AB_1 RF PA is 35 to 45 percent efficient; a class-AB_2 RF PA can approach 60 percent. Class-B amplifiers are 50 to 65 percent efficient, and class-C RF PA systems can be up to 75 percent efficient.

Drive and overdrive

In theory, class-A and class-AB_1 power amplifiers draw no power from a signal source to produce useful power output. Class-AB_2 amplifiers need some driving power to produce output. Class-B amplifiers require more drive than class-AB_2, and class-C amplifiers need still more. Whatever class of PA is used, it is important that the driving signal not be too strong. If *overdrive* takes place, problems occur.

In Fig. 12.7, the output signal waveshapes for power amplifiers are shown in various situations. In Fig. 12.7A and B, the amplifier operates in class-A or in class-B push-pull. The waveform in Fig. 12.7A shows the signal output from a properly driven amplifier, and the waveform in Fig. 12.7B shows the signal output from an overdriven amplifier. Figure 12.7C and D shows the output signal waveforms from properly driven and overdriven class-B RF PAs, respectively. Figure 12.7E and F shows the output waveforms from properly driven and overdriven class-C RF PAs, respectively. Note the *flat topping* that occurs with overdrive. This causes excessive harmonic emission, modulation-envelope distortion, and degraded efficiency.

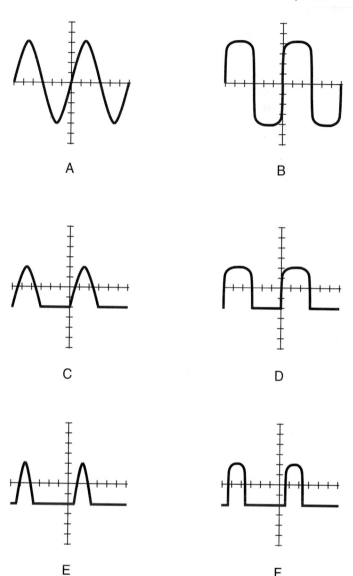

Figure 12.7 Output waveforms of power amplifiers: (A) properly driven class A; (B) overdriven class A; (C) properly driven class B; (D) overdriven class B; (E) properly driven class C; (F) overdriven class C.

AF Amplification

AF amplifiers are intended for use below approximately 100 kHz. Characteristics include adjustable gain, adjustable frequency response, and low distortion.

Frequency response

High-fidelity audio amplifiers work from a few hertz up to approximately 100 kHz. Audio amplifiers for voice communications cover about 300 Hz to 3 kHz. In digital communications, audio amplifiers work over a narrow range of frequencies, sometimes less than 100 Hz wide.

High-frequency amplifiers are equipped with *RC tone controls* that tailor the frequency response. The simplest amplifiers use a single knob to control the tone. More sophisticated systems have separate controls, one for bass and the other for treble. The most advanced systems have *graphic equalizers,* sets of several controls that affect the amplifier gain over various frequency spans.

Gain-versus-frequency curves for three hypothetical audio amplifiers are shown in Fig. 12.8. In Fig. 12.8A, a wideband, flat curve is illustrated. This is typical of hi-fi system amplifiers. In Fig. 12.8B, a voice communications response is shown. In Fig. 12.8C, a narrowband response curve, typical of audio amplifiers in digital wireless receivers, is illustrated.

Volume control

Audio amplifier systems usually consist of several stages. In an early stage, a *volume control* is used. An example is shown in Fig. 12.9. The ac output signal passes through C_1 and appears across R_1, a potentiometer. The wiper (indicated by the arrow) of the potentiometer "picks off" more or less of the ac output signal. Capacitor C_2 isolates the potentiometer from the dc bias of the following stage.

Volume control is usually done in a stage where the audio power level is low. This allows the use of a potentiometer rated for about 1 W.

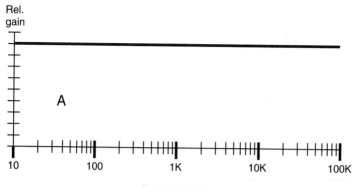

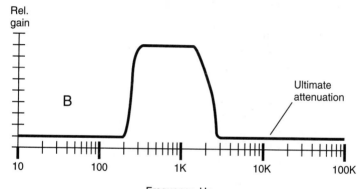

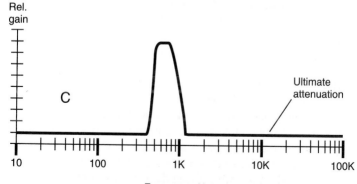

Figure 12.8 Audio amplifier frequency response curves: (A) typical hi-fi; (B) voice communications; (C) digital communications.

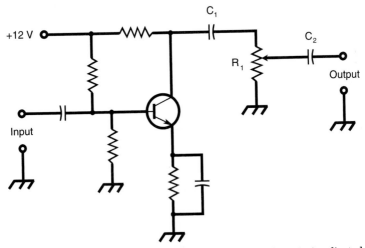

Figure 12.9 A simple audio-amplifier volume control. The gain is adjusted via R_1.

Transformer and tuned-circuit coupling

An example of *transformer coupling* is shown in Fig. 12.10. Capacitors C_1 and C_2 keep one end of the transformer primary and secondary at signal ground. Resistor R_1 limits the current through the first transistor, Q_1. Resistors R_2 and R_3 provide the proper base bias for transistor Q_2. This scheme provides optimum signal transfer between amplifier stages with minimum loss because of the impedance-matching ability of the transformer. The output impedance of Q_1 can be perfectly matched to the input impedance of Q_2.

In some amplifier systems, capacitors are added across the primary and/or secondary of the transformer. This results in *resonance* at a frequency determined by the capacitance and the transformer winding inductance. If the set of amplifiers is intended for a single frequency, *tuned-circuit coupling* enhances the efficiency. But care must be taken to ensure that the amplifier chain does not oscillate at the resonant frequency.

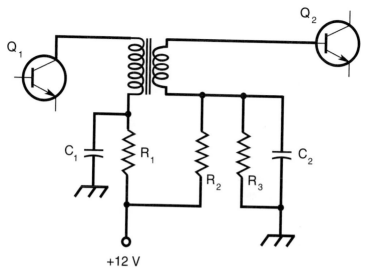

Figure 12.10 An example of transformer coupling.

RF Amplification

The RF spectrum begins at about 9 kHz and extends to over 300 GHz. At the low-frequency end of this range, RF amplifiers resemble AF amplifiers. As the frequency increases, amplifier design and characteristics change.

Weak-signal amplifiers

The *front end,* or first amplifying stage, of a radio receiver requires the most sensitive possible amplifier. Sensitivity is determined by two factors: *gain* and *noise figure.* Gain is measured in decibels. Noise figure is a measure of how well a circuit or system can amplify desired signals without injecting unwanted noise. All bipolar transistors or FETs create some noise because of the movement of charge carriers. In general, FETs produce less noise than bipolar transistors.

The higher the frequency for which a weak-signal amplifier is designed, the more important the noise figure becomes, because there is less atmospheric noise at the higher radio frequencies, as compared with the lower frequencies.

Weak-signal amplifiers almost always use resonant circuits. This optimizes the amplification at the desired frequency, while helping to cut out noise on unwanted frequencies. A tuned GaAsFET weak-signal RF amplifier is shown in Fig. 12.11. It is designed for approximately 10 MHz. At higher frequencies, the inductances and capacitances are smaller; at lower frequencies, the values are larger.

Broadband RF power amplifiers

The main advantage of a *broadband RF PA* is ease of operation, because the circuit does not require tuning. The operator need not worry about critical adjustments or bother to change settings when changing the frequency within a certain range. However, broadband RF PAs are slightly less efficient than their tuned counterparts.

A broadband PA will amplify any signal whose frequency lies within the design range, whether or not that signal is

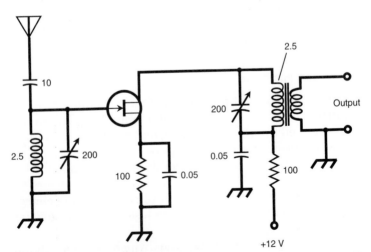

Figure 12.11 A tuned weak-signal RF amplifier for use at about 10 MHz. Resistances are in ohms. Capacitances are in microfarads if less than 1 and in picofarads if greater than 1. Inductances are in microhenrys.

intended for transmission. If some stage in a transmitter is oscillating at a frequency other than the intended signal frequency, and if this undesired signal falls within the design frequency range of a broadband PA, the signal will be amplified. The result will be *spurious emissions* from the transmitter.

A typical broadband PA circuit is diagrammed schematically in Fig. 12.12. The *NPN* bipolar transistor is a power transistor. The transformers are a critical part of this circuit; they must be designed to function efficiently over a wide range of frequencies, preferably 10:1 or more. This circuit is suitable for use on the amateur radio bands at wavelengths between 160 and 20 m.

Tuned RF power amplifiers

A *tuned RF PA* offers improved efficiency compared with broadband designs. Also, the tuning helps to reduce the chance of spurious signals being amplified and transmitted over the air. A tuned RF PA can work into a wide range of

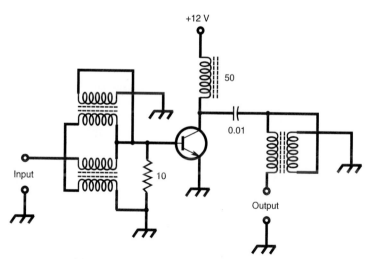

Figure 12.12 A broadband RF power amplifier, capable of producing a few watts output. Resistances are in ohms. Capacitances are in microfarads if less than 1 and in picofarads if greater than 1. Inductances are in microhenrys.

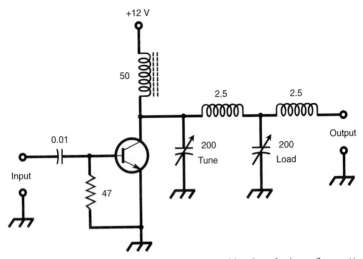

Figure 12.13   A tuned RF power amplifier, capable of producing a few watts output. Resistances are in ohms. Capacitances are in microfarads if less than 1 and in picofarads if greater than 1. Inductances are in microhenrys.

load impedances. In addition to a *tuning control,* or resonant circuit that adjusts the output of the amplifier to the operating frequency, there is a *loading control* that optimizes the signal transfer between the amplifier and the load (usually an antenna system).

The main drawbacks of a tuned PA are that adjustment can be time consuming and improper adjustment can result in damage to the amplifying device (bipolar transistor or FET). If the tuning and/or loading controls are improperly set, the efficiency of the amplifier will be near zero while the dc collector or drain power input is high.

A tuned RF PA, providing useful power output at approximately 10 MHz, is shown in Fig. 12.13. The tuning and loading controls must be adjusted for maximum RF power output as indicated on a wattmeter in the feed line going to the load.

13

Filters and Attenuators

A *filter* tailors the way an electronic circuit or system responds to energy at various frequencies. An *attenuator* reduces the amplitude of a signal to a specified extent over a range of frequencies.

Selectivity Curves

In electronic circuits, there are several types of *selectivity curves* or response patterns. The most common filters include bandpass, band-rejection, notch, high-pass, low-pass, and graphic-equalizer devices.

Bandpass filter

A circuit that discriminates against all frequencies except a specific frequency f_0, or a band of frequencies between two limiting frequencies f_0 and f_1, is a *bandpass filter*. In a parallel-tuned inductance-capacitance (LC) circuit (Fig. 13.1A), a bandpass filter exhibits high impedance at desired frequencies and low impedance at unwanted frequencies. In a series LC configuration (shown in Fig. 13.1B), the filter

has low impedance at desired frequencies and high impedance at unwanted frequencies.

Bandpass filters do not always consist of coils and capacitors. Quartz crystals can be used. Lengths of transmission
line, either short-circuited or open, are useful as bandpass
filters at the higher radio frequencies.

A bandpass filter can have a well-defined resonant frequency f_0 (Fig. 13.1C), or the response might be flat in the
passband, with two cutoff frequencies (Fig. 13.1D). The
bandwidth might be as narrow as a few dozen hertz, such as
with an audio filter designed for reception of frequency-shift

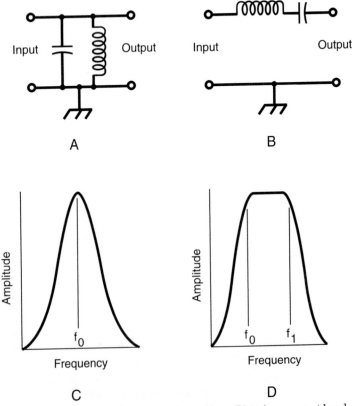

Figure 13.1   (A) Parallel-resonant bandpass filter; (B) series-resonant bandpass filter; (C) sharp bandpass response; (D) broad bandpass response.

keying (FSK). Or the bandwidth might be as wide as several megahertz, as in a helical filter designed for the front end of a VHF or UHF radio receiver.

Band-rejection filter

A *band-rejection filter,* also called a *bandstop filter,* is designed to pass energy at all frequencies except within a certain range. The attenuation is greatest at the resonant frequency f_0 or between two limiting frequencies f_0 and f_1. Figure 13.2A shows a simple parallel-resonant LC band-rejection filter; Fig. 13.2B shows a simple series-resonant LC band-rejection filter. A common example of a band-rejection filter is a *trap* in a multiband antenna. Another is a *parasitic suppressor* in a power amplifier.

Band-rejection filters, like bandpass filters, need not consist of coils and capacitors. Quartz crystals or lengths of transmission line are commonly used at radio frequencies.

All band-rejection filters have attenuation-versus-frequency responses characterized by low loss at all frequencies, except those frequencies within a prescribed range. Figure 13.2C and D shows two types of band-rejection response. A sharp, or peaked, response (Fig. 13.2C) occurs at or near a single resonant frequency. A rectangular response (Fig. 13.2D) exhibits low attenuation below a limit f_0 and above a limit f_1 and high attenuation between these limiting frequencies.

Notch filter

A *notch filter* is a narrowband-rejection filter that can reduce interference caused by strong, unmodulated carriers within the passband of a receiver. Notch-filter circuits are generally inserted in one of the IF stages of a superheterodyne receiver, where the bandpass frequency is constant. There are several different kinds of notch-filter circuit. One of the simplest is a parallel-tuned, band-rejection LC circuit in series with the signal path (Fig. 13.2A). More sophisticated filters employ multiple-tuned LC circuits or piezoelectric materials such as ceramics or quartz crystals.

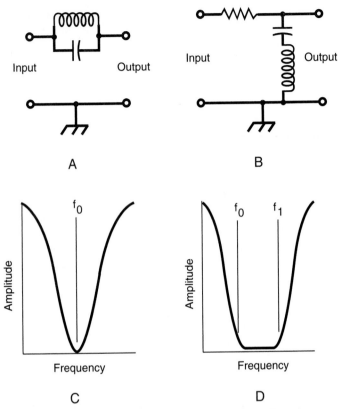

Figure 13.2 (A) Parallel-resonant band-rejection filter; (B) series-resonant band-rejection filter; (C) sharp band-rejection response; (D) broad band-rejection response.

High-pass filter

A *high-pass filter* exhibits high attenuation below its cutoff frequency and little or no attenuation above that frequency. At the cutoff, the power attenuation is 3 dB (and the voltage attenuation is 6 dB, assuming constant impedance) with respect to the attenuation at much higher frequencies.

A basic *LC* high-pass filter consists of parallel inductors and/or series capacitors, such as the circuits shown in Fig. 13.3. The filter in Fig. 13.3A is an *L-section high-pass filter*; that in Fig. 13.3B is a *T-section high-pass filter*. Resistors can be substituted for the inductors in audio applications

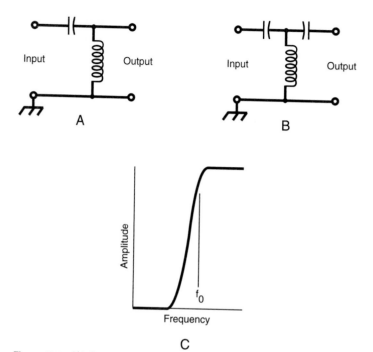

Figure 13.3 (A) L-section high-pass filter; (B) T-section high-pass filter; (C) typical high-pass response.

when active devices, such as operational amplifiers (op amps), are used. A *high-pass response* shows greater attenuation at lower frequencies than at higher frequencies. An example is shown in Fig. 13.3C.

A common use for an RF high-pass filter is at the input of a TV or FM broadcast receiver. The cutoff frequency is approximately 40 MHz. The filter reduces the susceptibility of the receiver to *electromagnetic interference (EMI)* from RF sources at lower frequencies, while allowing the TV signals, which are much higher than 40 MHz, to pass unhindered.

Low-pass filter

A *low-pass filter* exhibits high attenuation above its cutoff frequency and little or no attenuation below that frequency.

At the cutoff, the power attenuation is 3 dB (and the voltage attenuation is 6 dB, assuming constant impedance) with respect to the attenuation at much lower frequencies.

A basic *LC* low-pass filter consists of series inductors and/or parallel capacitors, such as the circuits shown in Fig. 13.4. The filter in Fig. 13.4A is an *L-section low-pass filter*; that in Fig. 13.4B is a *pi-section low-pass filter*. Resistors can be substituted for the inductors in audio applications when active devices, such as op amps, are used. A *low-pass response* shows greater attenuation at higher frequencies than at lower frequencies. A typical low-pass response is shown in Fig. 13.4C.

A low-pass filter can be inserted in the antenna feed line for an RF transmitter in the "shortwave" bands (frequencies below 30 MHz). The cutoff frequency is about 40 MHz. The filter reduces the risk, or the severity, of EMI to TV and FM broadcast receivers using independent antennas (rather than cable) in the vicinity.

Graphic equalizer

A *graphic equalizer* allows tailoring of the audio output of hi-fi sound equipment. There are several independent tone controls, each affecting a certain part of the audio spectrum. The controls are slide potentiometers that provide an intuitive graph, on the control panel, of the output response curve.

Figure 13.5 is a block diagram of a graphic equalizer with seven controls. The input is fed to an audio splitter that breaks the signal into paths of equal and independent impedance and prevents interaction among the circuits. The signals are fed to audio bandpass filters, each having its own gain control. The center frequencies of the filters in this example are 30, 100, 300, and 900 Hz and 2.5, 7, and 18 kHz. These are not standard frequencies; they are given here for illustrative purposes.

Selective Filter Characteristics

All selective filters have certain characteristics that are important to design engineers. The most common considerations are as follows.

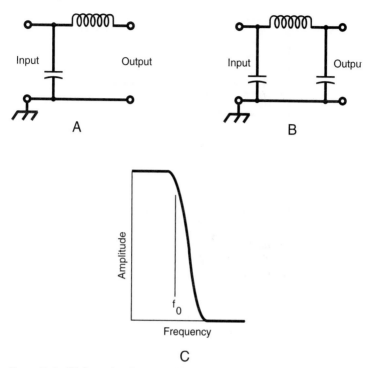

Figure 13.4 (A) L-section low-pass filter; (B) pi-section low-pass filter; (C) typical low-pass response.

Insertion gain

Insertion gain is a comparison of output signal amplitude with and without an active filter in the line. If the signal voltage output of a device is E_1 without the filter in the line and E_2 with the filter in the line, the insertion gain G, assuming constant and identical input and output impedances, is

$$G \text{ (dB)} = 20 \log_{10} \frac{E_2}{E_1}$$

If I_1 is the signal current output without the filter in the line and I_2 is the signal current output with the filter in the line, the insertion gain G, assuming constant and identical input and output impedances, is

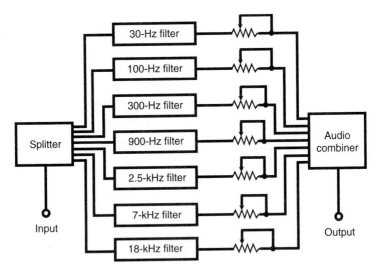

Figure 13.5 Example of a graphic equalizer for audio applications.

$$G \text{ (dB)} = 20 \log_{10} \frac{I_2}{I_1}$$

If P_1 is the signal output power without the filter in the line and P_2 is the signal output power with the filter in the line, the insertion gain G is

$$G \text{ (dB)} = 10 \log_{10} \frac{P_2}{P_1}$$

Insertion loss

Insertion loss is a comparison of output signal amplitude with and without a passive filter or attenuator in the line.

If the signal voltage output of a system is E_1 without an intermediary network in the line and E_2 with the network in the line, the insertion loss L, assuming constant and identical input and output impedances, is

$$L \text{ (dB)} = 20 \log_{10} \frac{E_1}{E_2}$$

If I_1 is the signal current output without a specific device in the line and I_2 is the signal current output with the device in the line, the insertion loss $L,$ assuming constant and identical input and output impedances, is

$$L \text{ (dB)} = 20 \log_{10} \frac{I_1}{I_2}$$

If P_1 is the signal output power without a device in the line and P_2 is the signal output power with the device in the line, the insertion loss L is

$$L \text{ (dB)} = 10 \log_{10} \frac{P_1}{P_2}$$

Bandwidth

Bandwidth is a quantitative measure of the spectrum space that a signal consumes. At the receiver, the selective filter should have a passband that corresponds to the bandwidth of the signal being received. The greater the bandwidth, the more data that can be transferred per unit time. The less the bandwidth, the more severe the constraints on data speed. This is true whether the signals are analog or digital and regardless of the modulation mode.

When displayed on a spectrum analyzer, a steady, unmodulated carrier is a single vertical pip (Fig. 13.6A). When a carrier has modulation, its bandwidth attains some finite value (Fig. 13.6B and C). The *signal bandwidth* is defined as the amount by which the 3-dB power-attenuation points (or the 6-dB voltage-attenuation points, assuming constant and identical input and output impedances) of the signal differ in frequency. The *filter bandwidth* is the difference in frequency between these points in the receiver system.

Half-power points

Figure 13.7 shows a hypothetical bandpass-filter response. The *half-power points* in this case indicate the bandwidth. For a bandpass filter, the half-power points are the frequencies at which the power from the filter drops 3 dB below the

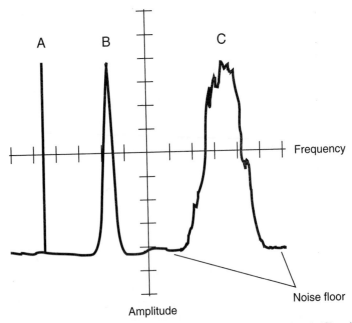

Figure 13.6   (A) An unmodulated carrier; (B) a narrowband signal; (C) a signal with wider bandwidth.

highest power output level in the passband. Half-power points can be specified for all types of selective filters: bandpass (as shown), high-pass, low-pass, and band-rejection.

The response characteristic of a bandpass filter can be described in terms of the half-power and the 30-dB power-attenuation points. This combined figure gives an indication of the skirt selectivity, or steepness of the cutoff, as well as the actual bandwidth. For example, a given filter might have a bandwidth of 2.700 kHz at the half-power points and 3.325 Hz at the 30-dB power-attenuation points. This would be suitable for single-sideband (SSB) reception.

Arithmetic symmetry

The term *arithmetic symmetry* refers to the shape of the spectrum of a signal or of a bandpass or band-rejection filter

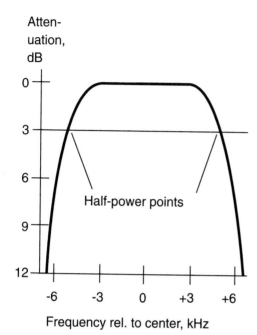

Figure 13.7 Half-power points for a passband filter.

response. Most digital signals exhibit this property, as do many analog signals. Some analog signals (such as SSB) are asymmetrical.

A theoretical example of arithmetic symmetry in a bandpass response is shown in Fig. 13.8. The curve is exactly symmetrical around the center frequency; the left-hand side (below center frequency) is a mirror image of the right-hand side (above center). In filter design, arithmetic symmetry represents a theoretical ideal.

Rectangularity

A bandpass or band-rejection filter is usually designed to have a nearly *rectangular response*. A theoretically perfect rectangular bandpass response would exhibit zero attenuation between the cutoff frequencies and infinite attenuation

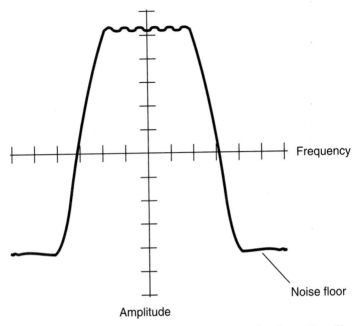

Figure 13.8 An example of arithmetic symmetry in a bandpass filter. The left and right halves of the response curve are exact mirror images.

outside the bandpass. A theoretically perfect rectangular band-rejection response would have infinite attenuation between the cutoff points and zero attenuation at all other frequencies.

In a radio receiver, a rectangular bandpass response provides the best possible *S/N ratio*. A single resonant circuit has a *peaked response,* whereas a group of tuned circuits or resonant devices, each having a slightly different natural frequency, can be connected together to produce a nearly rectangular response.

Specialized Filter Types

Several specific filter designs, commonly encountered in electronic equipment, are depicted in the following paragraphs and illustrations.

M-derived *LC* filter

An *m-derived LC filter* gets its name from the fact that the values of L and C are multiplied by a common factor, designated m. Figure 13.9A shows a typical T-section low-pass filter for unbalanced line. Figure 13.9B shows an *m*-derived filter with component values altered by the factor m. An additional inductor is placed in series with the capacitor in the *m*-derived filter.

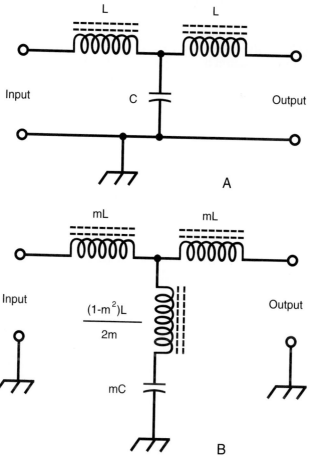

Figure 13.9 (A) A conventional T-section low-pass filter; (B) an *m*-derived, T-section low-pass filter.

The value of the factor m is always between 0 and 1. The optimum value for m in a given situation depends on the type of response and the cutoff frequency desired. A properly designed m-derived filter has a sharper cutoff than a simple LC filter for a given frequency.

Butterworth filter

A *Butterworth filter* is designed to have a flat response in its passband and a smooth roll-off characteristic (skirt). This type of filter can be designed to exhibit a low-pass, high-pass, bandpass, or band-rejection response.

Figure 13.10 shows ideal Butterworth responses for low-pass, high-pass, and bandpass filters. Examples of low-pass, high-pass, and bandpass Butterworth filters are shown in Fig. 13.11. The input and output impedances must be correctly chosen for proper operation. The values of the components depend on these impedances.

Chebyshev filter

A *Chebyshev filter* (also spelled Tschebyscheff or Tschebyshev) is similar to the Butterworth filter. It has a nearly flat response within its passband, nearly complete attenuation outside the passband, and a sharp cutoff response. The primary advantage over other types of filters is the extremely sharp roll-off, so the response is close to the theoretical rectangular ideal.

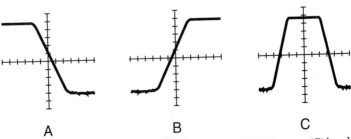

Figure 13.10   Butterworth responses: (A) low-pass; (B) high-pass; (C) bandpass. The horizontal scales show frequency; the vertical scales show amplitude.

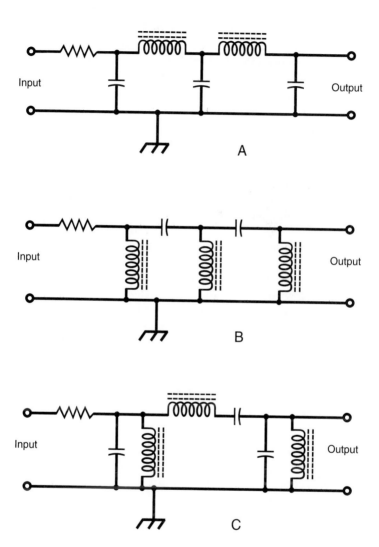

Figure 13.11 Butterworth *LC* filters: (A) low-pass; (B) high-pass; (C) bandpass.

Response curves for Chebyshev filters are similar to those shown in Fig. 13.10, but the skirts are steeper. The schematic diagrams are the same as those in Fig. 13.11, except the resistors are not used. The input and load impedances must be properly chosen for the best filter response. The values of the inductors and capacitors depend on the input and load impedances and also on the frequency response desired.

Mechanical filter

A *mechanical filter,* sometimes called an *ultrasonic filter,* has a bandpass response. The input signal is converted into acoustical vibration by an electromechanical transducer that is similar to a miniature loudspeaker. The vibrations travel through a set of resonant metallic disks to an output transducer that resembles a miniature microphone. The output transducer converts the vibrations back into an electrical signal.

A mechanical filter offers some advantages over electrical filters at low and medium frequencies. No adjustment is needed (or possible) because the resonant disks are of fixed size. Mechanical filters can be designed to have a nearly flat response within the passband and very steep skirts with excellent ultimate attenuation. A mechanical filter is sensitive to physical shock. Care must be exercised to ensure that it is not subjected to excessive vibration.

Piezoelectric filters

A *ceramic filter* makes use of piezoelectric materials to obtain a bandpass response. Ceramic disks resonate at the filter frequency. A ceramic filter is essentially the same as a mechanical filter in terms of construction; the only difference is the composition of the disk material.

A *crystal filter* is similar in construction to a ceramic filter, consisting of several separate piezoelectric crystals cut to slightly different resonant frequencies to obtain the desired bandwidth and selectivity characteristics.

Ceramic and crystal filters are used to provide selectivity in the IF sections of transmitters and receivers. When the filters are properly terminated at their input and output sections, the response is nearly rectangular.

Line Filters

A *line filter* is a device that can be inserted in the ac power-supply cord for an electronic system. Line filters generally consist of series inductors and/or parallel capacitors.

Basic line filter design

A simple ac line filter is shown in Fig. 13.12. The capacitors have values of about 0.1 µF and are rated at several hundred volts. The inductors are wound from wire heavy enough to deal with the current demanded by the load. Toroidal ferromagnetic cores offer high permeability and allow the inductors to be compact. The *LC* combination acts as a low-pass filter; the cutoff is somewhat above the 60-Hz line frequency.

Line filters are available from commercial sources. These filters typically have several outlets and are rated at 10 to 20 A for 117-V ac service. A circuit breaker protects the equipment in the event of a severe transient, such as the induced

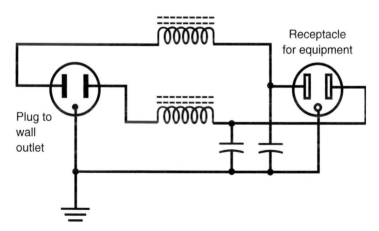

Figure 13.12 An ac line filter reduces the effects of transients and EMI.

voltage from a nearby lightning stroke. The circuit breaker also protects the components in the filter from damage in case a piece of equipment shorts out.

Uses for line filters

A line filter is useful for *transient suppression* in utility lines. Transients, also called "spikes" or (mistakenly) "surges," can cause some electronic equipment, particularly computers, to malfunction.

Line filters are helpful in reducing EMI that is sometimes conducted along ac power lines. Installed between a communications transmitter and the utility lines, such a filter can choke off RF currents and help keep the utility wiring from acting as an antenna. In the power cords of home entertainment equipment such as a stereo hi-fi amplifier, line filters can keep stray RF energy from entering the apparatus through its power supply.

Attenuators

An *attenuator* is a component, circuit, or device deliberately designed to reduce the strength of a signal by a specified amount. Some attenuators are designed for AF work, and others are intended to function at RF.

Basic attenuator design

Most attenuators are constructed with noninductive resistors, so they can function over a wide band of frequencies.

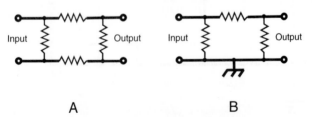

Figure 13.13   Basic attenuator circuits. (A) Balanced; (B) unbalanced.

Two simple examples are shown in Fig. 13.13. The circuit in Fig. 13.13A is for use with a system having a balanced input and a balanced output; the circuit in Fig. 13.13B is for an unbalanced input and an unbalanced output. The values of the resistors depend on the input impedance, the output impedance, and the extent of attenuation.

The circuits shown are fixed (nonvariable) attenuators, also known as *padders* or *pads*. If the extent of the attenuation must be adjustable, the design becomes complex if constant input and output impedance are to be maintained.

A frequency-sensitive attenuator for AF can employ an *RC* combination in conjunction with an op amp wired for gain less than unity. A frequency-sensitive attenuator for RF can use either an *RC* combination or a resistance-inductance-capacitance (*RLC*) combination.

All attenuators must be built to withstand the power applied to them. In receivers, this is not a problem; small resistors (1/4 or 1/8 W) can be used. In padders designed for the input circuits of linear amplifiers in RF transmitting installations, the components must dissipate considerable power. Large noninductive resistors, or series-parallel combinations of smaller resistors, are used to obtain the needed dissipation rating. The *duty cycle* must also be taken into consideration. For a given signal power, the component power-dissipation ratings must be higher for continuous modes such as FSK or FM, as opposed to variable-amplitude or intermittent modes such as ON-OFF keying or SSB.

Uses for attenuators

An attenuator can be useful in cases of *desensitization* ("desensing") in an RF communications receiver. If an extremely strong signal comes in, desensing causes the receiver gain to decrease. If the signal is intermittent, the receiver gain varies constantly, and this can make reception difficult over an entire frequency band. An attenuator, placed between the antenna and the first RF amplifier stage, can reduce the strength of the offending signal enough to prevent this. The attenuator is bypassed under normal conditions and used only when an interfering signal is present. Although the attenuator makes it impossible to hear the weakest signals, it allows effective reception of most signals.

In hi-fi audio systems, a graphic equalizer (described earlier in this chapter) can employ several attenuators, each designed to reduce the sound level within a certain range of audio frequencies. These attenuators are, in effect, frequency-specific volume controls. "Outboard" graphic equalizers are installed in the speaker and/or headphone outputs of a hi-fi amplifier.

Attenuators are sometimes used in power amplification, both for AF and RF. A variable attenuator can serve as a gain or volume control for such an amplifier. The attenuator can also prevent the exciter from overdriving the amplifier and causing distortion. This is especially important in a linear amplifier for SSB communication, because distortion unnecessarily increases the bandwidth of such a signal.

14

Telecommunications

The term *telecommunications* refers to the transfer of data between individuals, businesses, and/or governments. This can be done via landline (wired), by EM waves (wireless), or by a combination of both modes.

Networks

The *Internet,* also called simply the *Net,* is a worldwide network of computers. It was formed in the 1960s as *ARPAnet,* named after the Advanced Research Project Agency (ARPA) of the United States federal government.

Packets

When a data file or program is sent over the Net, it is divided into units called *packets* at the source, or transmitting computer. A packet consists of a header followed by a certain number of data bits or bytes. Each packet is routed individually. The packets are reassembled at the destination, or receiving computer(s), into the original message.

Figure 14.1 is a simplified drawing of Internet data transfer for a hypothetical file containing five packets transferred during a period of heavy usage. Intermediate computers in the circuits, called *nodes,* are the black dots surrounded by circles. The file or program cannot be completely reconstructed until all the packets have arrived and the destination computer has ensured that there are no errors.

The modem

The term *modem* is a contraction of *modulator / demodulator.* A modem interfaces a computer to a telephone line, TV cable system, or radio transceiver.

Figure 14.2 is a block diagram of a modem suitable for interfacing a home or business computer with the telephone line. The modulator, or D/A converter, changes outgoing digital data into audio tones. The demodulator, or A/D converter, changes incoming audio tones into digital data. The audio tones fall within the band of approximately 300 Hz to 3 kHz, the same as that used for most voice communications.

E-mail and newsgroups

For many computer users, communication via Internet *electronic mail* (*e-mail*) and/or *newsgroups* has practically replaced the postal service. Use of e-mail or newsgroups requires an e-mail address. An example is

stangib@aol.com

The part of the address before the @ symbol is the *username.* The characters after the @ sign and before the dot represent the *domain name.* The three-letter abbreviation after the dot is the *domain type.* Table 14.1 shows the most common domain-type abbreviations.

Internet conversations

Computer users can carry on "teletype" conversations with other computer users via the Internet. This is called *Internet relay chat* (*IRC*). It is also possible to digitize voice signals and transfer them via the Net. When Net traffic is light,

Destination data

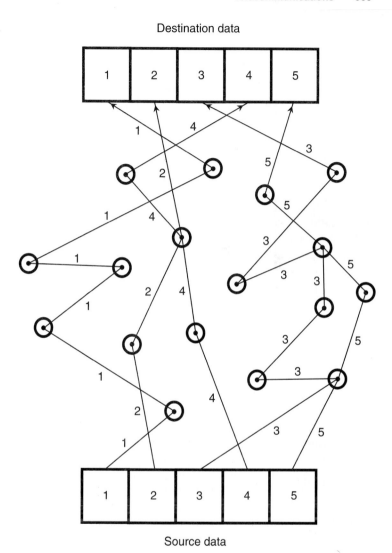

Source data

Figure 14.1 Internet data flows in packets from the source to the destination.

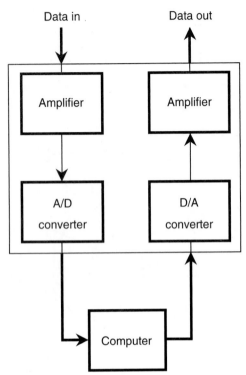

Figure 14.2 Block diagram of a telephone-line modem.

such connections can be almost as good as those provided by telephone companies. But when Net traffic is heavy, the quality is marginal to poor. When Net traffic is extreme, so-called Internet telephone connections might be impossible to establish or maintain between points separated by a large number of nodes.

Getting information

The Internet gets people in touch with millions of sources of information. Data is transferred among computers by means of a *file transfer protocol* (*FTP*). When you use FTP, the files or programs at the remote computer become available to you, exactly as if they were stored in your own computer.

TABLE 14.1   Common Internet Domain Abbreviations

Abbreviation	Domain type
.com	Commercial
.edu	Educational
.gov	Government
.mil	Military
.net	Network
.org	Organization

The *World Wide Web* (also called *WWW* or *the Web*) employs *hypertext,* a scheme of cross referencing. Certain words or phrases are highlighted and/or underlined. When you select one of these words or phrases, you are linked to another document dealing with the same or a related subject.

Local area networks

A *LAN* is a group of computers linked within a small region. In a *client-server LAN* (Fig. 14.3A), there is one central computer called a *file server,* to which smaller personal computers (labeled PC) are linked. In a peer-to-peer LAN (Fig. 14.3B), all of the computers are PCs with more or less equal computing power, speed, and storage capacity. A peer-to-peer LAN offers greater privacy and individuality than a client-server LAN, but it is slower when all users must share the same data.

Satellites

A *communications satellite* system is, in a certain sense, a gigantic cellular network in which the repeaters orbit the earth. The end users can be in fixed, mobile, or portable locations.

Geostationary satellites

At an altitude of 22,300 miles, a satellite in a circular orbit takes precisely one day to complete each revolution. If a satellite is placed in such an orbit over the equator, and if it

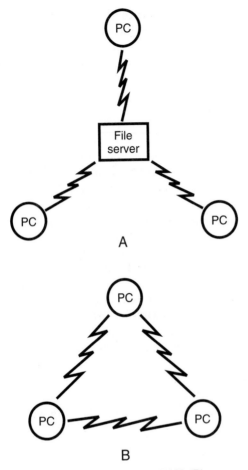

Figure 14.3 (A) A client-server LAN; (B) a peer-to-peer LAN.

revolves in the same direction as the earth rotates, it is a *geostationary satellite*. From the viewpoint of someone on the earth, a geostationary satellite stays in the same spot in the sky all the time. A single geostationary satellite can provide communications coverage over 40 percent of the earth's surface. Three such satellites spaced 120° apart provide coverage over the entire civilized world.

Earth-based stations can communicate via a single "bird" only when the stations are both on a line of sight with the satellite. If two stations are nearly on opposite sides of the planet, they must operate through two satellites (Fig. 14.4).

Low-earth-orbit satellites

A geostationary orbit requires constant adjustment. Geostationary satellites are expensive to launch and maintain. There is a signal delay because of the path length. It takes high transmitter power and a precisely aimed antenna to communicate reliably. These problems with geostationary satellites have given rise to the *low-earth-orbit* (LEO) concept.

In a LEO system, there are dozens of satellites spaced strategically around the globe in polar orbits a few hundred miles above the surface. The satellites relay messages among each other and to and from end users. If there are enough satellites, any two end users can maintain constant contact.

A LEO satellite link is easier to use than a geostationary-satellite link. A simple antenna will suffice, and it does not have to be aimed. The transmitter can reach the network using only a few watts of power. The propagation delay is less than 0.1 μs.

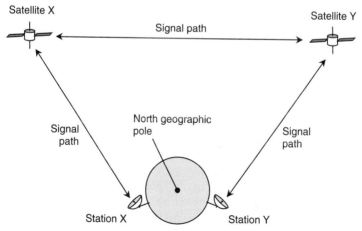

Figure 14.4 A communications link using two geostationary satellites.

Global Positioning System

The *Global Positioning System* (GPS) is a network of satellites that allows users to determine their exact latitude, longitude, and (if applicable) altitude. The satellites transmit signals in the microwave radio spectrum. The signals are modulated with timing and identification codes. A GPS receiver contains a computer that calculates the distances to four different satellites by timing the signals as they travel between the satellites and the receiver. A GPS receiver can give the user location data to within a few feet (for government and industrial subscribers) or a few hundred feet (for civilians).

Personal Communications Systems

Personal communications systems (PCS) include the use of cell phones, pagers, beepers, shortwave radio, citizens band (CB) radio, and amateur (ham) radio.

Cellular telecommunications

A *cellular telecommunications* system is a network of *repeaters,* also known as *base stations,* that allows portable or mobile radio transceivers to be used as telephone sets. A *cell* is the coverage zone of a base station.

If a cell phone set is in a fixed location, such as a restaurant or residence, communication takes place via a single cell. If the cell phone is in a moving vehicle such as a car or boat, it goes from cell to cell (Fig. 14.5A). All the base stations are connected to the telephone system by wires, microwave links, or fiber-optic cables. A computer can be connected to a cell phone set with a specialized modem (Fig. 14.5B) for access to the Internet.

Wireless local loop

A *wireless local loop* (WLL) is similar to cellular system. Telephone sets and data terminals are linked into the system via radio transceivers. An example is shown in Fig. 14.6. Heavy lines represent wire connections, and thin lines

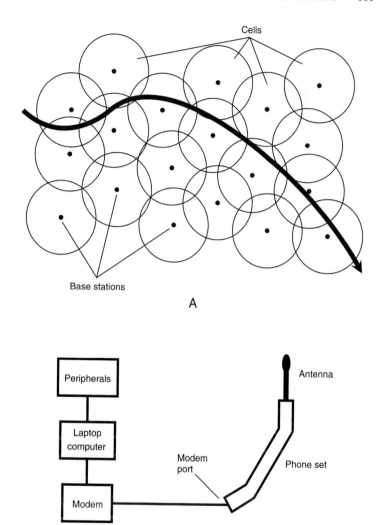

Figure 14.5 (A) Cellular telecommunications scheme; (B) connection of a laptop computer to a cell phone set.

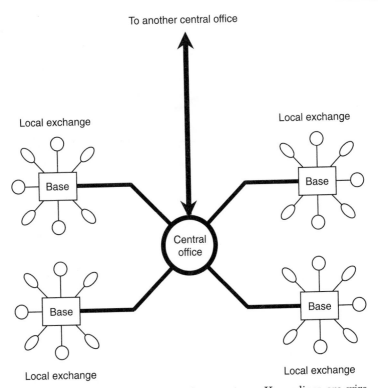

Figure 14.6 Wireless local loop telephone systems. Heavy lines are wire links; thin lines are wireless links. Circles are telephone sets; ellipses are data terminals.

represent wireless links. Ellipses and circles represent subscribers. Each subscriber can use a telephone set or a personal computer equipped with a modem. Several telephone sets and/or computers can be connected together by a system of wires confined to the house or business.

Pagers (beepers)

A simple *pager* or *beeper* employs a small, battery-powered radio receiver that picks up signals in the VHF or UHF radio spectrum. Transmitters are located in various places throughout a city, county, or telephone area code district.

The receiver picks up a simple encoded signal that causes the unit to display a series of numerals and/or letters of the alphabet. A pager equipped with *voice mail* allows the sender to leave a brief spoken message following the beep.

Wireless e-mail

A pager receiver equipped to receive e-mail resembles a handheld computer or small calculator. When the unit emits a sequence of beeps, the user looks at the screen to read messages. Messages can be stored for later retrieval or transfer to a laptop or desktop computer. Some pagers can send e-mail messages as well as receive them. This is done via a system similar to a cellular telephone network. The pager contains a small radio transmitter with an attached whip antenna.

Two notes of caution: (1) The use of wireless e-mail transmitters is generally forbidden on board aircraft, and (2) reading and composing e-mail should never be attempted while driving a motor vehicle.

Facsimile

Facsimile is a method of sending and receiving nonmoving images over telephone lines or radio. A fax machine has an *image transmitter* and an *image receiver.* Figure 14.7 shows the transmission of a wireless fax signal. Only the sending part of the fax machine is shown at the source, and only the receiving part is shown at the destination. A complete fax installation has a transmitter and a receiver, allowing for two-way exchange of fax messages.

To send a fax, a page of printed material is placed in an *optical scanner.* This device converts the image into binary digital signals (1 and 0). The output of the scanner is sent to a *modem* that converts the binary digital data into an analog signal suitable for transmission. At the destination, the analog signals are converted back into digital pulses like those produced by the optical scanner at the source. These pulses are routed to a printing device, computer, or terminal.

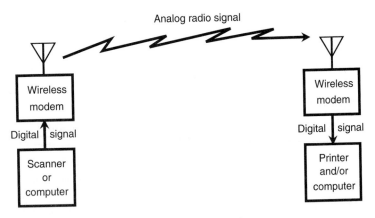

Figure 14.7   Transmission of a wireless fax signal.

Hobby Communications

Hobby communications includes *shortwave listening* (SWLing), the *citizens radio service,* and *amateur radio.*

Shortwave listening

A HF radio communications receiver is sometimes called a *shortwave receiver.* Most of these radios function at all frequencies from 1.5 through 30 MHz. Some also work in the standard AM broadcast band at 535 kHz to 1.605 Mhz, and a few can receive signals below 535 kHz. Most electronics stores carry shortwave receivers.

A shortwave listener in the United States need not obtain a license to receive signals. But, in general, a license is required if shortwave transmission is contemplated. Shortwave listeners often get interested enough in communications to obtain an amateur radio license.

Citizens radio service

The citizens radio service, also known as CB, is a radio communications/control service. The most familiar form is *class D,* which operates on 40 discrete channels near 27 MHz (11 m) in the HF radio spectrum.

A 40-channel, 12-W transceiver is the basic radio for class-D fixed-station operation. It employs the SSB voice mode and is operated from the standard 117-V utility circuit. Mobile transceivers typically run 12 W SSB as well and operate from the 13.7-V vehicle battery. The power connection should be made directly to the battery; "cigarette-lighter" adapters are not recommended.

Radio Emergency Associated Communications Teams (REACT) is a worldwide organization of radio communications operators. They provide assistance to authorities in disaster areas. On the class-D band, the emergency channel at 27.065 MHz (channel 9) is monitored by REACT operators. The *General Mobile Radio Service* (GMRS) operates at frequencies between 460 and 470 MHz using *class A* citizens band. The maximum communications range between two individual transceivers in class A is about 40 miles. Communications beyond 40 miles uses repeaters.

Some classes of CB operation require government licenses, but others do not. For the latest regulations, check with an electronics store that sells CB equipment.

Amateur radio

A fixed amateur (ham) radio station has several components (Fig. 14.8). A computer can be used to network via packet radio with other hams who own computers. The station can be equipped for on-line telephone (landline) services. The computer can control the antennas for the station and can keep a log of all stations that have been contacted. Most modern transceivers can be operated by computer, either locally or by remote control over the radio or landline.

Mobile amateur-radio equipment is operated in a moving vehicle such as a car, truck, train, boat, or airplane. Mobile equipment is generally more compact than fixed-station apparatus. In addition, mobile gear is designed to withstand large changes in temperature and humidity, as well as severe mechanical vibration.

Portable amateur-radio equipment is almost always battery-operated and can be set up and dismantled quickly. Some

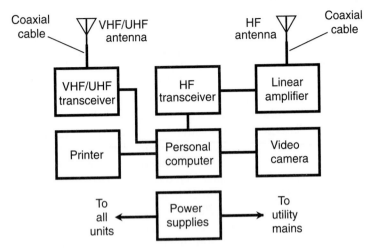

Figure 14.8 A basic amateur radio station.

portable equipment can be operated while being carried; an example is the *handy-talkie* (HT) or *walkie-talkie.* Portable equipment must withstand vibration, temperature and humidity extremes, and prolonged use.

All amateur radio operation requires licensing. For information, contact

American Radio Relay League (ARRL)
225 Main Street
Newington, CT 06111
http://www.arrl.org

Lightning

Lightning is a hazard to radio amateurs, CB operators, and shortwave listeners because of the outdoor antennas used in these systems.

The nature of lightning

A lightning surge, or *stroke,* lasts for a small fraction of a second. There are four types of lightning (see Fig. 14.9):

1. Intracloud lightning (within a single cloud)
2. Cloud-to-ground lightning
3. Intercloud lightning (between different clouds)
4. Ground-to-cloud lightning

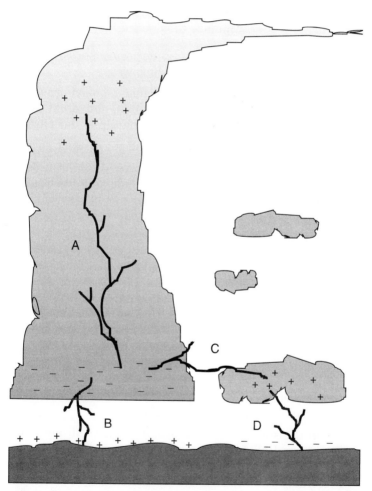

Figure 14.9 Four types of lightning stroke: (A) intracloud; (B) cloud-to-ground; (C) intercloud; (D) ground-to-cloud.

Types 2 and 4 present the greatest danger to electronics hobbyists and equipment, although types 1 and 3 can cause an EMP sufficient to damage sensitive apparatus.

Protecting yourself

The following precautions are recommended if lightning is taking place near you. These measures will not guarantee immunity, but they minimize the danger.

1. Stay indoors or inside a metal enclosure such as a car, bus, or train. Stay away from windows.

2. If it is not possible to get indoors, find a low-lying spot on the ground, such as a ditch or ravine, and squat down with your feet close together until the threat has passed.

3. Avoid lone trees or other isolated, tall objects such as utility poles or flagpoles.

4. Avoid electric appliances or electronic equipment that makes use of the utility power lines or has an outdoor antenna.

5. Stay out of the shower or bathtub.

6. Avoid swimming pools, either indoors or outdoors.

7. Do not use the telephone.

Protecting hardware

Precautions that minimize the risk of damage to electronic equipment (but not guarantee immunity) are as follows:

1. Never operate, or experiment with, a radio station when lightning is occurring near your location.

2. When the station is not in use, disconnect all antennas and ground all feed line conductors to a good electrical ground other than the utility power-line ground. Preferably the lines should be left entirely outside the building and connected to an earth ground several feet away from the building.

3. When the station is not in use, unplug all equipment from utility outlets.

4. When the station is not in use, disconnect and ground all rotator cables and other wiring that leads outdoors.

5. *Lightning arrestors* provide some protection from electrostatic-charge buildup, but they cannot offer complete safety and should not be relied upon for routine protection.

6. *Lightning rods* reduce the chance of a direct hit, but they should not be used as an excuse to neglect the other precautions.

7. Power line *transient suppressors* ("surge protectors") reduce computer "glitches" and can sometimes protect sensitive components in a power supply, but they should not be used as an excuse to neglect the other precautions.

8. The antenna mast or tower should be connected to an earth ground using heavy-gauge wire or braid. Several parallel lengths of AWG No. 8 ground wire, run in a straight line from the mast or tower to ground, form an adequate conductor.

9. Other secondary protection devices are advertised in electronics-related and radio-related magazines.

For more information, consult a competent communications engineer, or refer to the Lightning Protection Code, published by

National Fire Protection Association
Batterymarch Park
Quincy, MA 02269

Security and Privacy

People are concerned about the security and privacy of information exchanged via telecommunications systems because this information is sometimes confused, misused, or falsified by unauthorized persons.

Wireless versus wired

Wireless eavesdropping differs from *wiretapping* in two fundamental ways. First, eavesdropping operations are easier to carry out in wireless systems than in hard-wired ones. Second, eavesdropping of a wireless link is impossible to physically detect, but a tap can usually be found in a hard-wired system.

If any portion of a communications link is done via wireless, an eavesdropping receiver can be positioned within range of the RF transmitting antenna (Fig. 14.10) and the signals intercepted. The existence of a *wireless tap* has no effect on the electronic characteristics of any equipment in the system.

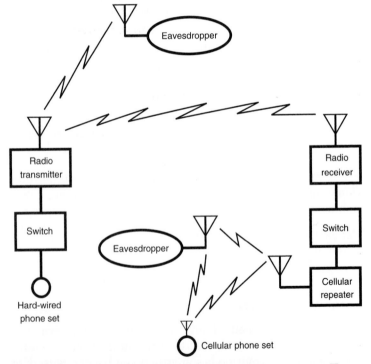

Figure 14.10 Eavesdropping on RF links in a telephone system. Heavy, straight lines represent wires or cables; zig-zags represent RF signals.

Levels of security

There are four levels of telecommunications security, ranging from zero (no security) to the most secure connections technology allows.

No security (level 0). In a *level-0-secure communications system,* anyone with the money and/or time to obtain the necessary equipment can eavesdrop on a connection at any time. Two examples of level-0 links are amateur radio and CB voice communications. The lack of privacy is compounded by the fact that if someone is eavesdropping, none of the communicating parties can detect the intrusion.

Wire-equivalent security (level 1). An end-to-end hard-wired connection requires considerable effort to tap, and sensitive detection apparatus can usually reveal the existence of any wiretap. A *level-1-secure communications system* must have certain characteristics to be effective and practical:

1. The cost must be affordable.

2. The system must be reasonably safe for transactions such as credit-card purchases.

3. When network usage is heavy, the degree of privacy afforded to each subscriber should not decrease, relative to the case when network usage is light.

4. Ciphers, if used, should be unbreakable for at least 12 months and preferably for 24 months or more.

5. Encryption technology, if used, should be updated at least every 12 months and preferably every six months.

Security for commercial transactions (level 2). Some financial and business data warrants protection beyond wire-equivalent. Many companies and individuals refuse to transfer money by electronic means because they fear criminals will gain access to an account.

In a *level-2-secure communications system,* the encryption used in commercial transactions should be such that it would take a hacker at least 10 years, and preferably 20 years or more, to break the cipher. The technology should be updated at least every 10 years but preferably every 3 to 5 years.

Mil-spec security (level 3). Security to military specifications (mil spec) involves the most sophisticated encryption available. Technologically advanced countries, and entities with economic power, have an advantage. However, as technology gains ever more (and arguably too much) power over human activities, aggressor nations and terrorists might injure powerful nations by seeking out, and striking at, the weak points in communications infrastructures.

The encryption in a *level-3-secure system* should be such that engineers believe it would take a hacker at least 20 years, and preferably 40 years or more, to break the cipher. The technology should be updated as often as economics allow.

Extent of encryption

Security and privacy are obtained by *digital encryption*. The idea is to render signals readable only to receivers with the necessary *decryption key*.

For level-1 security, encryption is required only for the wireless portion(s) of the circuit. The cipher should be changed at regular intervals to keep it "fresh." The block diagram in Fig. 14.11A shows wireless-only encryption for a hypothetical cellular telephone connection.

For security at levels 2 and 3, *end-to-end encryption* is necessary. The signal is encrypted at all intermediate points, even those for which signals are transmitted by wire or cable. Figure 14.11B shows this scheme in place for the same hypothetical cellular connection as depicted in Fig. 14.11A. The cipher might be changed at intermediate points, if this will offer greater security than the use of a single cipher from end to end.

Security with cordless phones

Most cordless phones are designed to make it difficult for unauthorized people to "pirate" a telephone line. Prevention of eavesdropping is a lower priority, except in expensive cordless systems. If there is concern about using a cordless phone in a particular situation, a hard-wired phone set should be used.

If someone knows the frequencies at which a cordless handset and base unit operate, and if that person is determined to eavesdrop on conversations that take place via that system, it is possible to place a *wireless tap* on the line. The conversation can be intercepted and recorded at a remote site (Fig. 14.12). It is more difficult to design and construct a wireless tap for a multiple-channel cordless set than for a single-channel set.

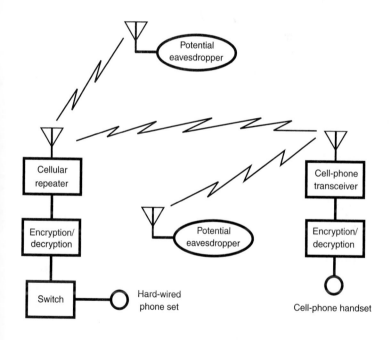

A

Figure 14.11 (A) Wireless-only encryption. Heavy, straight lines represent wires or cables; zig-zags represent RF signals.

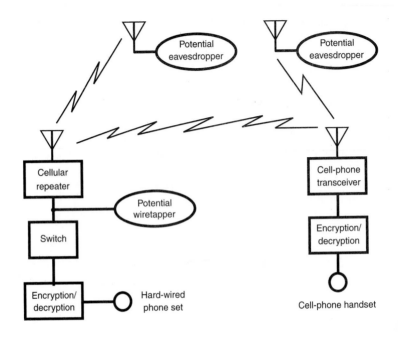

B

Figure 14.11 (*Continued*) (B) End-to-end encryption. Heavy, straight lines represent wires or cables; zig-zags represent RF signals.

Security with cell phones

Cellular telephones are, in effect, long-range cordless phones. The wider coverage increases the risk of eavesdropping and unauthorized use. In recent years, cell phone vendors have begun advertising their systems as "snoop proof" and "clone resistant." Some of these claims have more merit than others. The word *proof,* in particular, should be regarded with skepticism. Digital encryption is the most effective way to maintain privacy and security of cellular communications.

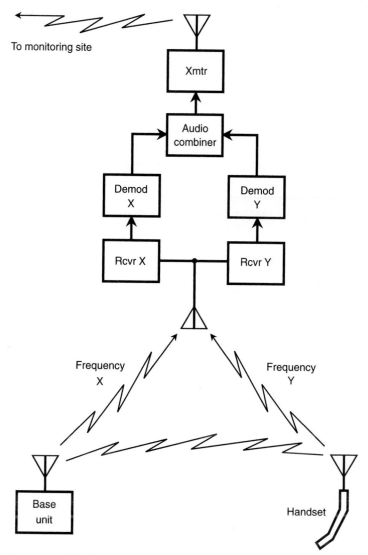

Figure 14.12 Wireless tapping of a cordless telephone.

Access and privacy codes, as well as data, must be encrypted if a cell-phone system is to be maximally secure. If an unauthorized person knows the code via which a cell phone set accesses the system (the "name" of the set), other sets can be programmed to fool the system into "thinking" the bogus sets belong to the user of the authorized set. This is *cell-phone cloning.*

In addition to digital encryption of data, *user identification* (user ID) must be employed. The simplest is a *personal identification number* (PIN). More sophisticated systems can employ *voice-pattern recognition,* in which the phone set functions only when the designated user's voice speaks into it. *Hand-print recognition* or *iris-print recognition* can also be employed.

Tone squelching

Subaudible tones or audible tone bursts can be used to keep a receiver from responding to unnecessary or unwanted signals. These schemes are known as *tone squelching.* For the receiver squelch to open in the presence of a signal, the signal must be modulated with a sine wave that has a certain frequency. Tone-burst systems are used by some police departments and business communications systems in the United States. Tone squelching is popular among radio amateurs as well.

Spread spectrum

Signals using *spread spectrum* require special receivers. A conventional receiver or scanner cannot intercept spread-spectrum signals unless the receiver follows exactly along with the changes in the signal frequency. The frequency-variation sequence can be complicated and in its most sophisticated form is almost as good as digital encryption for security purposes. For further information, see Chap. 15.

Audio scrambling

An amateur-radio SSB transmitter and receiver can demonstrate basic *audio scrambling* (Fig. 14.13). The transmitter

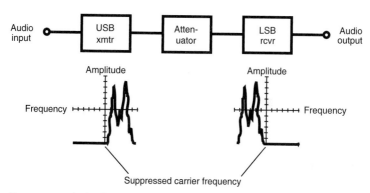

Figure 14.13   A simple voice scrambling and descrambling circuit.

is connected to a dummy antenna and is set for upper sideband (USB) operation. The receiver is set for lower sideband (LSB) reception and is tuned 3000 Hz higher than the transmitter.

Suppose the suppressed-carrier frequency of the USB transmitter is 3.800000 Mhz. This produces a spectral output similar to that shown in the graph on the left of the figure. Each horizontal division represents 500 Hz; each vertical division represents 5 dB. The receiver is tuned to 3.803000 Mhz. The signal energy from the transmitter therefore falls into the receiver passband, as shown by the graph on the right, but the audio frequencies are inverted within the 3000-Hz voice passband. To demonstrate unscrambling, the audio output is tape recorded, and the recorded signal is applied to the input. The "upside-down" signal is rendered "right-side-up" again.

Sophisticated audio scramblers split up the audio passband into two or more subpassbands, inverting some or all of the segments, and perhaps rearranging them according to frequency within the main passband. But audio scrambling is an analog mode and cannot provide the levels of protection available with digital encryption.

15

Wireless Receivers

A *wireless receiver* converts EM waves into the original messages sent by a distant *wireless transmitter*. In the broadest sense, a receiver is a form of transducer.

Simple Designs

In hobby radio and for instructional purposes, simple circuits can function effectively as EM signal receivers.

Crystal set

When an RF diode is connected in a circuit such as the one shown in Fig. 15.1, the result is a crude receiver capable of picking up AM signals. The diode, sometimes called a *crystal,* gave rise to the nickname *crystal set* for this low-sensitivity receiver. There is no source of power except the incoming signal, which comes from the antenna. The output is sufficient to drive a common earphone or headset if the transmitting station is within a few miles of the receiver and if the antenna is sufficiently large.

The diode acts as a *detector* to recover the modulating waveform from the signal. If the detector is to be effective, the diode must have low capacitance so that it works as a

Antenna

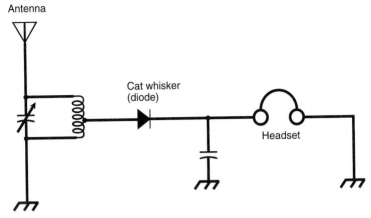

Figure 15.1 A crystal set.

rectifier at RF but not as a capacitor. *Point-contact diodes* are excellent for this purpose.

Direct-conversion receiver

A *direct-conversion receiver* derives its output by mixing incoming signals with the output of a variable-frequency *local oscillator* (LO). The received signal is fed into a *mixer,* along with the output of the LO. A diagram of a direct-conversion receiver is shown in Fig. 15.2.

For reception of CW radiotelegraphy signals, the LO is set slightly above or below the signal frequency. The audio output has a frequency equal to the difference between the LO and signal frequencies. For reception of AM or SSB signals, the LO is set to zero beat with the carrier frequency of the incoming signal.

A direct-conversion receiver does not provide optimum selectivity because signals on either side of the LO frequency interfere with one another. A selective filter can theoretically eliminate this. But such a filter must be designed for a fixed frequency if it is to have sufficiently steep skirts, and in a direct-conversion receiver, the RF chain works over a wide range of frequencies.

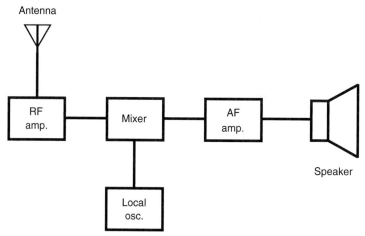

Figure 15.2 Block diagram of a direct-conversion receiver.

Superheterodyne receiver

A *superheterodyne receiver* uses one or more local oscillators and mixers to obtain a constant-frequency signal. A fixed-frequency signal is more easily processed than a signal that changes in frequency. The incoming signal is first passed through a tunable, sensitive amplifier called the *front end.* The output of the front end is mixed with the signal from a tunable, unmodulated LO. Either the sum or the difference signal is amplified. This is the first *IF,* which can be filtered to obtain a high degree of selectivity.

If the first IF signal is detected, the radio is a *single-conversion receiver.* Some receivers use a second mixer and second LO, converting the first IF to a lower-frequency *second IF.* This is a *double-conversion receiver.* The IF bandpass filter can be constructed for use on a fixed frequency, allowing superior skirt selectivity and facilitating adjustable bandwidth. The sensitivity is enhanced because fixed IF amplifiers are easy to keep in tune.

A superheterodyne receiver can intercept or generate unwanted signals. False signals external to the receiver are *images*; internally generated signals are *birdies.* If the LO frequencies are judiciously chosen, images and birdies are not a problem.

The Modern Receiver

Wireless communications receivers operate over specific ranges of the radio spectrum. The range covered depends on the application for which the receiver is designed.

Specifications

The *specifications* of a receiver indicate how well it can do the functions it is designed to perform.

Sensitivity. The most common way to express receiver sensitivity is to state the number of microvolts that must exist at the antenna terminals to produce a certain S/N ratio or *signal-plus-noise-to-noise ratio* (S + N/N) in decibels. Sensitivity is related to the gain of the front end, but the amount of noise this stage generates is more significant because subsequent stages amplify the front-end noise output as well as the signal output.

Selectivity. The passband of a receiver is established by a wideband *preselector* in the early RF amplification stages and is honed to precision by narrowband filters in later amplifier stages. The preselector makes the receiver optimally sensitive within a range of approximately ±10 percent of the desired signal frequency. The narrowband filter responds only to the emission band of a specific signal to be received; signals in nearby channels are rejected.

Dynamic range. The signals at a receiver input vary over several orders of magnitude in terms of absolute voltage. Dynamic range is the ability of a receiver to maintain a fairly constant output, and yet to maintain its rated sensitivity, in the presence of signals ranging from very weak to very strong. The dynamic range in a good receiver is in excess of 100 dB.

Noise figure. The less internal noise a receiver produces, in general, the lower the noise figure and the better is the

S/N ratio. This is paramount at VHF, UHF, and microwave frequencies. *Gallium-arsenide* FETs are well known for the low levels of noise they generate, even at quite high frequencies. Other types of FETs can be used at lower frequencies. *Bipolar transistors* tend to be rather noisy.

Overview

A block diagram of a basic single-conversion superheterodyne receiver is illustrated at Fig. 15.3. Individual receiver designs vary somewhat.

Front end. The front end consists of the first RF amplifier and often includes bandpass filters between the amplifier and the antenna. The dynamic range and sensitivity of a receiver are determined by the performance of the front end.

Mixer. One or two mixer stages convert the variable signal frequency to a constant IF. The output is either the sum or the difference of the signal frequency and the LO frequency.

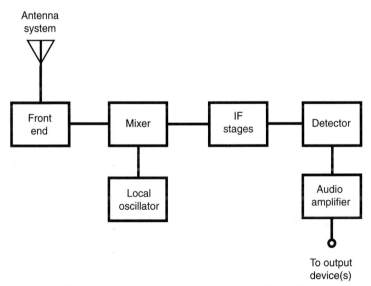

Figure 15.3 Block diagram of a single-conversion superheterodyne receiver.

IF stages. The IF stages are where most of the gain takes place. These stages are also where optimum selectivity is obtained. *Crystal-lattice* or *mechanical filters* are commonly used to obtain the desired bandwidth and response.

Detector. The detector extracts the information from the signal. Common circuits are the *envelope detector* for AM, the *product detector* for SSB, FSK, and CW, and the *phase-locked loop* and *ratio detector* for FM.

Postdetector stages. Following the detector, one or two stages of audio amplification are generally employed to boost the signal to a level suitable for listening with a speaker or headset. Alternatively, the signal can be fed to a printer, facsimile machine, picture tube, or computer.

Predetector Stages

The stages preceding the first mixer must be designed so that they provide reasonable gain but produce as little noise as possible.

Preamplifier

All preamplifiers operate in class A, and most employ FETs. An FET has a high input impedance that is ideally suited to weak-signal work. Figure 15.4 shows a simple RF preamplifier. Input tuning reduces noise and provides some selectivity. This circuit produces 5- to 10-dB gain, depending on the frequency and the choice of FET.

It is important that the preamplifier be linear. Nonlinearity results in *intermodulation distortion* (IMD) that can wreak havoc in a receiver to which the preamplifier is connected.

The front end

At low and medium frequencies, there is considerable atmospheric noise, and the design of a front-end circuit is simple. Above 30 MHz, atmospheric noise diminishes, and

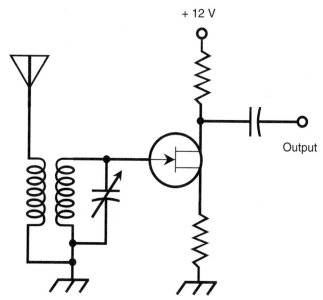

Figure 15.4 Schematic diagram of a basic preamplifier.

the main factor that limits the sensitivity is noise generated within the receiver. For this reason, front-end design becomes increasingly critical as the frequency rises through the VHF, UHF, and microwave spectra.

The front end, like a preamplifier, must be as linear as possible; the greater the degree of nonlinearity, the more susceptible the circuit is to the generation of mixing products. The front end should also have the greatest possible dynamic range.

Preselector

The preselector provides a bandpass response that improves the S/N ratio and reduces the likelihood of the receiver overloading by a strong signal far removed from the operating frequency. The preselector provides image rejection in a superheterodyne circuit. Most preselectors have a 3-dB bandwidth that is a small percentage of the received frequency.

A preselector can be tuned by means of *tracking* with the tuning dial, but this requires careful design and alignment. Some receivers incorporate preselectors that are adjusted independently.

Mixers

A mixer requires a nonlinear circuit element such as a diode or combination of diodes. This forms a *passive mixer*. The diode mixer does not require an external source of power, and there is some insertion loss. An *active mixer* employs one or more transistors or integrated circuits to produce gain.

The output of the mixer circuit can be tuned to either the sum frequency or the difference frequency, as desired.

IF chain

A high IF (several megahertz) is preferable to a low IF (less than 1 MHz) for image rejection. But a low IF is better for obtaining sharp selectivity. Double-conversion receivers have a comparatively high first IF and a low second IF to get the best of both worlds.

Intermediate-frequency amplifiers can be cascaded with tuned-transformer coupling. The amplifiers follow the mixer and precede the detector. Double-conversion receivers have two chains of IF amplifiers. The first IF chain follows the first mixer and precedes the second mixer, and the second IF chain follows the second mixer and precedes the detector.

The selectivity of the IF chain in a superheterodyne receiver can be expressed mathematically. The bandwidths are compared for two power-attenuation values, usually 3 and 30 dB. This gives an indication of the shape of the band-pass response. The ratio of the 30-dB selectivity to the 3-dB selectivity is the *shape factor*. A *rectangular response* is desirable in most applications. The smaller the shape factor, the more rectangular the response. Filters such as the ceramic, crystal-lattice, and mechanical types have small shape factors.

Detectors

Detection, also called *demodulation,* is the recovery of information such as audio, images, or printed data from a signal.

Detection of AM

The modulating waveform can be extracted from an AM signal by rectifying the carrier wave. A somewhat oversimplified view of this is shown in Fig. 15.5A. The rapid pulsations occur at the carrier frequency; the slower fluctuation is a duplication of the modulating intelligence. The carrier pulsations are smoothed out by passing the output through a capacitor large enough to hold the charge for one carrier current cycle, but not so large that it smoothes out the cycles of the modulating signal. This scheme is known as *envelope detection.*

Detection of CW

For detection of CW radiotelegraph signals, it is necessary to inject a signal into the receiver a few hundred hertz from the carrier. The injected signal is produced by a tunable *beat-frequency oscillator* (BFO). The BFO and desired CW signals are mixed to produce audio output at the difference frequency. To receive a keyed Morse-code CW signal, the BFO is tuned to a frequency that results in a comfortable listening pitch. For most people this is approximately 700 Hz. This is *heterodyne detection.*

Detection of FSK

Frequency-shift keying can be detected using the same method as CW detection. The carrier beats against the BFO in the mixer, producing an audio tone that alternates between two different pitches.

With FSK, the BFO frequency is set a few hundred hertz above or below both the mark and the space carrier frequencies. The *frequency offset,* or difference between the BFO and signal frequencies, determines the audio output frequencies and must be set so that standard tone pitches result. Unlike the situation with CW reception, there is little tolerance for BFO adjustment variation.

Detection of FM

Frequency-modulated signals can be detected in various ways. These methods also work for phase modulation.

Slope detection.   An AM receiver can be used to detect FM in a crude manner by setting the receiver frequency near, but not on, the FM unmodulated-carrier frequency. An AM receiver has a narrowband filter with a passband of a few kilohertz. This gives a selectivity curve such as that shown in Fig. 15.5B. If the FM unmodulated-carrier frequency is near the skirt of the filter response, modulation causes the signal to move in and out of the passband. This causes the receiver output to vary.

Phase-locked loop.   If an FM signal is injected into a PLL, the loop produces an error voltage that is a duplicate of the modulating waveform. A limiter can be placed ahead of the PLL so that the receiver does not respond to AM. Weak signals tend to appear and disappear, rather than fading, in an FM receiver that employs limiting.

Discriminator.   This FM detector produces an output voltage that depends on the instantaneous signal frequency. When the signal is at the center of the passband, the output voltage is zero. If the instantaneous signal frequency decreases, the output voltage becomes positive. If the frequency rises above center, the output becomes negative. A discriminator is sensitive to amplitude variations, but this can be overcome by a limiter.

Ratio detector.   This FM detector is a discriminator with a built-in limiter. The original design was developed by Radio Corporation of America (RCA) and is used in high-fidelity receivers and in the audio portions of TV receivers. A simple ratio detector circuit is shown in Fig. 15.5C.

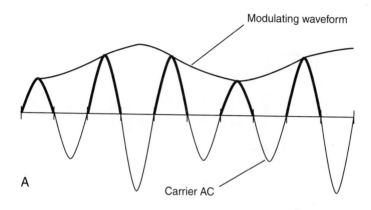

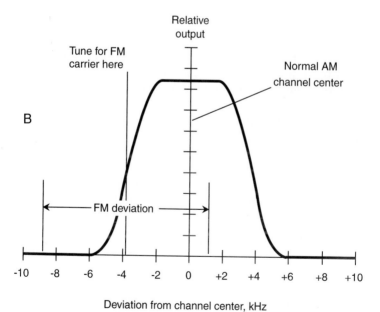

Figure 15.5A, B (A) Envelope detection of AM. (This rendition is oversimplified for clarity of illustration.) (B) Slope detection of FM.

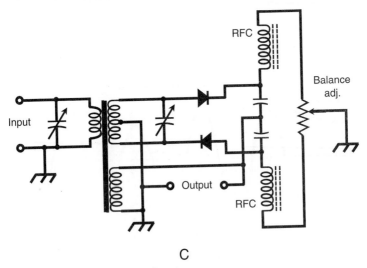

C

Figure 15.5C (C) An FM ratio detector.

Detection of SSB

For reception of CW, FSK, and SSB signals, a *product detector* is used. The incoming signal combines with the signal from an unmodulated LO, producing audio or video output. Product detection is done at a single frequency, rather than at a variable frequency as in direct-conversion reception. The single, constant frequency is obtained by mixing the incoming signal with the output of the LO.

Two product-detector circuits are shown in Fig. 15.5D and E. In Fig. 15.5D, diodes are used; there is no amplification. In Fig. 15.5E, a bipolar transistor is employed; this circuit provides some gain. The essential characteristic of either circuit is the nonlinearity of the semiconductor devices. This generates the sum and difference frequency signals that result in audio or video output.

Audio Stages

In a receiver, enhanced selectivity can be obtained by tailoring the audio response in stages following the detector.

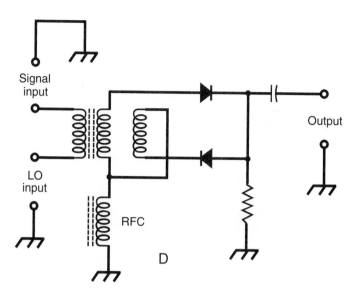

Signal input

Output

LO input

RFC

D

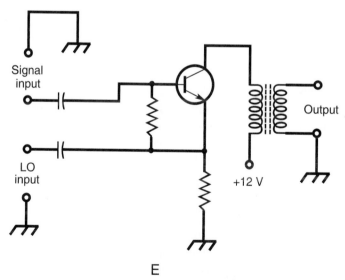

Signal input

Output

LO input

+12 V

E

Figure 15.5D, E (D and E) Product detectors for SSB.

Filtering

A voice signal requires a band of frequencies ranging from about 300 to 3000 Hz. An audio bandpass filter, with a passband of 300 to 3000 Hz, can improve the quality of reception with some voice receivers. An ideal voice audio filter has little or no attenuation within the passband range, and has high attenuation outside the range, with steep skirts (rectangular response).

A CW signal requires only a few hundred hertz of bandwidth to be clearly read. Audio CW filters can narrow the response bandwidth to as little as 100 Hz. Passbands narrower than about 100 Hz produce *ringing*, degrading the quality of reception. The center frequency of a CW audio filter is about 700 Hz.

An *audio notch filter* is a band-rejection filter with a sharp, narrow response. An interfering carrier that produces a tone of constant frequency in the receiver output can be nulled out. Audio notch filters are tunable from roughly 300 to 3000 Hz. Some tune wider ranges. Some sophisticated devices tune automatically; when a heterodyne appears and remains for a few tenths of a second, the notch is activated and centers itself on the frequency of the heterodyne.

Squelching

A *squelch* silences a receiver when no signal is present, allowing reception of signals when they appear. Most FM communications receivers use squelching systems. The squelch is normally closed when no signal is present. A signal opens the squelch if its amplitude exceeds the squelch threshold, which can be adjusted by the operator.

In some systems, the squelch will not open unless the signal has certain characteristics. The most common method of *selective squelching* uses subaudible-tone or tone-burst generators. This can prevent unauthorized transmissions from accessing repeaters or being picked up by receivers.

Television Reception

A TV receiver has a tunable front end, an oscillator and mixer, a set of IF amplifiers, a video demodulator, an audio demodulator and amplifier chain, a picture tube with associated peripheral circuitry, and a loudspeaker.

Fast-scan TV

A receiver for analog *fast-scan television* (FSTV) is shown in simplified block form in Fig. 15.6. In standard American TV, there are 525 lines per frame and 30 complete frames per second. High-definition TV (HDTV) and digital TV have more lines per frame and also may have more frames per second. As of this writing, the 525-line, 30-frame-per-second National Television Standards Committee (NTSC) scheme remains the de facto standard in the United States.

In the United States, wireless FSTV broadcasts are made on 68 different channels numbered from 2 through 69. Each channel is 6 MHz wide, including video and audio information. Channels 2 through 13 comprise the *VHF TV broadcast*

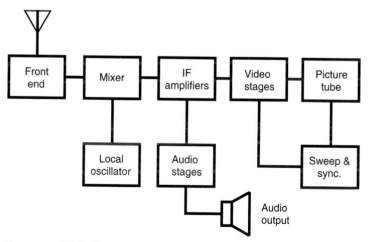

Figure 15.6 Block diagram of an FSTV receiver.

TABLE 15.1   VHF Television Broadcast Channels

Channel	Frequency, MHz
2	54–60
3	60–66
4	66–72
5	76–82
6	82–88
7	174–180
8	180–186
9	186–192
10	192–198
11	198–204
12	204–210
13	210–216

channels (Table 15.1). Channels 14 through 69 are the *UHF TV broadcast channels* (Table 15.2).

Slow-scan TV

A *slow-scan television* (SSTV) communications station needs a transceiver with SSB capability, a standard NTSC TV set or personal computer, a video camera, and a *scan converter*. The standard transmission rate is only eight frames per second. This reduces the required bandwidth to essentially the same as that of an SSB signal (the audio range of 300 to 3000 Hz).

The scan converter consists of two data converters (one for receiving and the other for transmitting), a memory, a tone generator, and a detector. Scan converters are commercially available. Computers can be programmed to perform this function. Amateur radio operators often build their own scan converters.

Specialized Wireless Modes

Some less common wireless communications techniques are effective under certain circumstances.

Dual-diversity reception

This is a scheme for reducing fading in radio reception via ionospheric propagation at high frequencies (approximately

TABLE 15.2 **UHF Television Broadcast Channels**

Channel	Frequency, MHz	Channel	Frequency, MHz
14	470–476	42	638–644
15	476–482	43	644–650
16	482–488	44	650–656
17	488–494	45	656–662
18	494–500	46	662–668
19	500–506	47	668–674
20	506–512	48	674–680
21	512–518	49	680–686
22	518–524	50	686–692
23	524–530	51	692–698
24	530–536	52	698–704
25	536–542	53	704–710
26	542–548	54	710–716
27	548–554	55	716–722
28	554–560	56	722–728
29	560–566	57	728–734
30	566–572	58	734–740
31	572–578	59	740–746
32	578–584	60	746–752
33	584–590	61	752–758
34	590–596	62	758–764
35	596–602	63	764–770
36	602–608	64	770–776
37	608–614	65	776–782
38	614–620	66	782–788
39	620–626	67	788–794
40	626–632	68	794–800
41	632–638	69	800–806

3 to 30 MHz). Two receivers are used; both are tuned to the same signal. They employ separate antennas, spaced several wavelengths apart. The outputs of the receivers are fed into a common audio amplifier, as shown in Fig. 15.7.

Dual-diversity tuning is critical, and the equipment is expensive. In some installations, three or more antennas and receivers are employed. This provides superior immunity to fading, but it compounds the tuning difficulty and further increases the expense.

Synchronized communications

Digital signals require less bandwidth than analog signals to convey a given amount of information per unit time.

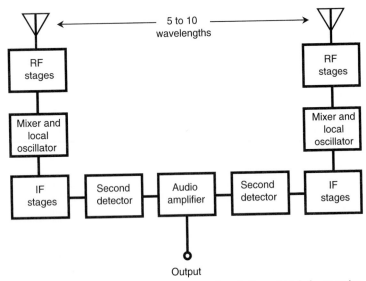

Figure 15.7 Dual-diversity reception can reduce fading at high frequencies.

Synchronized communications refers to a specialized digital mode in which the transmitter and receiver operate from a common time standard to optimize the amount of data that can be sent in a communications channel or band.

In synchronized digital communications, also called *coherent communications,* the receiver and transmitter operate in lock-step. The receiver evaluates each transmitted bit as a unit for a block of time lasting from the bit's exact start to its exact finish. This makes it possible to use a receiving filter that has an extremely narrow bandwidth. The synchronization requires the use of an external frequency/time standard. The broadcasts of the *National Bureau of Standards* stations WWV or WWVH can be used for this purpose. Frequency dividers are employed to obtain the necessary synchronizing frequencies. A tone or pulse is generated in the receiver output for a particular bit if, but only if, the average signal voltage exceeds a certain value over the duration of that bit. False signals, such as might be caused by filter ringing, sferics (radio noise produced by lightning), or other noise, are generally ignored because they rarely produce sufficient average bit voltage.

Experiments with synchronized communications have shown that the improvement in S/N ratio, compared with nonsynchronized systems, is several decibels at low to moderate data speeds. Further improvement can be obtained by the use of DSP in the receiver.

Digital signal processing

In DSP with analog modes such as SSB or SSTV, the signals are first changed into digital form by *A/D conversion*. Then the digital data is "cleaned up" so that the pulse timing and amplitude adhere strictly to the protocol for the type of digital data being used. Finally, the digital signal is changed back to the original voice or video via *D/A conversion*.

Digital signal processing can extend the workable range of a communications circuit because it allows reception under worse conditions than would be possible without it. Digital signal processing also improves the quality of fair signals, so the receiving equipment or operator makes fewer errors. In circuits that use only digital modes, A/D and D/A conversion are irrelevant, but DSP can still be used to clean up the signal. This improves the accuracy of the system and also makes it possible to copy data over many times (that is, to produce multigeneration duplicates).

The DSP circuit minimizes noise and interference in a received digital signal, as shown in Fig. 15.8. A hypothetical signal before DSP is shown at the top; the signal after processing is shown at the bottom. If the incoming signal is above a certain level for an interval of time, the DSP output is high (logic 1). If the level is below the critical point for a time interval, the output is low (logic 0).

Multiplexing

Signals in a communications channel or band can be intertwined in various ways. The most common methods are *frequency-division multiplexing* (FDM) and TDM. Multiplexing

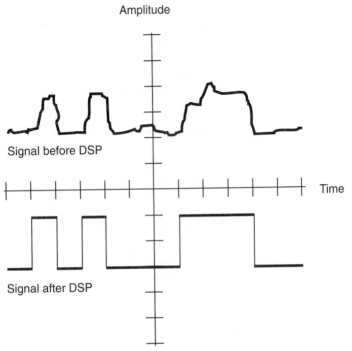

Figure 15.8 Digital signal processing "cleans up" a signal.

requires an encoder at the transmitter and a decoder at the receiver.

In FDM, the channel is broken down into subchannels. The carrier frequencies of the signals are spaced so that they do not overlap. Each signal is independent of the others.

In TDM, signals are broken into segments by time, and then the segments are transferred in a rotating sequence. The receiver must be synchronized with the transmitter. They can both be clocked from a time standard such as station WWV.

Spread spectrum

In *spread-spectrum communications,* the main carrier frequency is rapidly varied independently of signal modulation, and the receiver is programmed to follow. As a result, the probability of catastrophic interference, in which one strong interfering signal can obliterate the desired signal, is near zero. It is difficult for unauthorized people to eavesdrop on a communication in progress.

Frequency-spreading functions can be complex and can be kept secret. If the transmitting and receiving operators do not divulge the function to anyone, and if they do not tell anyone about the existence of their contact, no one else on the band will know the contact is taking place.

During a spread-spectrum contact between a given transmitter and receiver, the operating frequency might fluctuate over a range of kilohertz, megahertz, or tens of megahertz. As a band becomes occupied with an increasing number of spread-spectrum signals, the overall noise level in the band appears to increase. Therefore, there is a practical limit to the number of spread-spectrum contacts that a band can handle. This limit is about the same as it would be if all the signals were constant in frequency.

A common method of generating a spread spectrum is *frequency hopping*. The transmitter has a list of channels that it follows in a certain order. The receiver must be programmed with this same list, in the same order, and must be synchronized with the transmitter. The *dwell time* is the interval at which the frequency changes occur. It should be short enough so that a signal will not be noticed, and not cause interference, on any single frequency. There are numerous *dwell frequencies,* so the signal energy is diluted to the extent that, if someone tunes to any frequency in the sequence, the signal is not noticeable.

Another method of obtaining spread spectrum, called *frequency sweeping,* is to frequency-modulate the main transmitted carrier with a sine wave that guides it up and down over the assigned band. This FM is independent of signal intelligence. A receiver can intercept the signal if, but only if, its tuning varies via the same sine-wave function, over the same band, at the same frequency, and in the same phase as the transmitter.

16

Wireless Transmitters

A *wireless transmitter* converts data into EM waves intended for recovery by one or more *wireless receivers*. A wireless transmitter is therefore a form of *transducer*.

Oscillation and Amplification

A wireless transmitter employs one or more *oscillators* to generate an RF signal and *amplifiers* to generate the required power output. For information about oscillators, see Chap. 11. For information about amplifiers, see Chap. 12.

Some transmitters have *mixers* in addition to the oscillating and amplifying stages. In a transmitter, a mixer works in the same way as in a superheterodyne receiver: It combines signals having two different frequencies to obtain output at a third frequency, which is either the sum or the difference of the input frequencies.

Modulation

Modulation is the process of imprinting data onto an electric current or EM wave. The process can be done by varying the amplitude, frequency, or phase. Another method is to transmit a series of pulses, whose duration, amplitude, or spacing is made to vary.

The carrier

The heart of a wireless signal is a *sine wave,* usually of a frequency far above the range of human hearing. This is known as the *carrier.* The lowest carrier frequency used for radio communications is a few kilohertz. The highest frequency is in the hundreds of gigahertz. For efficient data transfer, the carrier frequency must be at least 10 times the highest frequency of the modulating signal.

ON/OFF keying

The simplest form of modulation is *ON/OFF keying,* usually accomplished in the oscillator of a CW radio transmitter. A block diagram of a simple CW transmitter is shown in Fig. 16.1. The use of Morse code in communications is archaic, but a CW transmitter is easy to build.

Morse code is a binary digital mode. The duration of a Morse-code *dot* is equal to the duration of 1 bit. A *dash* is 3 bits long. The space between dots and dashes within a *character* is 1 bit. The space between characters in a *word* is 3 bits. The space between words is 7 bits. A punctuation symbol is sent as a character attached to the preceding word. An amplitude-versus-time rendition of the Morse word *eat* is shown in Fig. 16.2. The key-down (full-carrier) condition is called *mark*; the key-up (no-signal) condition is called *space.*

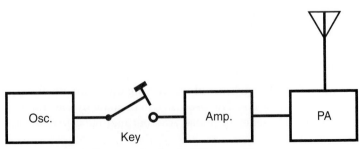

Figure 16.1   Block diagram of a simple CW transmitter.

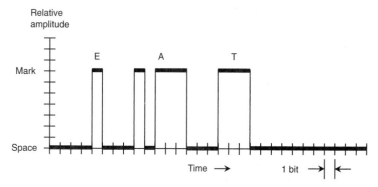

Figure 16.2 The Morse code word *eat* as sent via CW.

Morse code is one of the slowest known methods of data transmission. Human operators use speeds ranging from about five words per minute (wpm) to 40 or 50 wpm. Machines, such as computers and data terminals, function at many times this rate. These systems usually employ FSK rather than ON/OFF keying.

Frequency-shift keying

In FSK, the frequency of the signal is altered between two values nominally separated by a few hundred hertz. In some systems, the carrier frequency itself is shifted between mark and space conditions. In other systems, a two-tone AF sine wave modulates the main carrier. This latter mode is *audio-frequency-shift keying* (AFSK).

There are two common codes commonly used with FSK and AFSK: *Baudot* (pronounced baw-DOE) and *ASCII* (pronounced ASK-ee).

In RTTY FSK and AFSK systems, a *terminal unit* (TU) converts the digital signals into electrical impulses that operate a teleprinter or display characters on a computer screen. The TU also generates the signals necessary to send RTTY as an operator types on a keyboard. A frequency-versus-time graph of the word *eat,* sent using FSK with Morse code, is shown in Fig. 16.3. A block diagram of an AFSK transmitter is shown in Fig. 16.4.

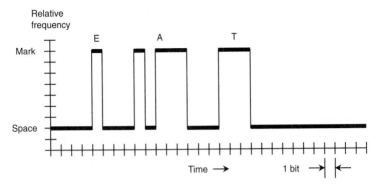

Figure 16.3 The Morse code word *eat* as sent via FSK.

The modem

There are three sets of standard tone frequencies used with AFSK: 1200 and 2200 Hz for general communications, 1070 and 1270 Hz for message origination, and 2025 and 2225 Hz for answering. These represent shifts of 1000 or 200 Hz. A device that sends and receives AFSK is known as a *modem*. A modem is basically the same as a TU.

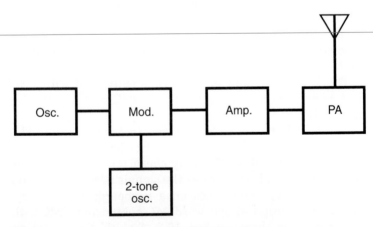

Figure 16.4 Block diagram of a simple AFSK transmitter.

Amplitude modulation

A voice signal is a complex waveform with frequencies mostly in the range between 300 Hz and 3 kHz. Some characteristics of a carrier can be varied, or modulated, by these waveforms, thereby transmitting voice information. Figure 16.5 shows a simple bipolar-transistor circuit for obtaining AM. This circuit works efficiently provided the AF input amplitude is not too great. If the AF input is excessive, distortion occurs, degrading signal intelligibility and increasing bandwidth beyond the minimum necessary to carry on communications.

Two complete AM transmitters are shown in block-diagram form in Fig. 16.6. In Fig. 16.6A, *low-level AM* is illustrated. In this mode, all the amplifier stages following the modulator must be linear. In some transmitters, AM is done in the final

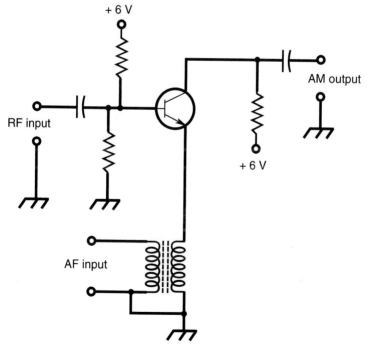

Figure 16.5 An amplitude modulator circuit.

PA as shown in Fig. 16.6B. The PA operates in class C; it is the modulator as well as the final amplifier. This is *high-level AM*.

The extent of AM is expressed as a percentage, from 0 (an unmodulated carrier) to 100 (full modulation). Increasing the modulation past 100 percent causes distortion of the signal and degrades the efficiency of data transmission. In an

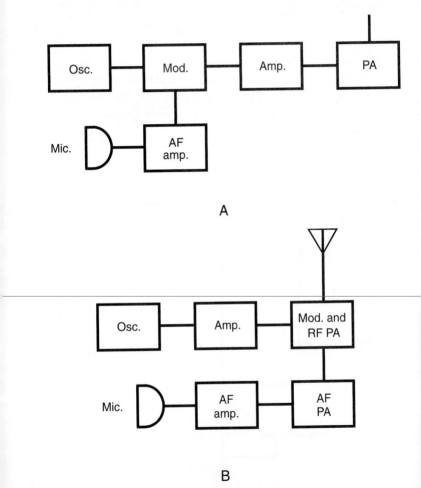

A

B

Figure 16.6 (A) Low-level AM; (B) high-level AM.

AM signal modulated 100 percent, one-third of the power is used to convey the data; the other two-thirds are consumed by the carrier wave.

In Fig. 16.7, the spectral display for an AM voice radio signal is illustrated. The horizontal scale is calibrated in increments of 1 kHz per division. Each vertical division represents 3 dB. The maximum (reference) amplitude is zero decibels relative to 1 mW (0 dBm). The AF components, containing the intelligence transmitted, appear as *sidebands* on either side of the carrier. The RF between −3 kHz and the carrier frequency constitutes the LSB; the RF from the carrier frequency to +3 kHz represents the USB. The *bandwidth* of the RF signal is the difference between the maximum and minimum sideband frequencies.

In an AM signal, the bandwidth is twice the highest audio modulating frequency. In the example of Fig. 16.7, the voice energy is at or below 3 kHz; thus the bandwidth of the

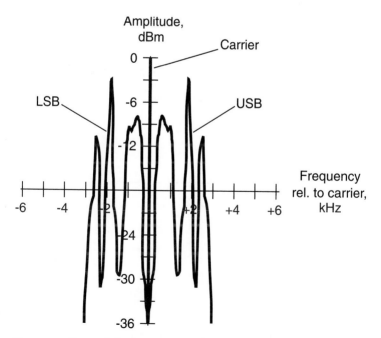

Figure 16.7 Spectral display of a typical AM voice signal.

complete signal is 6 kHz. This is typical of a communications signal. In AM broadcasting, the energy is often spread over a wider bandwidth.

Single sideband

In AM, most of the power is used up by the carrier. The two sidebands are mirror-image duplicates. If the carrier and one of the sidebands is eliminated, all of the available power goes into data transmission, and the bandwidth can be reduced by more than 50 percent. The remaining voice signal has a spectral display resembling Fig. 16.8. This is SSB transmission. Either the LSB or the USB can be used; they both work equally well.

An SSB transmitter employs a *balanced modulator*. This circuit works like an amplitude modulator, except the carrier

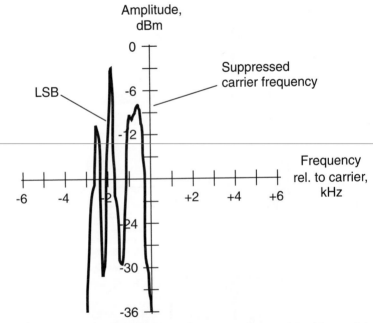

Figure 16.8 Spectral display of a typical voice SSB signal. In this case it is LSB.

is phased out (Fig. 16.9). This leaves only the sideband energy. One of the sidebands is removed by a *bandpass filter* in a later stage of the transmitter. A block diagram of a basic SSB transmitter is shown in Fig. 16.10.

High-level modulation cannot be used for SSB transmission. The balanced modulator is always placed in a low-power section of the transmitter. The RF amplifiers that follow any amplitude modulator must all be linear to avoid distortion and unnecessary spreading of signal bandwidth ("splatter"). The RF amplifiers following a balanced modulator generally work in class A, except for the PA, which functions in class AB or class B.

Frequency and phase modulation

In FM, the overall amplitude of the signal remains constant, and the instantaneous frequency or phase is made to change. Class-C power amplifiers can be used in FM transmitters without introducing distortion because the amplitude does not fluctuate, and linearity is of no concern.

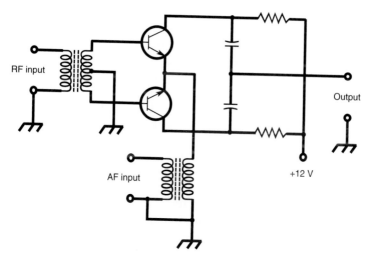

Figure 16.9 A balanced modulator circuit using bipolar transistors.

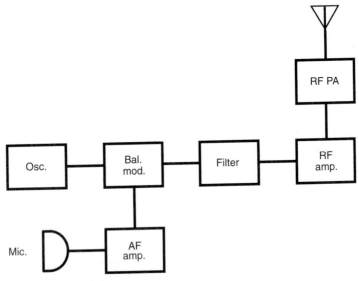

Figure 16.10   Block diagram of an SSB transmitter.

The most direct way to get FM is to apply the audio signal to a *varactor* in a tuned oscillator. An example of this scheme, known as *reactance modulation,* is shown in Fig. 16.11. The varying voltage across the varactor causes its capacitance to change in accordance with the audio waveform. The changing capacitance results in variation of the resonant frequency of the *LC* tuned circuit, causing a swing in the frequency generated by the oscillator.

Another way to get FM is to modulate the phase of the oscillator signal. This causes small fluctuations in the frequency as well, because any instantaneous phase change shows up as an instantaneous frequency change (and vice versa). When *phase modulation* is used, the audio signal must be processed, adjusting the amplitude-versus-frequency response of the audio amplifiers. Otherwise the signal will sound muffled in an FM receiver.

Deviation is the maximum extent to which the instantaneous carrier frequency differs from the unmodulated-carrier frequency. For most FM voice transmitters, the deviation is standardized at ±5.0 kHz (Fig. 16.12). The deviation

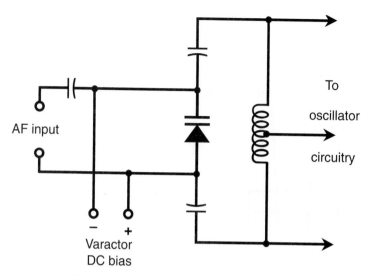

Figure 16.11   Reactance modulation to obtain FM.

obtainable by means of direct FM is greater, for a given oscillator frequency, than the deviation that can be obtained via phase modulation. Deviation can be increased by a *frequency multiplier.* When an FM signal is passed through a frequency multiplier, the deviation is multiplied along with the carrier frequency.

In FM hi-fi music broadcasting, and in some other applications, the deviation is much greater than ±5.0 kHz. This is called *wideband FM,* as opposed to *narrowband FM* discussed above. The deviation for an FM signal should be equal to the highest modulating audio frequency if optimum fidelity is to be obtained. Thus, ±5.0 kHz is more than enough for voice. For music, a deviation of at least ±15 or 20 kHz is needed.

The ratio of the frequency deviation to the highest modulating audio frequency is called the *modulation index.* Ideally this figure should be between 1:1 and 2:1. If it is less than 1:1, the signal sounds muffled or distorted, and efficiency is sacrificed. Increasing it beyond 2:1 broadens the bandwidth without providing significant improvement in intelligibility or fidelity.

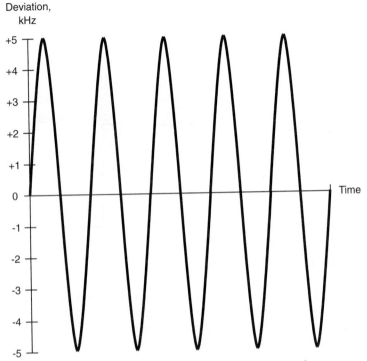

Deviation,
kHz

Figure 16.12   Frequency-versus-time rendition of an FM signal.

Pulse modulation

Several types of pulse modulation (PM) are briefly described below. They are diagrammed in Fig. 16.13 as amplitude-versus-time graphs.

In *pulse amplitude modulation* (PAM), the strength of each individual pulse varies according to the modulating waveform (Fig. 16.13A). In PAM, the pulses all have equal duration. *Pulse duration modulation* (PDM), also called *pulse width modulation* (PWM), is shown in Fig. 16.13B. *Pulse interval modulation* (PIM), also called *pulse frequency modulation* (PFM), is shown in Fig. 16.13C.

In *pulse-code modulation* (PCM), any of the aforementioned aspects—amplitude, duration, or frequency—of a pulse train can be varied. But rather than having infinitely

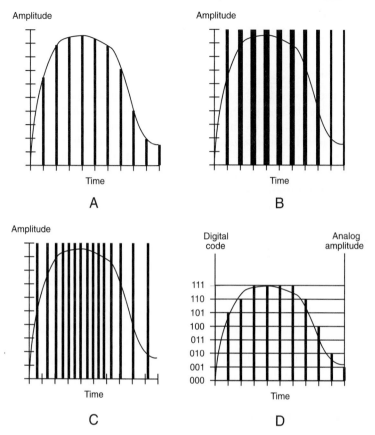

Figure 16.13 (A) Pulse amplitude modulation; (B) pulse width modulation; (C) pulse interval modulation; (D) pulse code modulation. Thin wavy lines represent analog input waveforms. Vertical bars represent pulses.

many possible states or levels, there are only a few. Usually this number is a power of 2, such as $2^3 = 8$ or $2^4 = 16$. Figure 16.13D shows an example of eight-level PCM in which the pulse amplitude is varied.

Analog-to-Digital Conversion

Pulse-code modulation, such as is shown in Fig. 16.13D, is one form of *A/D conversion*. A voice signal, or any continuously variable signal, can be digitized, or converted into a string of pulses, whose amplitudes can achieve only a finite number of states.

Resolution

In A/D conversion, the number of states is always a power of 2 so that it can be represented as a binary-number code. Fidelity improves as the exponent increases. The number of states is called the *sampling resolution,* or simply the *resolution.* A resolution of $2^3 = 8$ (as shown in Fig. 16.13D) is good enough for voice transmission and is the standard for commercial digital voice circuits. A resolution of $2^4 = 16$ is adequate for hi-fi music reproduction.

Sampling rate

The efficiency with which a signal can be digitized depends on the frequency at which sampling is done. In general, the *sampling rate* must be at least twice the highest data frequency. For an audio signal with components as high as 3 kHz, the minimum sampling rate for effective digitization is 6 kHz; the commercial voice standard is 8 kHz. For hi-fi digital transmission, the standard sampling rate is 44.1 kHz; the highest audible frequency is approximately 20 kHz.

Image Transmission

Nonmoving images can be sent within the same bandwidth as voice signals. For high-resolution, moving images, the necessary bandwidth is greater.

Wireless facsimile

Nonmoving images are transmitted by *fax.* If data is sent slowly enough, any amount of detail can be transmitted within a 3-kHz-wide band. This is how telephone fax works, for example.

To send an image by fax, a document or photo is wrapped around a *drum.* The drum is rotated at a slow, controlled rate.

A spot of light scans from left to right; the drum moves the document so a single line is scanned with each pass of the light spot. This continues, line by line, until the complete frame, or picture, has been scanned. The reflected light is picked up by a *photodetector.* Dark parts of the image reflect less light than bright parts, so the current through the photodetector varies. This current modulates a carrier in one of the modes described earlier, such as AM, FM, or SSB. Typically, black is sent as a 1.5-kHz audio sine wave, and white, as a 2.3-kHz audio sine wave. Gray shades produce audio sine waves that have frequencies between these extremes.

At the receiver, the scanning rate and pattern can be duplicated, and a CRT or special printer can be used to reproduce the image in gray scale.

Analog fast-scan television

To get a realistic impression of motion, it is necessary to transmit at least 20 complete *frames* (stationary images) per second, and the detail must be adequate. An FSTV system usually provides 30 frames per second. There are 525 or 625 lines in each frame in a conventional FSTV picture. In HDTV, there are more lines; in SSTV there are fewer lines. The lines run horizontally across the picture, which typically has a horizontal-to-vertical dimensional ratio, or *aspect ratio,* of 4:3. Each line contains shades of brightness in a gray-scale system and shades of brightness and hue in a color system. In FSTV broadcasting, the image is sent as an AM signal, and the sound is sent as an FM signal. In SSTV communications, SSB is the most often-used mode.

Because of the large amount of information sent, an FSTV channel is wide. A standard FSTV channel in the North American system takes up 6 MHz of spectrum space. All FSTV broadcasting is done at VHF and above for this reason. Figure 16.14 shows a typical frame in an FSTV signal.

An FSTV transmitter consists of a camera tube, an oscillator, an amplitude modulator, and a series of amplifiers for the video signal. The audio system consists of an input

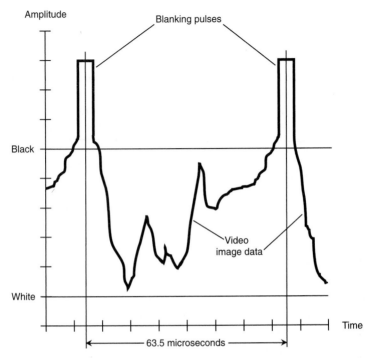

Figure 16.14 An FSTV signal, showing one line of data.

device (such as a microphone), an oscillator, a frequency modulator, and a feed system that couples the RF output into the video amplifier chain. There is also an antenna or cable output. A simplified block diagram of an analog FSTV transmitter is shown in Fig. 16.15.

Analog slow-scan television

If some aspects of the signal are compromised, it is possible to send a video image in a band much narrower than 6 MHz. In SSTV, this is done by greatly reducing the rate at which the frames are transmitted. The image resolution is compromised also; it is lower than with conventional FSTV.

An SSTV signal is sent within 3 kHz of spectrum, the same as needed by an SSB voice signal or fax signal. An SSTV signal typically contains one frame every 8 s. There are 120 lines per frame.

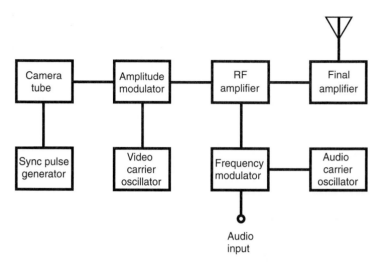

Figure 16.15 Block diagram of an analog FSTV transmitter.

The modulation for SSTV in radio communications is obtained by inputting audio signals into an SSB transmitter. An AF sine wave having a frequency of 1.5 kHz corresponds to black; a sine wave having a frequency of 2.3 kHz corresponds to white. Intermediate audio frequencies produce shades of gray. Synchronization signals are sent at 1.2 kHz. These are short bursts, lasting 0.030 s (30 milliseconds, or ms) for vertical synchronization and 0.005 s (5 ms) for horizontal synchronization.

Sometimes, SSTV is sent along with voice data. The video is sent on one sideband, and the audio is sent on the other sideband.

The differences between color and gray-scale SSTV are similar to the differences between color and gray-scale FSTV. Separate signals are sent for the red (R), green (G), and blue (B) primary colors. These are combined at the receiver according to the *RGB color model*.

High-definition television

The term *HDTV* refers to any of several similar methods for getting more detail into a TV picture and for obtaining better audio quality, compared with standard FSTV.

A standard FSTV picture has 525 lines per frame, but HDTV systems have between 787 and 1125 lines per frame. The image is scanned about 60 times per second. High-definition TV is a digital mode; this offers another advantage over conventional FSTV. Digital signals propagate better, are easier to deal with when they are weak, and can be processed in ways that analog signals cannot.

Some HDTV systems use *interlacing* in which two *rasters* are "meshed" together. This effectively doubles the image resolution without doubling the cost of the hardware. But it can cause "jittering" with fast-moving pictures.

Digital satellite TV

Until the early 1990s, a satellite television installation required a dish antenna several feet in diameter. Many such systems are still in use. The antennas are expensive, they attract attention (sometimes unwanted), and they are subject to damage from ice storms, heavy snows, and high winds. Digitization has changed this situation. In any communications system, digitization allows the use of smaller receiving antennas, smaller transmitting antennas, and/or lower transmitter power levels. Engineers have managed to get the diameter of the receiving dish down to about 2 ft.

A pioneer in digital TV was RCA, which developed the *Digital Satellite System* (DSS). Figure 16.16 is a simplified block diagram of a DSS link. The analog signal is changed into digital pulses at the transmitting station via A/D conversion. The digital signal is amplified and sent up to a geostationary satellite. The satellite has a *transponder* that receives the signal, converts it to a different frequency, and retransmits it toward earth. The return signal is picked up by a portable dish. A *tuner* selects the channel. *Digital signal processing* can be used to improve the quality of reception under marginal conditions. The digital signal is changed back into analog form, suitable for viewing on a conventional FSTV set, via *D/A conversion*.

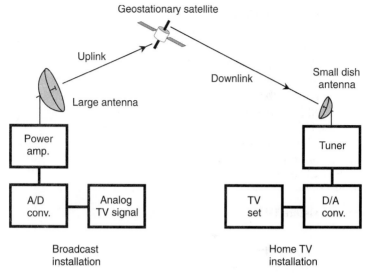

Figure 16.16 Block diagram of a DSS link.

17

Location, Navigation, and Control Systems

Electronic devices are employed to locate and track objects, people, and animals on the surface of the earth, under water, in the air, and in space.

Radar

The term *radar* is an acronym derived from the words "radio detection and ranging." Electromagnetic waves having certain frequencies reflect from various objects, especially if those objects contain metals or other electrical conductors. By ascertaining the direction(s) from which radio signals are returned, and by measuring the time it takes for an EM pulse to travel from the transmitter location to a target and back, it is possible to locate flying objects and to evaluate some weather phenomena.

A complete radar set consists of a transmitter, a highly directional antenna, a receiver, and an indicator or display. The transmitter produces microwave pulses that are propagated in a narrow beam. The EM waves strike objects at various distances. The greater the distance to the target, the longer the delay before the echo is received. The transmitting antenna is rotated so that all azimuth bearings (compass directions) can be observed.

A typical circular radar display is a CRT. The basic display configuration is shown in Fig. 17.1. The observing station is at the center of the display. Azimuth bearings are indicated in degrees clockwise from true north and are marked around the perimeter of the screen. The distance, or range, is indicated by the radial displacement of the echo. Airborne long-range radar can detect echoes from several hundred miles away under ideal conditions. A modest system with an antenna at low height can receive echoes from up to about 50 miles away.

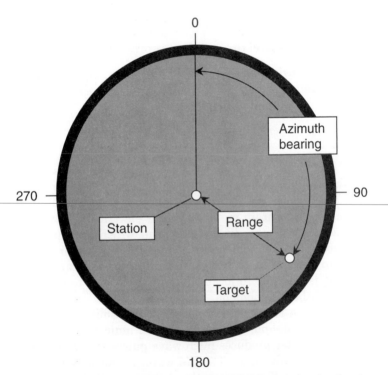

Figure 17.1 Simplified rendition of a typical radar display, showing the station and one target.

Some radar sets can detect changes in the frequencies of returned pulses, thereby allowing measurement of wind speeds in hurricanes and tornadoes. This is called *Doppler radar*. This type of radar is also employed to measure the speeds of approaching or receding targets.

Sonar

The term *sonar* is an acronym based on the words *sound detection and ranging*. An elementary sonar system consists of an acoustic pulse generator, an acoustic emitter, an acoustic pickup, a receiver, a delay timer, and an indicating device such as a numeric display, CRT, or pen recorder. The transmitter sends out acoustic waves through the medium, usually water or air. These waves are reflected by objects, and the echoes are picked up by the receiver. The distance to an object is determined on the basis of the echo delay, provided the speed of the acoustic waves in the medium is known.

A simple sonar system is diagrammed in Fig. 17.2A. A *computer map* can be generated on the basis of sounds returned from various directions in two or three dimensions. This system can be confused if the echo delay is equal to or longer than the time interval between pulses, as shown in Fig. 17.2B. To overcome this, a computer can instruct the generator to send pulses of various frequencies in a pseudorandom sequence. The computer keeps track of which echo corresponds to which pulse.

Direction Finding

At RF, location and navigation systems operate between a few kilohertz and the microwave region. Acoustic systems use frequencies between a few hundred hertz and a few hundred kilohertz.

Signal comparison

A machine or vessel can find its position by comparing the signals from two fixed stations whose positions are known,

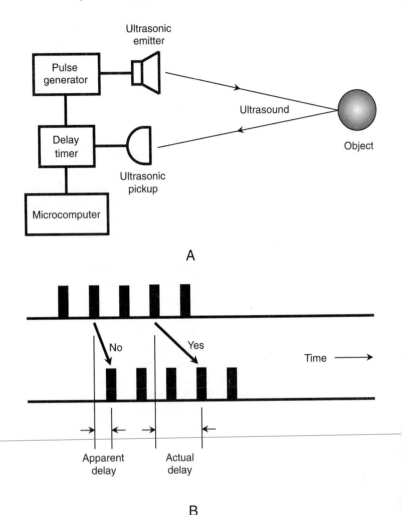

Figure 17.2 (A) A block diagram of a short-range sonar system; (B) a sonar fooled by long delays.

as shown in Fig. 17.3A. By adding 180° to the bearings of the sources X and Y, the machine or vessel (square) obtains its bearings as "seen" from the sources (dots). The machine or vessel can determine its direction and speed by taking two readings separated by a certain amount of time.

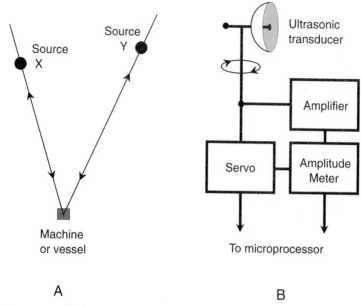

Figure 17.3   (A) A simple direction-finding scheme; (B) an ultrasonic direction finder.

Computers can assist in precisely determining, and displaying, the position and the velocity vector.

Figure 17.3B is a block diagram of an acoustic direction finder such as can be used by a mobile robot. The receiver has a signal-strength indicator and a *servo* that turns a directional ultrasonic transducer. There are two signal sources at different frequencies. When the transducer is rotated so that the signal from one source is maximum, a bearing is obtained by comparing the orientation of the transducer with some known standard such as a magnetic compass. The same is done for the other source. A computer uses triangulation to figure out the precise location of the robot.

Radio direction finding

A radio receiver, equipped with a signal-strength indicator and connected to a rotatable, directional antenna, can be used to determine the direction from which signals are coming. *Radio direction finding* (RDF) equipment aboard a mobile vehicle facilitates determining the location of a transmitter. RDF equipment can be used to find one's own position with respect to two or more transmitters operating on different frequencies.

In an RDF receiver, a loop antenna is generally used. It is shielded against the electric component of radio waves, so it picks up only the magnetic flux. The circumference is less than 0.1 wavelength. The loop is rotated until a null occurs in the received signal strength. When the null is found, the axis of the loop lies along a line toward the transmitter. When readings are taken from two or more locations separated by a sufficient distance, the transmitter can be pinpointed by finding the intersection point of the azimuth bearing lines on a map.

At frequencies above approximately 300 MHz, a directional transmitting/receiving antenna, such as a Yagi, quad, dish, or helical type, gives better results than a small loop. When such an antenna is employed for RDF, the azimuth bearing is indicated by a signal peak rather than by a null.

Ranging

Ranging is a method for a machine or vessel to navigate in its environment. It also allows a central computer to keep track of the whereabouts of mobile robots.

Loran

The term *loran* is an acronym derived from the words *long-range navigation*. Loran is one of the oldest electronic navigation schemes and is still used by some ships and aircraft. The system employs pulse transmission at a rate of about 20 pulses per second. Two transmitters are used for each coordinate determination; transmitter pairs are separated by 300 miles. The operating frequency is 100 kHz, corresponding to a wavelength of 3 km. Location is found by comparing the time delay in the arrival of the signal from the more distant transmitter, relative to the arrival time of the signal from the nearer transmitter.

Distance resolution

Distance resolution is the minimum radial separation (range difference) between two targets required for a machine to tell them apart. In Fig. 17.4, objects Y and Z are at almost exactly the same range from the machine or vessel. The delays, both long, are nevertheless almost identical. Depending on the precision of the system, the machine might or might not be able to distinguish between objects Y and Z on the basis of distance measurement alone. But a direction indicator would have no difficulty telling objects Y and Z apart. A range detector alone could easily differentiate between objects X and Y or objects X and Z, because the distance to object X is much different from the distance to objects Y and Z.

Sensing and plotting

Range sensing is used to measure distances in a single dimension. *Range plotting* is the creation of a graph of the distances to targets.

In *one-dimensional (1-D) range sensing,* a signal is transmitted, and the machine measures the time it takes for the echo to return. A 1-D range-sensing system could differentiate between objects X and Y, or between objects X and Z, in Fig. 17.4. But it might not be able to distinguish between

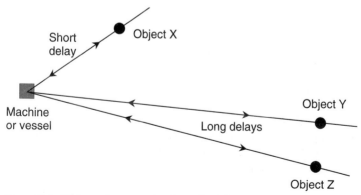

Figure 17.4 Distance is proportional to echo delay time.

objects Y and Z. Localized 1-D range sensing is called *proximity sensing*.

Two-dimensional (2-D) range plotting involves mapping the distances to various objects as a function of direction. The accuracy of the rendition depends on how many readings are taken around a 360° circle. A 2-D range plot can show objects in only one plane, usually a horizontal plane. Such a system could distinguish all three objects X, Y, and Z in Fig. 17.4.

Three-dimensional (3-D) range plotting is usually done in spherical coordinates: range (distance), azimuth (compass bearing), and elevation or altitude (vertical angle relative to the horizon). To obtain reasonable resolution, thousands of readings are necessary.

Global positioning system

The GPS is a network of radio location and radio navigation apparatus that operates on a worldwide basis. The system employs several satellites and allows determination of latitude, longitude, and altitude.

All GPS satellites transmit signals in the UHF radio spectrum. The signals are modulated with codes that contain timing information used by the receiving apparatus to make measurements. A GPS receiver determines its location by measuring the distances to four or more different satellites and using a computer to process the information received from the satellites. From this information the receiver can give the user an indication of position accurate to within a few meters (for government/industry subscribers) or a few hundred meters (for civilian subscribers).

Epipolar and Log-Polar Navigation

When the geometry of the environment is known, the position and course of a vessel can be determined by observations from a single point of reference.

Epipolar navigation

Epipolar navigation works by evaluating the way an image changes as viewed from a moving perspective.

Imagine piloting an aircraft over the ocean. The only land in sight is a small island. The on-board computer sees an image of the island that constantly changes shape. Figure 17.5 shows three sample sighting positions (A, B, C) and the size/shape of the island as seen by a machine-vision system in each case. The computer has the map data, so it knows the true size, shape, and location of the island. The

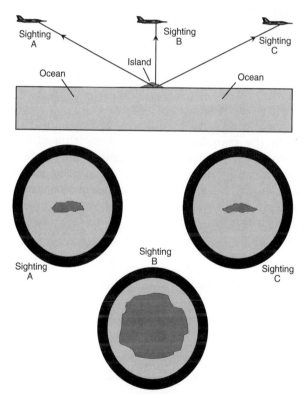

Figure 17.5 Epipolar navigation is the electronic analog of human spatial perception.

computer compares the shape/size of the image it sees at each point in time, from the vantage point of the aircraft, with the actual shape/size of the island from the map data. From this, the computer can ascertain the altitude of the aircraft, its speed, and direction of movement relative to the surface, its latitude, and its longitude.

Log-polar navigation

In *log-polar navigation,* a computer converts an image in polar coordinates to an image in rectangular coordinates. The principle is shown in Fig. 17.6. The polar map, with two object paths plotted, is shown in Fig. 17.6A. The rectangular equivalent, with the same paths shown, is in Fig. 17.6B. The polar radius (Fig. 17.6A) is mapped onto the vertical rectangular axis (Fig. 17.6B); the polar angle (Fig. 17.6A) is mapped onto the horizontal rectangular axis (Fig. 17.6B).

Radial coordinates are unevenly spaced in the polar map (Fig. 17.6A) but are uniform in the rectangular map (Fig. 17.6B). During the transformation, the logarithm of the radius is taken. The resolution is degraded for distant objects but is enhanced for nearby ones. A log-polar transform distorts the way a scene appears to human observers, but it translates images and motions into data that can be efficiently dealt with by a computerized scanning system.

Robot Guidance

There are several methods via which robots can use electronic devices to navigate in their surroundings.

Beacons

A tricorner reflector is a good example of a *passive beacon.* The robot requires a transmitting transducer such as a laser and a receiving transducer such as a photocell. The distance to each reflector can be determined by the time required for a visible-light or IR pulse to travel to the reflector and return to the robot.

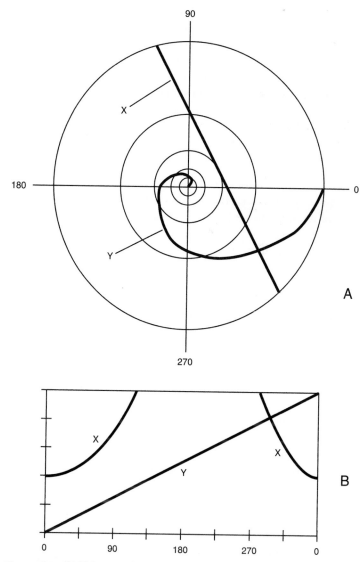

Figure 17.6 (A) Polar coordinates in real space; (B) coordinates after log-polar transform. Lines X and Y show hypothetical paths in each system.

An example of an *active beacon* is a radio transmitter. Several transmitters can be put in various places and their signals synchronized so that they are all exactly in phase. As the robot moves around, the relative phases of the signals change. Using an internal computer, the robot can determine its position by comparing the phases of the signals from the beacons. The system works like a small-scale version of the GPS.

Edge detection

Edge detection refers to the ability of a machine vision system to locate boundaries. It also refers to a robot's knowledge of what to do with respect to those boundaries. A robot automobile can use edge detection to see the edges of a road and to keep itself on the road (Fig. 17.7). A *personal robot,* of the type people often imagine doing the work of a maid or butler, must see the edges of a door before going through the door. It must be able to detect the presence of stairwells and ascertain the outlines of common household furnishings.

Embedded path

One common type of *embedded path* is a buried, current-carrying wire. The current produces a magnetic field that a robot can follow. This method of guidance has been suggested as a way to keep a car on a highway, even if the driver is not paying attention. The wire needs a constant supply of electric current.

Alternatives to wires, such as colored paints or tapes, do not require electric current. Robots follow these devices using edge detection; the brightly colored or reflective media assist the machine in keeping track of the boundary.

A large number of passive beacons, placed at regular intervals along a boundary, can serve as a hybrid beacon/edge/path system. Small, raised plastic "dots," designed to reflect light back toward the source and commonly found on highways, can serve this purpose. The dots must be kept in good condition and are not reliable in places where snow falls or sand and dust accumulate.

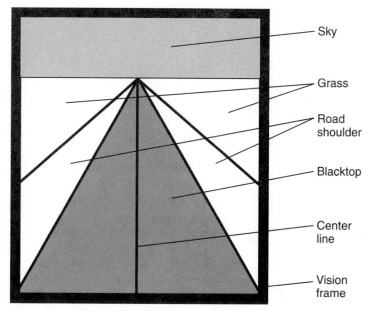

Figure 17.7 Edge detection. Boundaries between different objects or substances show up as well-defined lines.

Biased search

A *biased search* is a method of finding a destination or target by first looking off to one side. The robot controller deliberately takes an inaccurate course. The controller chooses its course on the basis of its general knowledge of the operating environment. Once a known landmark is spotted, the controller employs its database to calculate the position of, and set course for, the destination.

Figure 17.8 shows the technique of biased searching as a robot-controlled boat might apply it when approaching a dock in dense fog. When the shore comes into view, the boat turns and follows the shoreline at a safe distance until the dock is found. For this scheme to work, the sense of the error (to the left or right) must be correct. The initial error must be large enough so its sense is known with certainty.

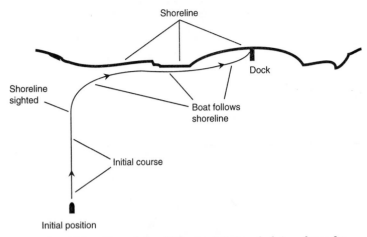

Figure 17.8 A biased search in which a boat finds a dock in a dense fog.

Fluxgate magnetometer

A *fluxgate magnetometer* is a magnetic compass for robot guidance. The device uses coils to sense changes in an artificially generated magnetic field. Navigation within a room, for example, can be done by checking the orientation and intensity of magnetic lines of flux generated by electromagnets in the walls. There is a mathematical one-to-one correspondence between magnetic flux orientation/intensity and the points inside the room. The robot controller is programmed with this information, allowing the robot to pinpoint its position in the room.

Gyroscope

A *gyroscope,* or *gyro,* is an inertial guidance system. Gyros are used when it is inconvenient or impossible to obtain bearings in any other way. A massive, rotating disk is mounted in a set of bearings. The disk is driven by a motor. A common application of a gyro is to keep track of the direction of travel (bearing or heading) of a robot or vessel without having to rely on external signals, fields, or points of reference. A gyroscope must be realigned at regular intervals.

Binaural machine hearing systems

In a *binaural machine hearing system,* two acoustic trans-ducers are positioned on either side of a robot or vehicle. The robot controller compares the relative phase and intensity of the signals from the two transducers. This lets the robot "know," with certain limitations, the direction from which the sound is coming. If the robot is confused, it can turn until ambiguity is eliminated and a meaning-ful bearing is obtained. If the positions of various sound sources are known, the robot can navigate its way via this directional sense of machine hearing. For the scheme to work, the robot controller must be programmed with information concerning the locations of the sound sources.

Machine Vision

One of the most advanced specialties in electronics is the field of *vision systems,* also called *machine vision.* There are sever-al different schemes. The optimum form of machine vision depends on the intended use. Vision systems are especially important in high-level computer and robotics applications.

Components

A visible-light vision system employs a camera tube such as a vidicon or charge-coupled device (CCD). In bright light an image orthicon can be used. The camera produces an analog video signal that is changed into digital form by A/D con-version. The digital signal is clarified by DSP. The resulting data goes to a computer or robot controller. A block diagram of this scheme is shown in Fig. 17.9.

A functional block diagram of a color-sensing system is shown in Fig. 17.10. Three gray-scale cameras are used. Each camera has a color filter. One filter is red (R), another is green (G), and another is blue (B). All possible hues, brightness levels, and saturation levels are made up of these three colors in various ratios. The signals from the three cameras are processed by a microcomputer, and the result is fed to the computer or robot controller.

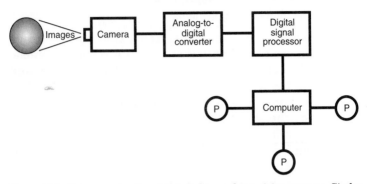

Figure 17.9 Components of a visible-light machine-vision system. Circles labeled P are computer peripherals.

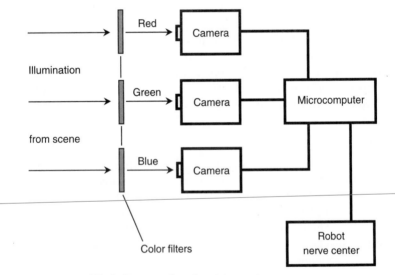

Figure 17.10 Block diagram of a color vision system.

Computer mapping

An autonomous robot must have a map of its environment. A computer map resembles a blueprint. An example is shown in Fig. 17.11. This is a dining room. The robot's assignment is to set the table before the people sit down. The robot can

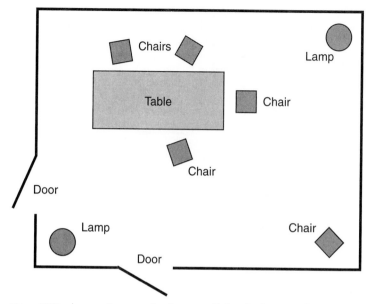

Figure 17.11 A computer map of a plane parallel to the floor in a dining room.

envision, relative to the table and chairs, where to put the plates. Then it can follow its programming and place the napkins, forks, knives, spoons, and glasses in their proper places. Note that one chair has been moved into the corner. This might confuse the robot and result in its setting an extra place at the table and/or dropping utensils on the carpet in the corner. Whether or not such problems occur depends on the quality of the map-interpretation program.

Remote Control

An example of a sophisticated remote-control system, in this case for a home computer, is shown in Fig. 17.12.

Features

There are various remote-control programs available, some with basic features only, and others with sophisticated capabilities. Common features are as follows:

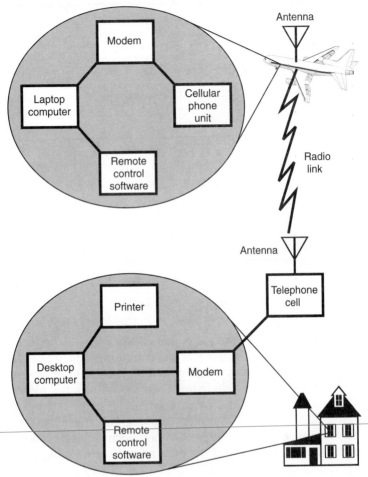

Figure 17.12 A home computer can be remotely operated from a portable computer.

- *Operating system flexibility.* Remote-control programs are available for all operating systems.

- *Networking.* Remote control is an asset when running on networks. A wide-area network (WAN) can be assembled, with one host computer and numerous remote machines.

- *Multiple hosts.*   High-end remote-control programs allow operation with several screens, in rotating sequence.

- *Modem specifications.*   Performance is directly proportional to the speed of the slowest modem in a system.

- *Unattended host.*   It is possible to power-up, operate, and power-down a host computer even when there is no one attending it.

- *Chat mode.*   This allows the use of on-line services or communication with someone attending the host computer. It can also allow the users of remote computers to communicate with each other through the host.

- *Timing out.*   Some remote-control systems, especially those with multiple remotes, allow for a time limitation. After a certain preprogrammed length of time, the connection will terminate.

- *Printing.*   It is possible to have the host computer print data if there is an automatic (and reliable) paper feed.

- *Call logging.*   Every time the host computer is accessed from a remote unit, the date, time, unit identity, and duration of the call can be logged.

- *Hierarchical password protection.*   Also called *multilevel password protection,* this feature prevents unauthorized use of the host computer.

- *Password retry limitation.*   This feature prevents hackers from making repeated guesses at passwords in an attempt to break into the system.

- *On-line help.*   Most sophisticated software packages have a provision for calling a toll-free number and getting assistance on-line.

In robotics

A simple example of a robotic remote-control system is the control box for a TV set. Other examples are the transmitter used to fly a model airplane and an electronic garage-door opener. Most TV controls employ IR beams to carry the data; the model airplane and the garage door receive their commands via radio signals.

Robots are used to explore the bottoms of oceans and lakes. These machines are operated via electric or fiber-optic cable links. The range is limited because it is impractical to have a cable longer than a few miles. Wireless laser links are possible in clear waters over short distances. Visible red and IR lasers propagate fairly well for short distances underwater.

Autonomous robots

An *autonomous robot,* also called a *smart robot,* is a self-contained machine with its own built-in computer system. Figure 17.13 shows one scheme in which a fleet of autonomous robots might work together. Robots are shown as squares and computers, as solid black dots. The computers are connected by radio links (straight lines) in a peer-to-peer LAN. Each robot can move around by rolling

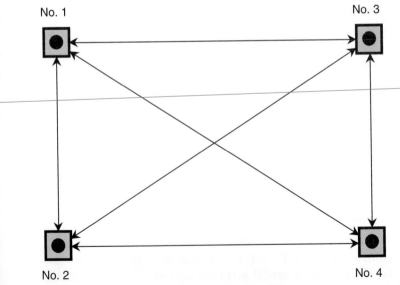

Figure 17.13 Autonomous robots in a peer-to-peer wireless LAN.

on wheels or on a track drive. Each robot controller in the fleet has its own programming. Robots share data with each other, just as computers share information in a LAN.

Insect robots

"Stupid robots," all under the absolute control of a single central computer, are called *insect robots* because the whole system functions as a unit, much like an anthill or beehive. This is the robotic counterpart of a client/server LAN.

Ants or bees are like idiot robots. But an anthill or beehive is an efficient system, run by the collective mind of all its members. Such systems are "smart in the generality." Rodney Brooks, who pioneered the insect-robot concept, saw this fundamental difference between autonomous and collective intelligence. He designed robot colonies consisting of many units under the control of a central computer.

Brooks envisions microscopic insect robots that might live in a home or office building, coming out at night to clean floors and counter tops. "Antibody robots" of even tinier proportions could be injected into a person infected with some heretofore incurable disease. Controlled by a central microprocessor, they could seek out the disease bacteria or viruses and destroy them. The communication between the central controller and the robots would be via wireless, probably at microwave radio frequencies.

18

Antenna Systems

Antennas can be categorized into two major classes: receiving and transmitting. Most transmitting antennas can function effectively for reception. Some receiving antennas can efficiently transmit EM signals; others cannot.

Radiation Resistance

When RF ac flows in an electrical conductor, some EM energy is radiated into space. If a resistor is substituted for the antenna, in combination with a capacitor or inductor to mimic any inherent reactance in the antenna, the transmitter behaves in the same manner as when connected to the actual antenna. For any antenna operating at a specific frequency, there is a specific resistance, in ohms, for which this can be done. This is known as the *radiation resistance* (R_R) of the antenna at the frequency in question. Radiation resistance is specified in ohms.

Determining factors

If a thin, straight, lossless vertical antenna is placed over perfectly conducting ground, R_R is a function of the vertical-antenna conductor height in wavelengths (Fig. 18.1A). If a thin, straight, lossless wire is placed in free space and fed at the center, R_R is a function of the overall conductor length in wavelengths (Fig. 18.1B).

Antenna efficiency

Efficiency is rarely crucial to the performance of a receiving antenna system, but it is important in a transmitting antenna system. A transmitting antenna ideally has a large R_R because radiation resistance appears in series with *loss resistance* (R_L). The antenna efficiency, Eff, is given by

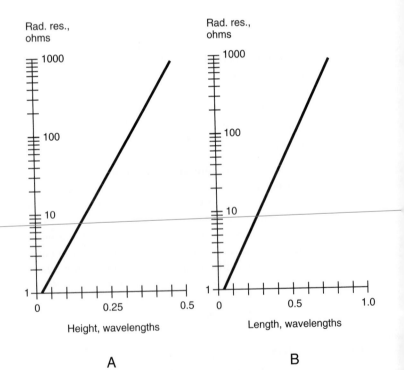

Figure 18.1 Approximate values of radiation resistance for vertical antennas over perfectly conducting ground (A) and for center-fed antennas in free space (B).

$$\text{Eff} = \frac{R_R}{R_R + R_L}$$

which is the ratio of the radiation resistance to the total antenna system resistance. As a percentage,

$$\text{Eff} = \frac{100\,R_R}{R_R + R_L}$$

Some short radiators, widely used for receiving, can function efficiently for transmitting if the earth is kept out of the antenna circuit, and if the antenna is set up in a favorable location. The design of such systems must minimize R_L.

Half-Wave Antennas

A half wavelength in free space is given by the equation

$$L = \frac{492}{f}$$

where L is the linear distance in feet, and f is the frequency in megahertz. A half wavelength in meters is given by

$$L = \frac{150}{f}$$

For ordinary wire, the results as obtained above should be multiplied by a *velocity factor* v of 0.95 (95 percent). For tubing or large-diameter wire, v can range down to approximately 0.90 (90 percent).

Open dipole

An *open dipole* or *doublet* is a half-wavelength radiator fed at the center. Each "leg" of the antenna is one-quarter wavelength long (Fig. 18.2A). For a straight wire radiator, the length L, in feet, at a design frequency f, in megahertz, for a practical half-wave dipole is approximately

$$L = \frac{467}{f}$$

The length is meters is approximately

$$L = \frac{143}{f}$$

These values assume $v = 0.95$. In free space, the impedance at the feed point is a pure resistance of approximately 73 Ω. This represents the radiation resistance and the absence of reactance at the resonant frequency.

Folded dipole

A *folded dipole antenna* is a half-wavelength, center-fed antenna constructed of two parallel wires with their ends

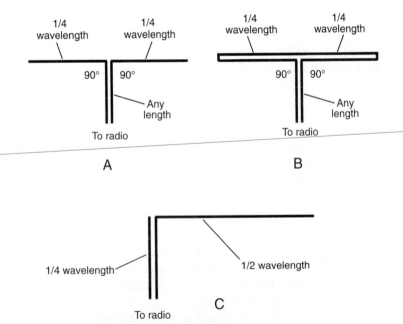

Figure 18.2 Basic half-wave antennas. (A) Dipole antenna; (B) folded-dipole antenna; (C) zepp antenna.

connected together (Fig. 18.2B). The feed-point impedance of the folded dipole is a pure resistance of approximately 290 Ω. This makes the folded dipole ideal for use with high-impedance, parallel-wire transmission lines in applications where gain and directivity are not especially important.

Zepp

A *zeppelin antenna,* also called a *zepp,* is a half-wavelength radiator, fed at one end with a quarter-wavelength section of open-wire line (Fig. 18.2C). The impedance at the feed point is an extremely high, pure resistance. At the transmitter end of the line, the impedance is a low, pure resistance. A zeppelin antenna can operate well at all harmonics of the design frequency. If a transmatch is available to tune out reactance, the feed line can be of any length. Feed-line radiation should be minimized by carefully cutting the radiator to one-half wavelength at the design frequency and by using the antenna only at this frequency or one of its harmonics.

Quarter-Wave Antennas

A *quarter wavelength* in free space is related to frequency by the equation

$$L = \frac{246}{f}$$

where L represents a quarter wavelength in feet, and f represents the frequency in megahertz. If L is expressed in meters, the formula is

$$L = \frac{75}{f}$$

If v is the velocity factor in a given medium other than free space,

$$L = \frac{246v}{f}$$

if L is expressed in feet. It is

$$L = \frac{75v}{f}$$

if L is expressed in meters. For a typical wire conductor, $v = 0.95$ (95 percent); for metal tubing, v can range down to approximately 0.90 (90 percent). For a vertical antenna constructed from aluminum tubing such that $v = 0.92$, for example, the formulas are

$$L = \frac{226}{f}$$

if L is expressed in feet and

$$L = \frac{69}{f}$$

if L is expressed in meters.

A quarter-wave antenna must be operated against an RF ground. The feed-point impedance over perfectly conducting ground is approximately 37 Ω. This represents radiation resistance in the absence of reactance and provides a reasonable impedance match to most coaxial transmission lines.

Ground-mounted vertical antenna

The simplest vertical antenna is a quarter-wave radiator mounted at ground level. The radiator is fed with a coaxial cable. The center conductor is connected to the base of the radiator, and the shield is connected to a ground system.

At some frequencies, the height of a quarter-wave vertical is unmanageable unless *inductive loading* is used to reduce the physical length of the radiator. A quarter-wave vertical antenna can be made resonant on several frequencies by the use of multiple loading coils or by inserting traps at specific points along the radiator.

Unless an extensive *ground radial* system is installed, a ground-mounted vertical antenna is inefficient. Inductive loading worsens this situation. Another problem is the fact that vertically polarized antennas receive more human-made noise than horizontal antennas. In addition, the EM fields from ground-mounted transmitting antennas are more likely to cause interference to electronic devices than are the EM fields from antennas installed at a height.

Ground plane

A *ground-plane antenna* is a vertical radiator operated against a system of quarter-wave radials and elevated at least one-quarter wavelength above the earth's surface. The radiator itself can be any length but should be tuned to resonance at the desired operating frequency.

When a ground plane is elevated at least 90 electrical degrees above the surface, only three or four radials are necessary to obtain a low-loss RF ground system. The radials extend outward from the base of the antenna at an angle between 0 and 45° with respect to the horizon. Figure 18.3A illustrates a typical ground-plane antenna.

A ground-plane antenna should be fed with coaxial cable. The feed-point impedance of a ground-plane antenna having a quarter-wave radiator is about 37 Ω if the radials are horizontal; the impedance increases as the radials droop, reaching about 50 Ω at a droop angle of 45°.

The radials in a ground-plane antenna can extend straight downward, in the form of a quarter-wave cylinder

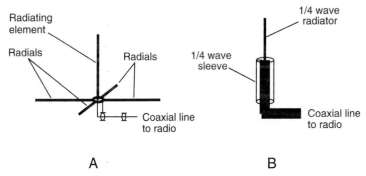

Figure 18.3 (A) Basic ground-plane antenna; (B) coaxial antenna.

concentric with the feed line. In this case, the feed-point impedance is approximately 73 Ω. This comprises a *coaxial antenna* (Fig. 18.3B).

Loops

Any receiving or transmitting antenna, consisting of one or more turns of wire forming a dc short circuit, is a *loop antenna*. Loop antennas can be categorized as either small or large.

Small loop antenna

A *small loop antenna* has a circumference of less than 0.1 wavelength and is suitable for receiving but generally not for transmitting. A small loop is the least responsive along its axis and is most responsive in the plane perpendicular to its axis. A capacitor can be connected in series or parallel with the loop to provide a resonant response. An example of such an arrangement is shown in Fig. 18.4.

Small loops are useful for RDF and also for reducing interference caused by human-made noise or strong local signals. The null along the axis is sharp and deep.

Loopsticks

For receiving applications at frequencies up to approximately 20 MHz, a *loopstick antenna* is sometimes used. This antenna, a variant of the small loop, consists of a coil wound on a rod-shaped, powdered-iron core. A series or parallel capacitor, in conjunction with the coil, forms a tuned circuit.

A loopstick displays directional characteristics similar to those of the small loop antenna shown in Fig. 18.4. The sensitivity is maximum off the sides of the coil, and a sharp null occurs off the ends. This null can minimize interference from local signals and from human-made sources of noise.

Large loop antennas

If a loop has a circumference greater than 0.1 wavelength, it is classified as a *large loop antenna*. A large loop usually

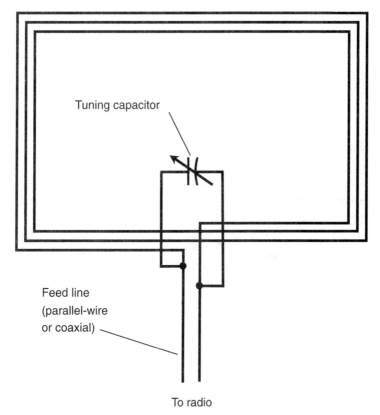

Figure 18.4 A small loop antenna with a capacitor for adjusting the resonant frequency.

has a circumference of either 0.5 or 1 wavelength and is self-resonant without the need for a tuning capacitor. The half-wavelength loop presents a high impedance at the feed point, and maximum radiation/response occurs in the plane of the loop. The full-wavelength loop presents an impedance of about 100 Ω at the feed point, and the maximum radiation/response occurs along the axis.

A large loop can be used for transmitting or receiving. The half-wavelength loop exhibits a slight power loss relative to a dipole; the full-wavelength loop shows a slight gain over a dipole in its favored directions.

Ground Systems

Unbalanced antenna systems, such as a quarter-wave vertical antenna, require low-loss RF ground systems. Most balanced systems, such as a half-wave dipole, do not. A low-loss dc ground, also called an *electrical ground,* is advisable for any antenna system.

Electrical versus RF ground

Electrical grounding is important for personal safety. It can help protect equipment from damage if lightning strikes nearby. It minimizes the chances for EMI to and from the equipment.

There's an unwritten commandment in electrical and electronics engineering: Never touch two grounds at the same time. This refers to the tendency for a potential difference to exist between supposedly neutral points, in apparent defiance of logic. Even experienced people can get severe electrical shocks because they forget this rule.

Some appliances have three-wire cords. One wire is connected to the "common" or "ground" part of the hardware and leads to a D- or U-shaped prong in the plug. This ground prong should never be defeated because such modification can result in dangerous voltages appearing on exposed metal surfaces.

A good *RF ground* system can help minimize EMI. Figure 18.5 shows a proper RF ground scheme (Fig. 18.5A) and an improper one (Fig. 18.5B). In a good RF ground system, each device is connected to a common *ground bus,* which in turn runs to the earth ground via a single conductor. This conductor should be as short as possible. A poor ground system contains *ground loops* that increase the susceptibility of equipment to EMI.

Radials and the counterpoise

Ground radials generally measure one-quarter wavelength or more. They run outward from the base of a surface-mounted vertical antenna and are connected to the shield of the coaxial feed line. The radials can be installed on the sur-

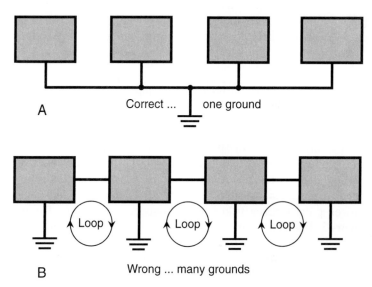

Figure 18.5 (A) The correct method for grounding multiple units; (B) an incorrect method creates RF ground loops.

face or buried under the earth. The greater the number of radials of a given length, the better the antenna will work. Also, the longer the radials for a given number, the better.

Radials can be used to obtain a good RF ground for any type of EM communications or broadcast installation. The ground rod should be located as close to the station as possible, and the wire from the rod to the station should be as straight, and as large in diameter, as possible.

A *counterpoise* is a means of obtaining an RF ground or ground plane without a direct earth-ground connection. A grid of wires, a screen, or a metal sheet is placed above the surface and oriented horizontally to provide capacitive coupling to the earth. This minimizes RF ground loss. Ideally, the radius of a counterpoise should be at least one-quarter wavelength at the lowest operating frequency.

Gain and Directivity

The *power gain* of a transmitting antenna is the ratio of the maximum *effective radiated power* (ERP) to the actual RF power applied at the feed point. Power gain is expressed in *decibels.* If the ERP is P_{ERP} watts and the applied power is P watts,

$$\text{Power gain (dB)} = 10 \log_{10} \frac{P_{\text{ERP}}}{P}$$

Power gain is always measured in the favored direction(s) of an antenna, that is, at the center(s) of the main lobe(s).

For power gain to be defined, a *reference antenna* must be chosen with a gain assumed to be unity, or 0 dB. This reference antenna is usually a half-wave dipole in free space. Power gain figures taken with respect to a dipole (in its favored directions) are expressed in units called dBd. The reference antenna for power-gain measurements can also be an *isotropic radiator,* which theoretically radiates and receives equally well in all directions in three dimensions. In this case, units of power gain are called dBi. For any given antenna, the power gains in dBd and dBi are different by approximately 2.15 dB:

$$\text{Power gain (dBi)} = 2.15 + \text{power gain (dBd)}$$

Directivity plot

Antenna radiation and response patterns are represented by plots such as those shown in Fig. 18.6. The location of the antenna is assumed to be at the center (origin) of the coordinate system. The greater the radiation or reception capability of the antenna in a certain direction, the farther from the center the points on the chart are plotted.

A dipole antenna, oriented horizontally so that its conductor runs in a north-south direction, has a horizontal-plane (H-plane) pattern similar to that in Fig. 18.6A. The elevation-plane (E-plane) pattern depends on the height of the antenna above effective ground at the viewing angle. With the dipole oriented so that its conductor runs perpendicular to the page, and the antenna one-quarter wavelength above effective ground, the E-plane antenna pattern resembles the graph shown in Fig. 18.6B.

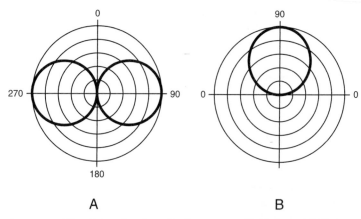

Figure 18.6 Directivity plots for a dipole antenna. (A) H-plane; (B) E-plane.

Forward gain

Forward gain is expressed as the level, in decibels, of the ERP in the main lobe of a unidirectional antenna to the ERP from a reference antenna, usually a half-wave dipole, in its favored directions. A typical two-element phased array has a forward gain of 3 dBd. As more in-phase elements are added, the gain increases. A typical two-element Yagi antenna has a gain of about 5 dBd. Some antennas have forward gain figures exceeding 25 dBd. At microwave frequencies, large dish antennas can have forward gain upward of 35 dBd. In general, as the wavelength decreases (the frequency gets higher), it becomes easier to obtain high forward gain figures.

Front-to-back ratio

The *front-to-back ratio* of a unidirectional antenna, abbreviated f/b, is an expression of the concentration of radiation/response in the main lobe, relative to the direction 180° opposite the center of the main lobe. Figure 18.7 shows a hypothetical directivity plot for a unidirectional antenna pointed north. The outermost circle depicts the field strength in the direction of the center of the main lobe and represents 0 dB. The next smaller circle represents a field strength 5 dB down with respect to the main lobe.

Continuing inward, circles represent 10, 15, and 20 dB down. The center, or origin, of the graph represents 25 dB down and also shows the location of the antenna. The f/b ratio is found, in this case, by comparing the signal levels between north (0° azimuth) and south (180° azimuth).

Front-to-side ratio

The *front-to-side ratio* of a directive antenna, abbreviated f/s, is another expression of the directivity of an antenna system. The term applies to unidirectional and bidirectional antennas. The f/s ratio is expressed in decibels. The EM field strength in the favored direction is compared with the field strength at right angles to the favored direction. An example is shown in Fig. 18.7. The f/s ratios are found, in this case, by comparing the signal levels between north and east (right-hand f/s) or between north and west (left-hand f/s). The right- and left-hand f/s ratios are usually the same in theory, although they might differ slightly in practice.

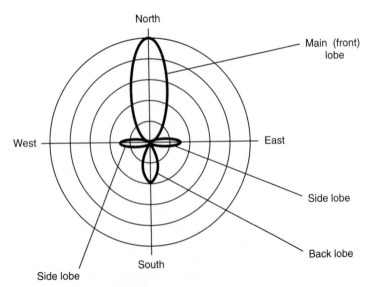

Figure 18.7 Directivity plot for a hypothetical antenna. Front-to-back and front-to-side ratios can be determined from such a graph.

Phased Arrays

A *phased array* uses two or more driven elements to produce gain in some directions at the expense of other directions.

Concept

Two simple pairs of phased dipoles are depicted in Fig. 18.8. In Fig. 18.8A, the dipoles are spaced one-quarter wavelength apart and are fed 90° out of phase, resulting in a unidirectional pattern. A bidirectional pattern can be obtained by spacing the dipoles one wavelength apart and feeding them in phase, as shown in Fig. 18.8B.

Phased arrays can have fixed directional patterns, or they might have rotatable or steerable patterns. The pair of phased dipoles shown in Fig. 18.8A can, if the wavelength is short enough to allow construction from metal tubing, be mounted on a rotator for 360° directional adjustability. With phased vertical antennas, the relative signal phase can be varied and the directional pattern thereby adjusted.

Longwire antenna

A wire antenna measuring a full wavelength or more, and fed at a high-current point or at one end, comprises a *longwire antenna*. A longwire antenna offers gain over a half-wave dipole. As the wire is made longer, the main lobes get more nearly in line with the antenna, and their amplitudes increase. The gain is a function of the length of the antenna: the longer the wire, the greater the gain. A longwire antenna must be as straight as possible for proper operation.

Broadside array

Figure 18.9 shows the geometric arrangement of a *broadside array*. The driven elements can each consist of a single radiator, as shown in the figure, or they can consist of Yagi antennas, loops, or other systems with individual directive properties. If a reflecting screen is placed behind the array of dipoles in Fig. 18.9, the system becomes a *billboard antenna*.

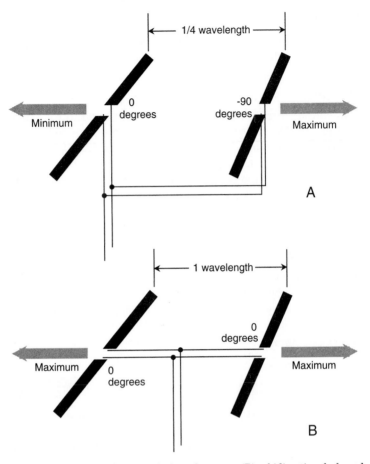

Figure 18.8 (A) A unidirectional phased system; (B) a bidirectional phased system. These are both examples of end-fire arrays.

The directional properties of a broadside array depend on the number of elements, whether or not the elements have gain themselves, and the spacing among the elements. In general, the larger the number of elements, the greater the gain.

End-fire array

A typical *end-fire array* consists of two parallel half-wave dipoles fed 90° out of phase and spaced one-quarter wave-

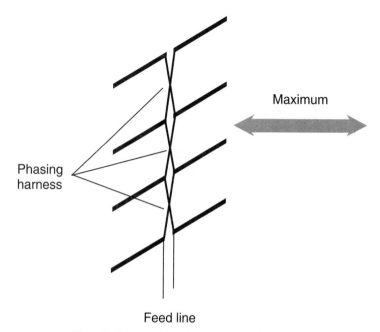

Figure 18.9 A broadside array. The elements are all fed in phase. Maximum radiation and response occur perpendicular to the plane containing the elements.

length apart. This produces a unidirectional radiation pattern. Or, the two elements might be driven in phase and spaced at a separation of one wavelength. This results in a bidirectional radiation pattern. In the phasing system, the two branches of transmission line must be cut to precisely the correct lengths, and the velocity factor of the line must be known and be taken into account. The phased dipole systems in Fig. 18.8 are both end-fire arrays.

Parasitic Arrays

Parasitic arrays are used at frequencies ranging from approximately 5 Mhz into the microwave range for obtaining directivity and forward gain. Examples include the *Yagi* and the *quad antenna.*

Concept

A *parasitic element* is an electrical conductor that comprises an important part of an antenna system but is not directly connected to the feed line. Parasitic elements operate via EM coupling to the driven element. When gain is produced in the direction of the parasitic element, the element is a *director*. When gain is produced in the direction opposite the parasitic element, the element is a *reflector*. Directors are generally a small percentage shorter than the driven element; reflectors are a small percentage longer.

Yagi antenna

The *Yagi antenna,* sometimes called a "beam," is an array of parallel, straight elements. A two-element Yagi can be formed by placing a director or a reflector parallel to, and a specific distance away from, a single driven element. The optimum spacing for a driven-element/director Yagi is 0.1 to 0.2 wavelength, with the director tuned 5 to 10 percent higher than the resonant frequency of the driven element. The optimum spacing for a driven-element/reflector Yagi is 0.15 to 0.2 wavelength, with the reflector tuned 5 to 10 percent lower than the resonant frequency of the driven element. The gain of a well-designed *two-element Yagi* is approximately 5 dBd.

A Yagi with one director and one reflector, along with the driven element, increases the gain and f/b ratio compared with a two-element Yagi. An optimally designed *three-element Yagi* has approximately 7 dBd gain. An example is shown in Fig. 18.10. (This is a conceptual drawing; it should not be used as an engineering blueprint.)

The gain and f/b ratio of a Yagi increase as elements are added. This is usually done by placing extra directors in front of a three-element Yagi. Each director is slightly shorter than its predecessor. The design and construction of Yagi antennas having numerous elements is a sophisticated business; as the number of elements increases, the optimum antenna dimensions become more and more intricate.

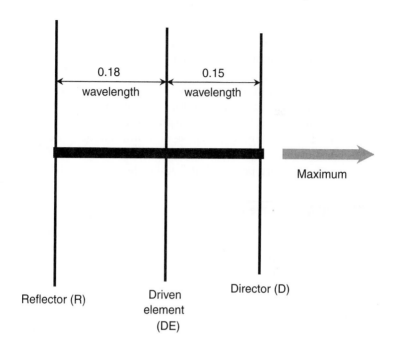

$$\text{Length DE} = 470/f$$
$$\text{Length D} = 425/f$$
$$\text{Length R} = 510/f$$

where f = operating frequency, MHz

Figure 18.10 A three-element Yagi antenna. Dimensions are discussed in the text.

Quad antennas

A *quad antenna* operates according to the same principles as the Yagi, except full-wavelength loops are used instead of straight half-wavelength elements.

A *two-element quad* can consist of a driven element and a reflector, or it can have a driven element and a director. A *three-element quad* has one driven element, one director, and one reflector. The director has a perimeter of 0.95 to 0.97 electrical wavelength, the driven element measures exactly one electrical wavelength around, and the reflector has a perimeter of 1.03 to 1.05 electrical wavelength.

Additional directors can be added to the basic three-element quad design to form quads that have any desired numbers of elements. The gain increases as the number of elements increases. Each succeeding director is slightly shorter than its predecessor. The additional directors might be half-wave, straight elements similar to those of a multielement Yagi. In this instance, the antenna is called a *quagi*. Long quad and quagi antennas are practical at frequencies above 100 Mhz.

UHF and Microwave Antennas

At UHF and microwave frequencies, high-gain antennas are reasonable in size because the wavelengths are short.

Dish antenna

A *dish antenna* must be correctly shaped and precisely aligned. The most efficient shape is a paraboloidal reflector. The feed system consists of a coaxial line or waveguide from the receiver and/or transmitter and a horn or helical driven element at the focal point of the reflector. *Conventional dish feed* is shown in Fig. 18.11A. *Cassegrain dish feed* is shown in Fig. 18.11B.

The larger the diameter of the reflector in wavelengths, the greater the gain and the narrower the main lobe. A dish antenna must be at least several wavelengths in diameter for proper operation. The reflecting element can be sheet metal, a screen, or a wire mesh. If a screen or mesh is used, the spacing between the wires must be a small fraction of a wavelength.

Helical antenna

A *helical antenna* is a circularly polarized, high-gain, unidirectional antenna. A typical helical antenna is shown in Fig. 18.12. The reflector diameter should be at least 0.8 wavelength at the lowest operating frequency. The radius of the helix should be approximately 0.17 wavelength at the center

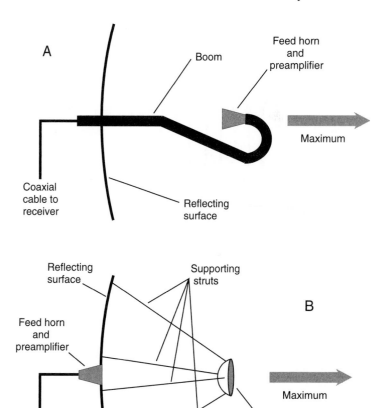

Figure 18.11   (A) Dish antennas with conventional feed; (B) Cassegrain feed.

of the intended operating frequency range. The longitudinal spacing between turns should be approximately 0.25 wavelength in the center of the operating frequency range. The overall length of the helix should be at least one wavelength at the lowest operating frequency.

A helical antenna can provide about 15 dBd forward gain. Bays of helical antennas are common in space communications systems.

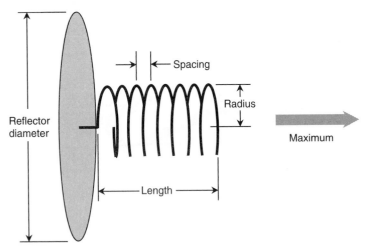

Figure 18.12 A helical antenna with reflecting element.

Corner reflector

A *corner reflector,* employed with a half-wave driven element, is illustrated in Fig. 18.13. This provides some gain over a dipole. The reflector is wire mesh, screen, or sheet metal. The flare angle of the reflecting element is approximately 90°. Corner reflectors are widely used in TV reception and satellite communications. Several half-wave dipoles can be fed in phase and placed along a common axis with a single, elongated reflector, forming a *collinear corner-reflector array.*

Horn antenna

There are several different configurations of the *horn antenna;* they all look similar. Figure 18.14 is a representative drawing. This antenna provides a unidirectional radiation and response pattern, with the favored direction coincident with the opening of the horn. The feed line is a waveguide that joins the antenna at the narrowest point (throat) of the horn.

Horns are not often used by themselves; they are commonly used to feed large dish antennas. The horn is aimed back at the center of the dish reflector. This minimizes extraneous radiation and response to and from the dish.

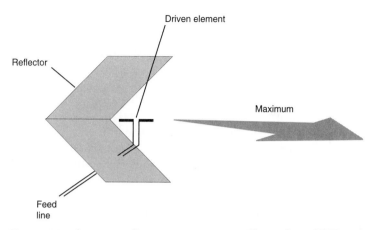

Figure 18.13 A corner-reflector antenna, generally used at UHF and microwave frequencies.

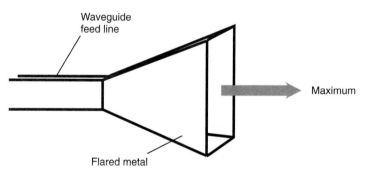

Figure 18.14 A horn antenna, used with a waveguide transmission line at microwave frequencies.

Feed Lines

A *feed line,* also called a *transmission line* in transmitting applications, transfers energy between wireless communications hardware and an antenna.

Basic characteristics

Feed lines can be unbalanced or balanced. Unbalanced lines include single-wire line, coaxial cable, and waveguides.

Balanced lines include open wire, ladder line, tubular line, ribbon line, and four-wire line.

All feed lines exhibit a *characteristic impedance* (Z_0) that depends on the physical construction of the line. This value, expressed in ohms, is the ratio of the RF voltage (in volts) to the RF current (in amperes) when there are no standing waves on the line—that is, when the current and voltage are in the same ratio at all points along the line.

Coaxial cable typically has Z_0 between 50 and 100 Ω. *Twinlead* is available in 75-Ω and 300-Ω Z_0 values. *Open-wire line* has Z_0 between 300 and 600 Ω, depending on the spacing between the conductors and also on the type of dielectric employed to keep the spacing constant between the conductors.

Waveguide

A *waveguide* is used at UHF and microwave frequencies. It is a hollow metal pipe, usually having a rectangular or circular cross section. To efficiently propagate an EM field, a *rectangular waveguide* must have sides measuring at least 0.5 wavelength and preferably more than 0.7 wavelength. A *circular waveguide* should be at least 0.6 wavelength in diameter and preferably 0.7 wavelength or more.

The Z_0 of a waveguide varies with the frequency. In this sense, it differs from coaxial or parallel-wire lines, whose Z_0 values are independent of the frequency.

Standing waves

Standing waves are voltage and current variations that exist on an RF transmission line when the load (antenna) impedance differs from the characteristic impedance of the line.

In a transmission line terminated in a pure resistance having a value equal to the characteristic impedance of the line, no standing waves occur. When standing waves are present, a nonuniform distribution of current and voltage exists. The greater the impedance mismatch, the greater the nonuniformity. The ratio of the maximum voltage to the minimum voltage, or the maximum current to the minimum current, is the SWR.

A large SWR can significantly increase the loss in a transmission line. If a high-power transmitter is used, the currents and voltages on a severely mismatched line can become large enough to cause physical damage.

Safety

Antennas should never be placed in such a way that they can fall or blow down on power lines. Also, it should not be possible for power lines to fall or blow down on an antenna.

Wireless equipment that has an outdoor antenna should not be used during thundershowers or when lightning is anywhere in the vicinity. Antenna construction and maintenance should never be undertaken when lightning is visible, even if a storm appears to be many miles away. Ideally, antennas should be disconnected from electronic equipment and connected to a substantial earth ground at all times when the equipment is not in use.

Tower and antenna climbing is a job for professionals. Under no circumstances should an inexperienced person attempt to climb such a structure.

Indoor transmitting antennas expose operating personnel to EM field energy. The extent of the hazard, if any, posed by such exposure has not been firmly established. However, there is sufficient concern to warrant checking the latest publications on the topic.

For detailed information concerning antenna safety, consult a professional antenna engineer and/or a comprehensive text on antenna design and construction.

19

The EM Spectrum

An *EM field* is produced when charge carriers, such as electrons in a conductor, accelerate.

Basic Properties

All EM fields contain electric and magnetic lines of flux. The electric lines of flux are perpendicular to the magnetic lines of flux at every point in space. The direction in which the EM field propagates (travels) is perpendicular to both sets of flux lines.

Frequency

Frequency is the rate at which the cycle of a periodic disturbance repeats. The fundamental unit of frequency is the *hertz,* formerly called *cycles per second* (cps). Frequencies are often specified in larger units: *kilohertz, megahertz, gigahertz,* and *terahertz.* Thus,

$$1 \text{ kHz} = 1000 \text{ Hz}$$

$$1 \text{ MHz} = 1000 \text{ kHz} = 10^6 \text{ Hz}$$

$$1 \text{ GHz} = 1000 \text{ MHz} = 10^9 \text{ Hz}$$

$$1 \text{ THz} = 1000 \text{ GHz} = 10^{12} \text{ Hz}$$

If the *period,* or the time required for one complete cycle, is equal to T seconds, the frequency f, in hertz, is equal to $1/T$.

Wave disturbances in the *EM spectrum* can have frequencies ranging from less than 1 Hz to trillions of terahertz. The *radio spectrum* extends from a few kilohertz up to hundreds of gigahertz.

Free-space wavelength

Any EM field has a *wavelength* that depends on the frequency of the disturbance and also on the speed with which it propagates. Wavelength is inversely proportional to frequency and is directly proportional to the propagation speed.

In a periodic disturbance, the wavelength is defined as the distance between identical points in adjacent wave cycles. These points can be wave *crests* (maxima), *troughs* (minima), or any other specifically defined points. In radio communications, EM wavelengths are measured in meters, centimeters, or millimeters. At higher frequencies, nanometers are commonly specified. These units are related according to the following equations:

$$1 \text{ cm} = 0.01 \text{ m} = 10^{-2} \text{ m}$$

$$1 \text{ mm} = 0.1 \text{ cm} = 10^{-3} \text{ m}$$

$$1 \text{ nm} = 10^{-6} \text{ mm} = 10^{-7} \text{ cm} = 10^{-9} \text{ m}$$

The general formula for the relationship between wavelength s, frequency f, and speed c for a wave disturbance is

$$s = \frac{c}{f}$$

For an EM wave in free space, c is approximately 300,000,000 (3.00×10^8) meters per second, the same as the speed of light. Therefore,

$$s = \frac{3.00 \times 10^8}{f}$$

Electrical wavelength

In any medium carrying RF signals, the *electrical wavelength* is the distance between a specific point in one cycle or wave and the corresponding point in the adjacent cycle or wave. This is usually expressed as the separation between points where the instantaneous amplitude of the electric field is zero and increasing positively (Fig. 19.1). The electrical wavelength depends on the *velocity factor v* of the medium through, or along, which the field propagates. The electrical wavelength also depends on the frequency f.

Along a single wire conductor, the electrical wavelength at a given frequency is somewhat shorter than the wavelength in free space. The velocity factor in a single-wire conductor is approximately 0.95, or 95 percent. Therefore the

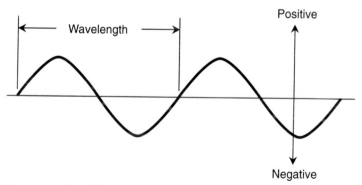

Figure 19.1 A wavelength is the distance between identical points on adjacent cycles.

above formula for free-space wavelength must be modified to approximately

$$s = \frac{3.00 \times 10^8 \times 0.95}{f} = \frac{2.85 \times 10^8}{f}$$

In general, in a transmission line with a velocity factor v (given as a fraction rather than as a percentage),

$$s = 3.00 \times 10^8 \times \frac{v}{f}$$

Often, this formula is modified for values of f expressed in megahertz rather than in hertz. The equation in this case becomes

$$s = 300 \times \frac{v}{f}$$

The electrical wavelength of a signal in an RF transmission line is always less than the wavelength in free space. In practice, the value of v ranges from approximately 0.66 (66 percent) in a coaxial cable with a solid polyethylene dielectric to 0.95 (95 percent) for parallel-wire line with an air dielectric.

Depiction of the spectrum

The EM spectrum can be graphically displayed on a logarithmic scale. A simplified rendition is shown in Fig. 19.2A, labeled for wavelength in meters. To find the frequencies in megahertz, divide 300 by the wavelength shown. For frequencies in hertz, use 300,000,000 (3.00×10^8) instead of 300. For kilohertz, use 300,000 (3.00×10^5); for gigahertz, use 0.300.

Radio waves fall into a subset of the EM spectrum at frequencies between approximately 3 kHz and 300 GHz. This corresponds to wavelengths of 100 km and 1 mm. The *radio spectrum,* which includes TV and microwaves, is "blown up" in Fig. 19.2B and is labeled for frequency. To find wavelengths in meters, divide 300 by the frequency in megahertz. For frequencies in hertz, use 3.00×10^8; for kilohertz,

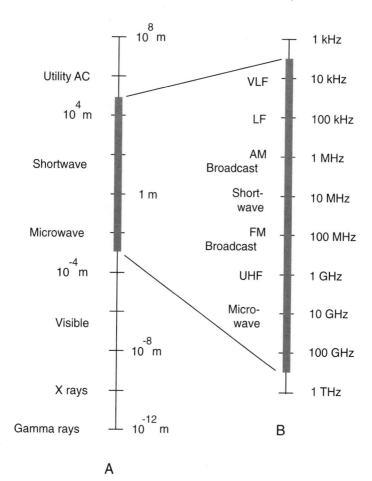

Figure 19.2 (A) The electromagnetic spectrum from 10^8 m (300,000 km) to 10^{212} m (0.001 nm). (B) The radio spectrum from 1,000 Hz (1 kHz) to 10^{12} Hz (1 THz).

use 3.00×10^5; for gigahertz, use 0.300. The radio spectrum is divided into eight ranges or bands (Table 19.1).

Wave Polarization

The *polarization* of an EM field is defined as the orientation of the electric flux lines. This orientation might

TABLE 19.1   Bands in the Radio Spectrum

Frequency designation	Frequency range	Wavelength range
Very low (VLF)	3–30 kHz	100–10 km
Low (LF)	30–300 kHz	10–1 km
Medium (MF)	300 kHz–3 MHz	1 km–100 m
High (HF)	3–30 MHz	100–10 m
Very high (VHF)	30–300 MHz	10–1 m
Ultra high (UHF)	300 MHz–3 GHz	1 m–100 mm
Super high (SHF)	3–30 GHz	100–10 mm
Extremely high (EHF)	30–300 GHz	10–1 mm

remain constant as a function of time, or it might change with time.

Horizontal polarization

In radio communications, *horizontal polarization* has certain advantages and disadvantages at various wavelengths.

At the low and very-low frequencies (below about 300 kHz), horizontal polarization is not often used. This is because the *surface wave,* an important factor in propagation at these frequencies, is more effectively transferred when the electric field is oriented vertically.

In the high-frequency part of the radio spectrum (3 to 30 MHz), horizontal polarization becomes practical. The surface wave is of lesser importance; the *sky wave* is the primary mode of propagation.

In the very-high, ultra-high, and microwave frequency ranges (above 30 MHz), either vertical or horizontal polarization can be used. Horizontal polarization provides better noise immunity and less fading than vertical polarization.

Vertical polarization

Vertical polarization is a condition in which the electric lines of flux of an EM wave are vertical, or perpendicular, to the surface of the earth.

At low and very-low radio frequencies, vertical polarization is ideal, because surface-wave propagation, the major mode of propagation at these wavelengths, requires a vertically polarized field. At very high and ultra-high frequencies, vertical polarization is used mainly for mobile communications and also in repeater installations.

The main disadvantage of vertical polarization is the fact that most human-made *electrical noise* is vertically polarized. Thus, a vertical antenna usually picks up more human-made interference than a horizontal antenna. At very-high and ultra-high frequencies, vertical polarization results in more "flutter" in mobile communications, compared with horizontal polarization.

Elliptical polarization

If the orientation of the electric lines of flux rotates as the signal is propagated from the transmitting antenna, the signal is said to have *elliptical polarization.* An elliptically polarized EM field can rotate either clockwise or counterclockwise as it approaches an observer. This variable is known as the polarization *sense.*

Elliptical polarization allows reception of signals having variable polarization with a minimum of fading. Ideally, the transmitting and receiving antennas should both have elliptical polarization, although signals with *linear polarization* (usually either vertical or horizontal) can be received with an elliptically polarized antenna. If the transmitted signal has elliptical polarization in a sense contrary to that of a receiving antenna, there is substantial loss.

Circular polarization

Uniformly rotating elliptical polarization is called *circular polarization.* The orientation of the electric flux completes one rotation for every cycle of the wave, with constant angular speed corresponding to the frequency of the signal. The flux rotation is accomplished by electrical means.

Circular polarization is compatible with linear polarization, but there is a 3-dB (2:1) power loss. If a circularly polarized signal arrives with opposite sense from that of the receiving antenna, the loss is approximately 30 dB (1000:1) compared with matched rotational sense.

Surface and Ionospheric Propagation

At long wavelengths, EM fields can travel for great distances in direct contact with the earth's surface. The ionosphere returns EM waves to the earth at some wavelengths.

The ground wave

In radio communication, the *ground wave* consists of three distinct components: the *direct wave* (also called the *line-of-sight wave*), the *reflected wave,* and the *surface wave.*

The direct wave travels in a straight line. At most radio frequencies, EM fields pass through objects such as trees and frame houses with little attenuation. Concrete-and-steel structures cause some loss in the direct wave at higher frequencies. Earth barriers such as hills and mountains block the direct wave.

A radio signal can be reflected from the earth or from structures containing metal. The reflected wave combines with the direct wave (if any) at the destination. Sometimes the two waves arrive out of phase, in which case the received signal is extremely weak.

The surface wave travels in contact with the earth, and the earth forms part of the circuit. This occurs only with vertically polarized fields at frequencies up to a few megahertz. Below approximately 300 kHz, the surface wave can travel thousands of miles.

The ionosphere

Ionization in the upper atmosphere takes place at three or four distinct layers. The lowest region is called the *D layer.* It exists at an altitude of about 30 miles (50 km) and is ordinarily present only on the daylight side of the planet. This layer absorbs radio waves at some frequencies, impeding long-distance ionospheric propagation. The *E layer,* about 50 miles (80 km) above the surface, also exists mainly during the day, although nighttime ionization is sometimes observed. The E layer can provide medium-range radio communication at certain frequencies. The uppermost layers are called the *F1* and the *F2 layers.* The F1 layer, normally present only on the daylight side of the earth, forms at about 125 miles (200 km) altitude; the F2 layer exists at about 180 miles (300 km) over most, or all, of the earth. Sometimes the distinction between the F1 and F2 layers is ignored, and they are spoken of together as the *F layer.* Communication via F-layer propagation can usually be accomplished between any two points on the earth at some frequencies between 5 and 30 MHz.

Figure 19.3 illustrates these ionospheric layers and their relative distances above the surface. The altitudes vary somewhat, and the layers are not sharply defined. Sometimes ionization occurs in patches or "clouds," particularly in the E layer.

Solar activity

The average number of sunspots varies dramatically from year to year (Fig. 19.4). This fluctuation is called the *sunspot cycle.* It has a period of 10 to 11 years. The sunspot cycle affects ionospheric radio-wave propagation conditions at frequencies up to about 200 MHz.

The sunspot cycle affects F-layer propagation to a significant extent and with predictable regularity. When there are few sunspots, the *maximum usable frequency* (MUF) for the F layer is relatively low, sometimes dropping below 5 MHz. At or near the time of a sunspot peak, the MUF is much higher on the average and can occasionally reach 70 MHz.

A solar flare is a sudden "storm" on the surface of the sun. Such an event emits high-speed subatomic particles that arrive at the earth within a few hours. Because the particles are electrically charged, they are accelerated by the geomagnetic field. This causes a sudden deterioration of ionospheric radio-wave propagation conditions at some

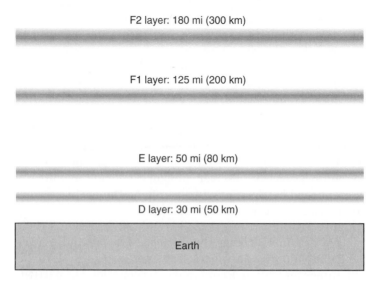

Figure 19.3   Layers of the ionosphere and their approximate altitudes.

frequencies. Solar flares can occur at any time, but they take place most often near the peak of the sunspot cycle.

Sporadic-E propagation

At certain radio frequencies, the E layer occasionally returns signals to the earth. This effect is intermittent, and conditions can change rapidly. For this reason, it is known as *sporadic-E propagation*. It is most likely to occur at frequencies between 20 and 150 MHz. Occasionally it is observed at frequencies as high as 200 MHz. The obtainable communications range is normally several hundred miles, but communication is sometimes possible over distances of up to 1500 miles.

The standard FM broadcast band can be affected by sporadic-E propagation. The same is true of the lowest TV broadcast channels, especially channels 2 and 3. Sporadic-E propagation often occurs on the amateur-radio bands at 21 through 148 MHz and is sometimes mistaken for propagation effects that take place in the lower atmosphere, independently of the ionosphere.

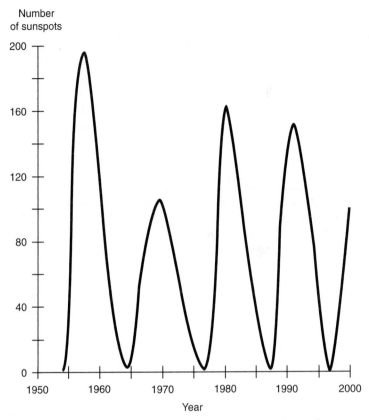

Number
of sunspots

Figure 19.4 Recent sunspot cycles, showing approximate observed sunspot numbers as a function of time.

Auroral propagation

In the presence of unusual solar activity, the aurora can return radio waves to the earth. This is *auroral propagation*. The aurora (also called "northern lights" and "southern lights") occur at altitudes of 40 to 250 miles. Theoretically, auroral propagation is possible, when the aurora are active, between any two points on the surface from which the same part of the aurora lie on a line of sight. Auroral propagation seldom occurs when one end of the circuit is at a latitude less than 35° north or south of the equator.

Auroral propagation is characterized by rapid and deep fading. This almost always renders analog voice and video signals unintelligible. Digital modes are most effective for communication via auroral propagation, but the carrier is often spread out over several hundred hertz as a result of phase modulation induced by auroral motion. This severely limits the maximum data transfer rate. Auroral propagation is often accompanied by deterioration in ionospheric propagation via the E and F layers.

Meteor-scatter propagation

Meteors produce ionized trails that persist for approximately 0.5 s up to several seconds, depending on the size of a particular meteor, its speed, and the angle at which it enters the atmosphere. This is not sufficient time for the transmission of much data; but during a meteor shower, multiple trails can result in almost continuous ionization for a period of hours. Such ionized regions reflect radio waves at certain frequencies. This is *meteor-scatter propagation*. It can take place at frequencies considerably above 30 MHz.

Meteor-scatter propagation is mainly of interest to experimenters and radio amateurs. This mode is sometimes used in amateur packet radio because the messages (packets) are of short duration. Meteor-scatter propagation occurs over distances ranging from just beyond the horizon up to about 1500 miles, depending on the altitude of the ionized trail and the relative positions of the trail, the transmitting station, and the receiving station.

Ionospheric propagation forecasting

Ionospheric propagation conditions vary from day to day, month to month, and year to year. To some extent, these changes are predictable. The sun is constantly monitored for fluctuations in the intensity of radiation at various EM frequencies. The *solar flux,* as it is called, provides a good indicator of propagation conditions for the ensuing several hours.

A sudden increase in the solar flux means that a solar flare is occurring. Within a short time, propagation conditions usually become disturbed following a flare. Flares sometimes wipe out F-layer communications completely, but the E layer might become sufficiently ionized to allow propagation over moderate distances. At higher latitudes, auroral propagation can usually be observed after a solar flare.

Daily and seasonal variations are typical for F-layer propagation. At frequencies below about 10 MHz, ionospheric propagation is generally better at night than during the day. Above about 10 MHz, this situation is reversed. The winter months tend to be better below 10 MHz, whereas the summer months bring better conditions above 10 MHz.

Propagation-forecast bulletins are regularly transmitted by the National Bureau of Standards' time-and-frequency stations, WWV and WWVH. Long-term propagation forecasts are also issued each month in amateur-radio magazines and other publications for the electronics hobbyist and professional.

Tropospheric Propagation

The lowest part of the earth's atmosphere, in which weather disturbances occur, is the *troposphere*. It extends from the surface up to about 10 miles. This region has an effect on radio-wave propagation at some frequencies. At wavelengths shorter than about 15 m (frequencies above 20 MHz), refraction and reflection can take place within and between air masses of different density. The air also produces some scattering of EM fields at wavelengths shorter than about 3 m (frequencies above 100 MHz). All of these effects are collectively known as *tropospheric propagation.*

Tropospheric bending

A common type of tropospheric propagation occurs when radio waves are refracted in the lower atmosphere. This takes place to a certain extent all the time but is most dramatic near weather fronts, where warm, relatively light air lies above cool, dense air. The cool air has a higher index of refraction than the warm air, causing radio waves to be bent downward at a considerable distance from the transmitter. This is called *tropospheric bending* (Fig. 19.5).

Bending is often responsible for anomalies in reception of FM and TV broadcast signals. On TV, a station hundreds of miles away might suddenly appear on a channel that is normally vacant. Unfamiliar stations might be received on a hi-fi FM radio tuner. Sometimes two or more distant stations come in on a single channel, interfering with each other. In the case of TV, the effects are sometimes mistaken for EMI.

Duct effect

The *duct effect,* also called *tropospheric ducting,* takes place at approximately the same frequencies as bending. A duct forms when a layer of cool air becomes sandwiched between two layers of warmer air. This is common along and near weather fronts in the temperate latitudes. It also takes place frequently above water surfaces during the daylight hours and over land surfaces at night. Cool air, being more

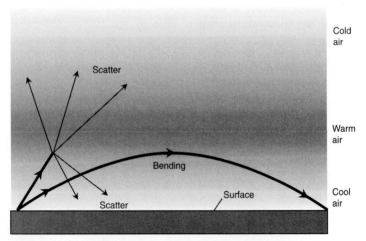

Figure 19.5 Tropospheric bending and scatter propagation.

dense than warmer air at the same humidity level, exhibits a higher index of refraction for radio waves of certain frequencies. Total internal reflection takes place inside the region of cooler air, in much the same way as light waves are trapped inside an optical fiber.

For ducting to provide communications, the transmitting and receiving antennas must both be located within the same duct, and this duct must be unbroken and unobstructed between the two locations. A duct might measure only a few feet from top to bottom but nevertheless cover many thousands of square miles parallel to the surface. Ducting often allows over-the-horizon communication of exceptional quality, over distances of hundreds of miles, at VHF and UHF.

Troposcatter

At frequencies above about 100 MHz, the atmosphere has a scattering effect on EM fields. Scattering allows over-the-horizon communication at VHF, UHF, and microwave frequencies. This mode of propagation is called *tropospheric scatter,* or *troposcatter* (Fig. 19.5). Dust and clouds in the air increase the scattering effect, but some troposcatter occurs regardless of the weather. Troposcatter takes place mostly at low altitudes where the air is the most dense.

Communication via troposcatter requires the use of high-gain antennas. The transmitting and receiving antennas are aimed at the same parcel of air, which is ideally located midway between the two stations. The maximum obtainable range depends on the gain of the antennas used for transmitting and receiving and on their elevation above the surface. The higher the antennas, the greater the range. The terrain also affects the range. Flat terrain is best. Hills, mountains, and large buildings impede troposcatter propagation by blocking low-angle radiation and reception at the transmitter and receiver locations. High-power transmitters and sensitive receivers are advantageous.

Propagation Characteristics by Frequency

The following several sections discuss EM wave propagation
in the earth's atmosphere from the lowest radio frequencies
through microwaves and beyond.

Very-low frequencies (below 30 kHz)

At *very-low frequencies* (VLF), propagation takes place mainly via the waveguide effect between the earth and the ionosphere. Surface-wave propagation also occurs for considerable
distances. With high-power transmitters and large antennas, communications can be realized over distances of several thousand miles. The earth-ionosphere waveguide has a
low-end cutoff frequency of approximately 9 kHz. For this
reason, signals much below 9 kHz suffer severe attenuation
and do not propagate well.

Antennas for VLF transmitting must be vertically polarized. Otherwise, surface-wave propagation will not take
place, because the earth short-circuits horizontally polarized EM fields at long radio wavelengths.

Propagation at VLF is remarkably stable; there is very little fading. Solar flares occasionally disrupt communication
in this frequency range by raising the cutoff frequency of the
earth-ionosphere waveguide.

Low frequencies (30 to 300 kHz)

At *low frequencies* (LF), propagation takes place in the surface-wave mode and also as a result of ionospheric effects.
Toward the low end of the LF band, wave propagation is
similar to that in the VLF range. As the frequency increases, surface-wave propagation becomes less efficient. A surface-wave range of more than 3000 miles is common at 30
kHz, but it is unusual for the range to be greater than a few
hundred miles at 300 kHz.

Ionospheric propagation at LF usually occurs via the E
layer. This increases the useful range during the nighttime
hours, especially toward the upper end of the band.
Intercontinental communication is possible with high-pow-

er transmitters. Solar flares disrupt conditions at LF. Following a flare, the D layer becomes absorptive, preventing ionospheric propagation.

Medium frequencies (300 kHz to 3 MHz)

Propagation at *medium frequencies* (MFs) occurs by means of the surface wave and by E- and F-layer ionospheric modes. Near the low end of the MF band, surface-wave communications is common up to distances of several hundred miles. As the frequency is raised, the surface-wave attenuation increases. At 3 MHz, the range of the surface wave is limited to about 150 miles.

Ionospheric propagation at MF is almost never observed during the daylight hours, because the D layer prevents radio waves from reaching the higher E and F layers. During the night, ionospheric propagation takes place mostly via the E layer in the lower portion of the band and primarily via the F layer in the upper part of the band. The range increases as the frequency is raised. At 3 MHz, worldwide communication is sometimes possible over nighttime paths.

Medium-frequency communication is severely affected by solar flares. The sunspot cycle and the season of the year also affect propagation at MF. Propagation is usually better in the winter than in the summer. This is partly because darkness prevails over a greater proportion of the hemisphere in the winter than in the summer and partly because there is less interference from *sferics* (radio noise produced by lightning) in the winter than in the summer.

High frequencies (3 to 30 MHz)

Propagation at HF exhibits widely variable characteristics. Effects are much different in the lower-frequency part of this band than in the upper-frequency part. Consider the lower portion as the range from 3 to 10 MHz and the upper portion as the range from 10 to 30 MHz.

Some surface-wave propagation occurs in the lower part of the HF band. At 3 MHz, the maximum range is approximately 150 miles; at 15 MHz it decreases to 15 or 20 miles.

Above 15 MHz, surface-wave propagation is essentially nonexistent.

Ionospheric communication at HF occurs mainly via the F layer. In the lower part of the band, there is very little daytime ionospheric propagation because of D-layer absorption. At night, worldwide communication can be realized, because the D layer thins out sufficiently to let signals reach the F layer. In the upper part of the band, signals penetrate the D layer, allowing worldwide communication during the day; conditions often deteriorate at night because F-layer ionization is not dense enough to return the waves to the surface.

Communications in the lower part of the HF band are generally better during the winter months than during the summer. In the upper part of the band, this situation is reversed. Sporadic-E propagation is occasionally observed in the upper part of the HF band. This can occur even when the F-layer MUF is below the communications frequency. At the extreme upper-frequency end of the HF band near 30 MHz, there is often no ionospheric propagation. This is especially true when the sunspot cycle is at or near a minimum.

Solar flares cause dramatic changes in conditions in the HF band. Ionospheric communications can be wiped out within minutes, or even seconds, following the arrival of the subatomic particles ejected by a large solar flare.

Very-high frequencies (30 to 300 MHz)

Propagation at VHF takes place via line-of-sight and tropospheric modes. Ionospheric F-layer propagation is rarely observed, although it can occur at times of sunspot maxima at frequencies up to about 70 MHz. Sporadic-E propagation can occur at frequencies up to approximately 200 MHz.

Meteor-scatter and auroral propagation are observed in the VHF band. The range for meteor scatter is a few hundred miles; auroral propagation can provide communications over distances of up to about 1500 miles. Digital modes must usually be employed for auroral communications.

Even then, the maximum data speed is limited because the aurora introduce phase modulation.

Repeaters are used extensively at VHF to extend the range of mobile communications equipment. A repeater at a high-elevation site such as a hilltop can provide coverage over thousands of square miles (Fig. 19.6). In the illustration, station X transmits signals to the repeater, which in turn retransmits them to stations Y and Z.

At VHF, active communications satellites provide world-wide coverage on a reliable basis. Active satellites are, in effect, orbiting repeaters. The moon is sometimes used as a passive communications satellite, mostly by amateur radio operators. This mode is called *earth-moon-earth* (EME) or *moonbounce*.

Ultra-high frequencies (300 MHz to 3 GHz)

Propagation at UHF occurs mainly via the line-of-sight mode and via satellites and repeaters. "Shortwave-type" ionospheric propagation does not take place. Auroral, mete-or-scatter, and tropospheric propagation are occasionally observed in the lower-frequency portion of this band. Ducting can result in propagation over distances of several hundred miles.

A major asset of the UHF band is its sheer size: 2.7 GHz of spectrum space. Because UHF energy has nothing to do

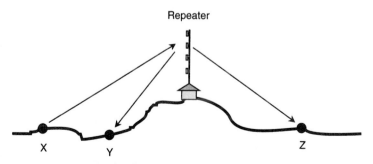

Figure 19.6 A repeater facilitates communication over greater distances than the line of sight allows.

with the ionospheric D, E, and F layers, disruptions in the ionosphere rarely have any effect. An extremely intense solar flare can sometimes interfere with UHF circuits, as a result of EM noise caused by rapid fluctuations in the geomagnetic field.

The UHF band was once regarded as an infinite spectral resource, but in recent decades this has changed. Digital communications is occurring among an increasing number of people, at ever-higher data speeds. The numbers of signals, and their bandwidths, are skyrocketing. As a result, competition for UHF spectrum space has become intense among various services, especially in metropolitan areas.

Microwaves (above 3 GHz)

Microwaves travel essentially in straight lines through the atmosphere, are rarely affected by temperature inversions or tropospheric scattering, and are not refracted or reflected by the ionosphere. The primary mode of propagation in the microwave range is line of sight. This facilitates repeater and satellite communications.

At some microwave frequencies, atmospheric attenuation becomes a consideration. Rain, fog, and other weather effects can increase path loss as the wavelength becomes comparable with the diameter of water droplets and dust particles. Transmitter and receiver designs become critical as the frequency increases, especially above several hundred gigahertz. Although a megawatt-power transmitter can be built for use at VLF, LF, MF, or HF, such a thing is unheard of in the microwave range. However, a device called the *maser,* which is in fact a microwave laser, can concentrate energy into narrow, coherent beams at microwave frequencies. This offers some possibilities for long-distance links, especially in outer space.

Beyond microwaves

Electromagnetic waves whose frequencies are higher than those of the radio microwaves always travel in straight lines through the atmosphere. The only factor that varies is the path loss.

Water vapor causes severe attenuation in the *IR spectrum* between the wavelengths of approximately 4500 and 8000 nm. Carbon dioxide (CO_2) gas interferes with the transmission of IR at wavelengths ranging from about 14,000 to 16,000 nm. Rain, snow, fog, and dust also interfere with the propagation of IR.

The *visible spectrum* extends from about 750 nm (red light) to 390 nm (violet light). Visible light is transmitted fairly well through the atmosphere at all wavelengths. Scattering increases toward the short-wavelength end of the spectrum (blue and violet). For free-space laser communications, red is the preferred color. Helium-neon (He-Ne) lasers produce red light and are available at reasonable prices for experimenters. Rain, snow, fog, and dust interfere with the transmission of visible light through the air.

In the *UV spectrum,* energy at the longer wavelengths penetrates the air with comparative ease, although some scattering takes place. Ozone (O_3) pollution increases the loss. At shorter UV wavelengths, attenuation increases. Rain, snow, fog, and dust interfere with UV propagation in the same way that they interfere with the propagation of visible energy.

Energy in the *x-ray spectrum* and the *gamma-ray spectrum* (wavelengths shorter than those of UV energy) do not propagate well for long distances through the air. This is because the sheer mass of the air, over paths of any appreciable distance, is sufficient to block radiation at these wavelengths. In addition to this problem, effective transmitters are almost impossible to construct. It is doubtful that x and gamma rays will ever be routinely employed for communications purposes, although they might be used in short-range wireless monitoring or control systems.

20

Noise and Interference

Noise is unwanted EM energy that comes from noncommunications sources. *Interference* is unwanted EM energy that comes from communications equipment, sometimes of the same type and used for the same purpose as the device(s) being interfered with.

External Noise

In RF systems, noise that comes from outside the receiver is *external noise*.

Cosmic noise

Noise from outer space is called *cosmic noise*. It primarily affects satellite and space communications systems. At some frequencies, the earth's atmosphere prevents cosmic noise from reaching the surface; at other frequencies, cosmic noise arrives at the surface with little or no attenuation. Cosmic noise correlates with the plane of the galaxy. The strongest *galactic noise* comes from the direction of the constellation Sagittarius, because this part of the sky lies on a line between our solar system and the center of the galaxy. Galactic noise contributes, along with noise from the sun, the planet Jupiter, and a few other celestial objects, to most of the cosmic radio noise arriving at the earth.

Solar flux

The amount of radio noise emitted by the sun is called the *solar radio-noise flux,* or simply the *solar flux.* The level of solar flux increases abruptly when a solar flare occurs. The solar flux is monitored at a wavelength of 10.7 cm, or a frequency of 2.8 GHz. At this frequency, the troposphere and ionosphere have no effect on radio waves, so the energy reaches the surface at full strength. The solar flux is correlated with the *sunspot cycle.* On the average, the solar flux is higher near the peak of the sunspot cycle and lower near sunspot minima.

Sferics

Electromagnetic noise is generated in the atmosphere of our planet, mostly by lightning discharges in thundershowers. This noise is called *sferics.* In a radio receiver, sferics produce a faint background hiss or roar, punctuated by bursts of "static." An example of sferics, as they might appear on the display of an oscilloscope connected to the IF section of a radio receiver, is shown in Fig. 20.1.

An individual lightning stroke produces a burst of RF energy from VLF through the microwave region. Energy is also produced in the IR, visible, and UV portions of the EM spectrum. Sferics propagate for hundreds of miles at VLF; this distance decreases to a few miles in the microwave region.

Precipitation static

Precipitation static is caused by electrically charged water droplets or ice crystals as they strike objects. This is especially likely to happen in snow and dust storms. Precipitation

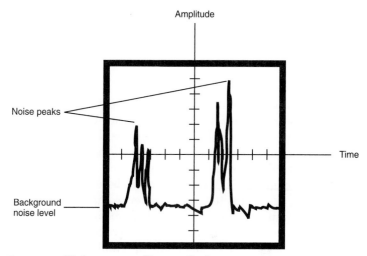

Figure 20.1   Sferics on an oscilloscope display.

static can be severe, especially at the very low, low, and medium frequencies (wavelengths greater than approximately 100 m). A *noise blanker* or *noise limiter* can reduce the effects of precipitation static. A means of facilitating discharge, such as an inductor between the antenna and dc ground, can be helpful. Improvement can also be obtained by blunting any sharp points in the antenna elements.

Corona

When the voltage on an electrical conductor exceeds a certain value, the air around the conductor ionizes. The result is *corona,* also called *Saint Elmo's fire.* This ionization causes broadband RF noise that can render a wireless communications receiving system ineffective. Corona is sometimes observed between the plates of air-variable capacitors handling large voltages. A pointed object, such as the end of a whip antenna, is more likely to produce corona than a flat or blunt surface. Corona can be caused by electrostatic-charge buildup during thunderstorms. In darkness, corona can be seen as a blue-white or blue-violet glow. Some inductively loaded transmitting antennas have metal disks or spheres at the ends to minimize corona.

Impulse noise

Any sudden, high-amplitude voltage pulse will generate an RF field. The result is *impulse noise*. It can be produced by household appliances such as vacuum cleaners, hair dryers, electric blankets, thermostat mechanisms, and fluorescent-light starters. Impulse noise is most severe at frequencies below approximately 1 MHz, but it can be observed at frequencies up to several tens of megahertz. Figure 20.2 is an example of impulse noise as it might appear on an oscilloscope connected in the IF chain of a radio receiver.

Impulse noise can be picked up by hi-fi audio systems. The greater the number of external peripherals (such as tape players, microphones, and speakers) there are in a system,

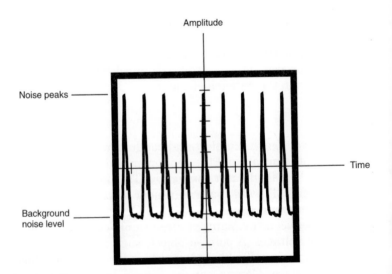

Figure 20.2   Impulse noise on an oscilloscope display.

the greater is the susceptibility to this type of noise. Wireless microphones and headsets are especially vulnerable.

Impulse noise can be reduced by the use of a good dc ground system. A noise blanker or noise limiter can be helpful. A radio receiver should be set for the narrowest response bandwidth consistent with the mode of reception. In hi-fi systems, interconnecting cables should be shielded.

Noise in mobile systems

Ignition noise is a wideband form of impulse noise, generated by electric arcing in the spark plugs of internal combustion engines. Ignition noise is radiated from most automobiles, trucks, lawn mowers, and gasoline-engine-driven generators. Ignition noise can be reduced in a receiver by means of a noise blanker. Sometimes, special spark plugs, called *resistance plugs,* can be installed to reduce the ignition noise. An excellent vehicle-chassis ground connection is imperative in any mobile wireless installation.

Ignition noise is not the only source of trouble in mobile wireless communications. Noise can be generated by the friction of the vehicle tires against pavement at freeway speeds. High-tension power lines often radiate significant wideband RF noise. A vehicle alternator can cause noise in the form of a whine that changes pitch as the car accelerates and decelerates.

Power-line noise

Utility lines, in addition to carrying the 60-Hz ac that they are intended to transmit, carry higher-frequency components that produce *power-line noise.* The currents usually occur because of electric arcing at some point in the circuit. The arcing can originate in appliances connected to terminating points, it can take place in faulty or obstructed transformers, and it can occur in high-tension lines as a corona discharge into humid air.

Power-line noise sounds like a buzz or hiss when picked up by a radio receiver. Some types of power-line noise can be attenuated by means of a noise blanker. In other cases, a limiter must be used to obtain improvement in reception. Phase cancellation, in which the noise picked up by an auxiliary antenna is used to null out the noise from the main receiving antenna, can sometimes reduce received power-line noise.

Internal Noise

Some noise is produced by the active components within electronic equipment. This is *internal noise.* Low-internal-noise receiver design is of major concern at VHF, UHF, and microwave frequencies.

Thermal noise

The level of *thermal noise* in any material is proportional to the absolute temperature. The noise decreases to near zero as the absolute temperature approaches zero degrees Kelvin (0 K).

Thermal noise imposes a limit on the sensitivity that can be obtained with electronic receiving equipment. This noise can be minimized by placing a *preamplifier* circuit in a bath of liquid gas, such as helium or nitrogen. Helium has a boiling point of only a few degrees Kelvin. Such cold temperatures cause the atoms and electrons to move at relatively slow speed, reducing the thermal noise compared with the level at room temperature.

Shot noise

In any current-carrying medium, charge carriers cause noise impulses as they move from atom to atom. This is called *shot effect,* and the resulting noise is known as *shot noise.* This noise is produced in the front end of a radio receiver and is amplified by succeeding stages along with the desired signals.

The amount of shot noise that a device produces is roughly proportional to the current that it carries. Low-current solid-state devices, such as the GaAsFET have been developed to optimize the sensitivity of a receiver front end by minimizing shot noise.

Hiss

Audio-frequency noise, in which the amplitude is peaked near the midrange or treble regions, is called *hiss*. It is generated as the result of random electron or hole movement in components. The hiss produced in early stages is amplified by subsequent stages. In a radio receiver, hiss can be generated in the IF stages and in the front end, mixers, and oscillators. Some hiss even originates outside the receiver, in the form of thermal noise in the antenna conductors and atmosphere.

Conducted Noise

When noise in a communications system comes through the power supply, it is said to be *conducted noise*. In a fixed wireless station, this noise can be suppressed by inserting RF chokes in series with all power leads from the station to utility outlet(s). In a mobile station, conducted noise comes from the alternator and spark plugs, through the dc power leads, and into the radio. Conducted noise in a mobile station can be minimized by connecting the power leads to the battery—never through the cigarette lighter. Filtering the alternator leads using capacitors of about 0.1 µF can minimize alternator whine. Resistance wiring in the ignition system sometimes reduces impulse noise generated by spark plugs.

Noise Reduction

In communications systems, there are various ways to reduce the level of noise compared with the level of the desired signals.

Random noise

Some noise exhibits no orderly relationship between amplitude and time; this is called *random noise*. Examples are sferics, thermal noise, and shot noise. Random noise is more difficult to suppress than noise that has an identifiable waveform or pattern.

A limiting circuit, which allows a desired signal to compete with noise, provides some defense against strong random noise. The use of narrowband emission, in conjunction with a narrow receiver bandpass, is helpful in dealing with strong random noise. The use of FM can give better results than AM. Directional or noise-canceling antenna systems can sometimes improve communications in the presence of random noise.

Noise versus mode

Radio modulation can be broadly classified as *analog* or *digital*. Analog modes include conventional AM and FM broadcast, SSB, conventional FSTV, and most SSTV. Digital modes include ON/OFF keying, FSK, digital television, and pulse modulation. In general, digital-modulation systems offer better noise immunity than analog-modulation systems.

Noise is characterized by fluctuations in amplitude, with the frequency ill defined and spread out. Because of this, FM offers better noise immunity, all other factors being equal, than AM, provided the FM receiver is equipped with a limiter to nullify variations in amplitude. This fact, in addition to the difference in frequency, accounts for the substantial difference in received sferics levels between the standard AM broadcast band and the standard FM broadcast band.

Noise versus bandwidth

Noise is wideband in nature. Signals, in contrast, usually occupy a narrow band of frequencies. The proper choice of receiver selectivity (bandwidth) affects the extent to which the receiver will discriminate between a narrowband signal and the noise.

Figure 20.3 shows two simplified spectral illustrations of a hypothetical signal as it might be received using two different *bandpass filters*. In both situations, the total energy contained in the noise is represented by the area of the noise rectangle, and the total energy in the signal is represented by the area of the signal rectangle. The signal energy is the same in both instances, but the noise energy is lower in the case shown in Fig. 20.3B, in which the receiver passband is narrower. The ratio of signal energy to noise energy is higher in Fig. 20.3B, and the signal is therefore clearer.

Noise blankers

A *noise blanker* discriminates between short, intense bursts (typical of some types of noise) and the more uniform characteristics of a desired signal. The noise blanker is usually installed in one of the IF amplifiers of a superheterodyne receiver or just prior to the detector stage in a direct-conversion receiver.

Under low-noise conditions, a noise blanker has no effect on signals passing through the IF stage in which it is installed. But when a sudden, high-amplitude, short-duration pulse occurs, the noise blanker cuts off the stage. The receiver output thus contains a brief moment of silence (in which no signal or noise is received) instead of a loud pop or click.

Noise blankers are highly effective against impulse noise, in which the interfering pulses are regular and of short duration. Noise blankers can also work against *key clicks* caused by an improperly operating, ON/OFF-keyed wireless transmitter. Noise blankers are less effective against sferics and power-line hiss.

Noise limiters

A *noise limiter*, also called a *noise clipper*, is a circuit that prevents externally generated noise from exceeding a certain amplitude. The circuit can consist of a pair of diodes with

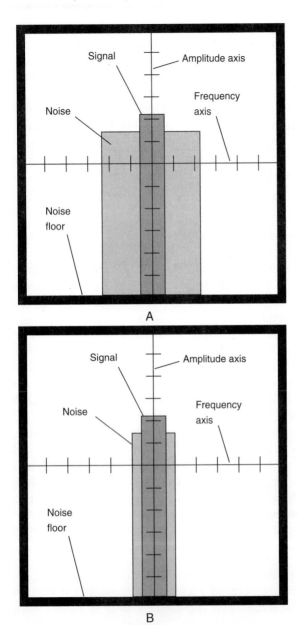

Figure 20.3 (A) The passband is much wider than the signal; (B) the passband is slightly wider than the signal.

variable bias for control of the clipping level (Fig. 20.4). The bias is adjusted until clipping occurs at the signal amplitude. Noise pulses then cannot exceed the signal amplitude. This makes it possible to receive (with some difficulty) a signal that would otherwise be obscured by the noise. A noise limiter is typically installed between two IF stages of a superheterodyne receiver. In a direct-conversion receiver, the best place for the noise limiter is immediately ahead of the detector.

A noise limiter can use a circuit that sets the clipping level automatically, according to the strength of an incoming signal. This is a useful feature, because it relieves the receiver operator of the necessity to continually readjusting the clipping level as the signal fades. Such a circuit is called an *automatic noise limiter* (ANL).

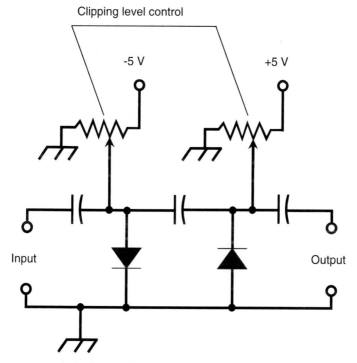

Figure 20.4. A simple noise limiter.

Limiters do not work as well as blankers against impulse noise. But a limiter can provide some relief from interference caused by sferics or transformer arcing when a noise blanker is ineffective.

Line filters

A *line filter* is a passive circuit that is placed in series with the ac power cord of an electronic device. The line filter allows 60-Hz ac to pass unaffected, but higher-frequency noise is suppressed. A typical line filter is in effect a lowpass filter, consisting of a capacitor or capacitors in parallel with the power leads and an inductor or inductors in series. The component values are chosen for a cutoff frequency somewhat above 60 Hz. Line filters are effective only against noise that enters equipment via the power line.

Special antennas

Any receiving antenna with a sharp directional null can be used to reduce the level of received human-made noise. Such an antenna can also suppress strong, unwanted local signals. A *ferrite loopstick* or *loop antenna* is especially useful for this purpose. When the null is oriented in the direction of a noise source, the noise level drops. When tuned to resonance by means of a variable capacitor, such an antenna exhibits narrow bandwidth, further suppressing broadband noise. The feed line must be properly balanced or shielded if the benefits of these antennas are to be realized.

The noise rejection of a small loop antenna can be enhanced by the addition of a *Faraday shield,* also called an *electrostatic shield,* around the loop element. The shield discriminates against the electric component of an EM field, while allowing the magnetic component to penetrate unaffected. Many noise signals are locally propagated by electrostatic coupling. When the antenna is insensitive to these fields, the signal-to-noise ratio is improved at the receiver input.

The overall noise susceptibility of an antenna system can be minimized by careful balancing or shielding of the feed line, maintaining an adequate RF ground, and minimizing the bandwidth. In general, at the medium and high frequencies, a horizontally polarized antenna is less susceptible to human-made noise than is a vertically polarized antenna.

Phase cancellation

Human-made noise can be reduced at frequencies below about 100 kHz by using two antennas that receive the noise at equal amplitudes but that receive the desired signals at different amplitudes. The two antennas are connected together, so the noise signals cancel. An example of this technique is illustrated in Fig. 20.5.

The signal antenna and the noise antenna must be reasonably near each other so that they pick up local human-made

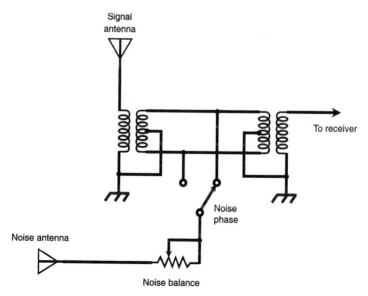

Figure 20.5 A noise-canceling antenna system.

noise whose characteristics are as nearly identical as possible. But they should not be too close together, or signals will be canceled along with the noise. The signal antenna should be mounted in a location favorable for the reception of radio signals, but the noise antenna should be placed so it will receive as much of the unwanted noise as possible and relatively little of the desired signal.

The potentiometer serves to adjust the level of signals and noise from the noise antenna, while having a minimal effect on the overall signal level from the two antennas combined. The phase switch ensures injection of noise-antenna energy out of phase with the noise from the signal antenna. After choosing the correct switch position by trial and error, the potentiometer is adjusted for minimum received noise.

Synchronization

In *synchronized digital communications,* the receiver follows the transmitter bit by bit in lockstep, so the receiver hears and evaluates each bit individually. This makes it possible to use a receiving filter that has an extremely narrow bandwidth. The synchronization requires the use of an external, common frequency or time standard. The broadcasts of station WWV or WWVH can be used for this purpose. Frequency dividers are used to obtain the necessary synchronizing frequencies. An output signal is generated in the receiver for a particular bit if (but only if) the average input signal voltage exceeds a certain value over the duration of that bit. False signals, such as might be caused by filter ringing, sferics, or other noise, are usually ignored, because they rarely result in sufficient average voltage.

Wireless Interference

Wireless interference is the presence of unwanted, human-made RF signals that impede reception in communications, monitoring, and control systems.

Electromagnetic Interference

Electronic devices sometimes interfere with one another's operation at short range. For example, a computer can interfere with amateur-radio reception. Television receivers are notorious for the spurious signals (actually harmonics of the scan signal) they generate at very-low, low, and medium frequencies. These phenomena are known collectively as EMI.

Good hardware engineering is the most effective defense against EMI. Common-sense measures include the shielding of wires that carry RF signals, RF grounding of all electronic equipment, and limitation of transmitter power output to the minimum necessary to maintain effective communications.

It should be noted that in most cases of EMI, the transmitting equipment is not at fault. It is merely performing its intended function. Usually, EMI results from poor engineering or installation of radio receivers or consumer electronic equipment. This is especially true in cases of EMI that occur in conjunction with amateur, citizens band, commercial two-way, and commercial broadcast wireless transmitters.

Adjacent-channel Interference

When a radio receiver is tuned to a particular frequency and interference is received from a signal on a nearby frequency, the effect is *adjacent-channel interference.* It can be reduced by proper engineering in transmitters and receivers. Transmitter audio amplifiers, modulators, and RF amplifiers should produce as little distortion as the state of the art permits. Receivers should employ selective filters of the proper bandwidth for the signals to be received, and the *adjacent-channel response* should be as low as possible. A relatively flat response in the passband and a steep drop-off in sensitivity outside the passband are characteristics of good receiver design. An example of a good response curve for narrowband FM reception is shown in Fig. 20.6.

Splatter

The colloquialism *splatter* is used to describe undesirable effects of wireless signals that have excessive bandwidth. Splatter can be prevented by

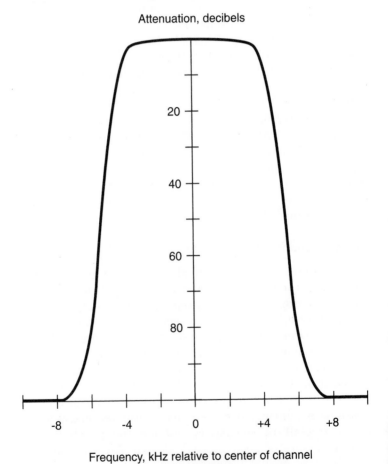

Figure 20.6   Proper attenuation-versus-frequency response in a receiver can minimize adjacent-channel interference.

- Avoiding overmodulation

- Operating audio amplifier circuits with minimum distortion

- Ensuring proper bias and drive for all transmitter amplifiers

- Ensuring that external linear amplifiers are operating properly

- Operating speech-processing circuits properly

Key clicks are a form of splatter from a transmitter that uses ON/OFF keying. The clicks are in fact sidebands, resulting from overly rapid signal rise and/or decay time. In a properly adjusted transmitter using ON/OFF keying, the rise and decay time should be such that the signal amplitude-versus-time function appears similar to Fig. 20.7A. If the decay or rise time is too short, as shown in Fig. 20.7B and C, clicks occur that can interfere with communications on nearby frequencies.

Harmonics

Any signal contains energy at integral multiples of its frequency, in addition to energy at the desired frequency. The lowest-frequency component of a signal is called the *fundamental frequency*; all integral multiples are *harmonics*. The signal having a frequency of twice the fundamental is the *second harmonic,* the signal having a frequency of three times the fundamental is the *third harmonic,* and so on.

A nearly perfect sine wave has very little harmonic energy. The sawtooth wave, square wave, and all distorted periodic oscillations contain large amounts of energy at harmonic frequencies. When a sine-wave signal passes through a nonlinear circuit, harmonic energy is produced.

Harmonic output from wireless transmitters is usually undesirable. Methods of minimizing harmonic emissions from a wireless transmitter include

- Tuning the final amplifier properly
- Avoiding overdrive of the final amplifier

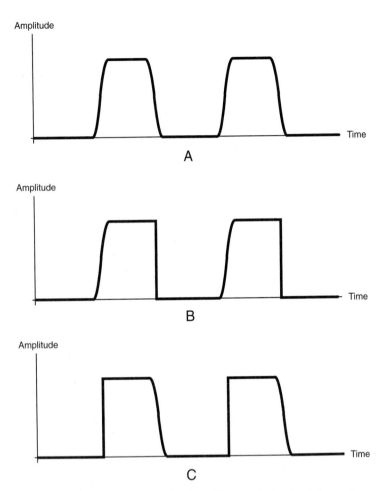

Figure 20.7 (A) Proper ON/OFF keying; (B) excessively short decay time; (C) excessively short rise time.

- Maintaining a good impedance match between the final amplifier and the antenna system
- Ensuring that all connections in the antenna system are sound
- Installing a low-pass filter in the antenna feed line

- Using a transmatch (antenna tuner) in the antenna feed line
- Using bandstop filters (traps) in the antenna feed line to suppress energy at specific harmonic frequencies

Spurious emissions

Wireless radio transmitters, when poorly designed and/or improperly operated, can radiate nonharmonic signals at frequencies other than the desired transmission frequency. Such signals can propagate for thousands of miles and cause widespread interference to other services. The most common cause of such *spurious emissions* is oscillation in the final amplifier circuit. Such oscillations are called *parasitics*. They can, in some cases, be eliminated by placing small inductors in series with the collector, drain, or plate leads of the final amplifier transistor, FET, or vacuum tube. In stubborn cases, the only effective cure is to redesign and rebuild the amplifier.

Spurious responses

Spurious responses take place when a strong signal appears at the *image frequency* of a superheterodyne receiver.

Suppose a single-conversion receiver has an IF of 9 MHz. Further suppose that it is set to receive a signal at 25 MHz. The local oscillator within the receiver is tuned to 16 MHz; this mixes with the incoming 25-MHz signal to produce an output at 9 MHz ($25 - 16 = 9$), as shown in Fig. 20.8A. Now suppose a strong signal comes in at 7 MHz. This mixes with the 16-MHz local oscillator signal to produce an output at 9 MHz ($16 - 7 = 9$), as shown in Fig. 20.8B. As a result, the 7-MHz signal interferes with the desired 25-MHz signal.

There are various ways to deal with situations of this kind. The use of one or more high-selectivity, tuned RF amplifiers ahead of the mixer will enhance the desired signal and suppress image signals. In especially severe cases, for example, when a nearby broadcast station happens to be transmitting at a receiver image frequency, the use of a *bandstop filter* (also called a *trap*) in the antenna system can help.

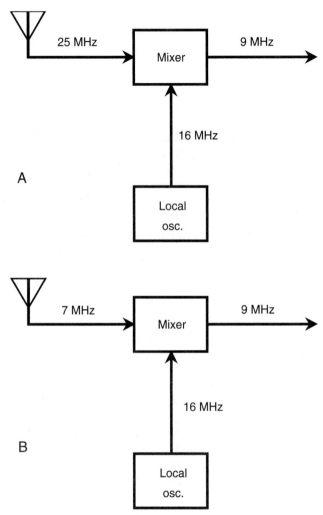

Figure 20.8 Normal operation (A) and abnormal operation (B) of a single-conversion wireless receiver.

Intermodulation

Another form of interference occurs with *intermodulation* ("intermod"). Nonlinear electronic components inside a receiver can cause unwanted mixing among external signals. Poor electrical connections external to the receiver can also cause this problem. In the downtown areas of some large cities, there are so many radio transmitters operating simultaneously that intermodulation is observed in all but the most sophisticated radio receivers. Intermodulation problems can be dealt with by the same means as spurious responses in general.

Non-RF Systems

Noise and interference can degrade the performance of optical, IR, and acoustic wireless systems.

Optical and IR systems

Bright daylight can bias a photocell and reduce its sensitivity, but in a well-designed system, this need not be a problem. By narrowing the receiver's aperture in a line-of-sight optical or IR communications system, most background illumination can be eliminated. The aperture can be narrowed by using a telescope with the photoreceptor placed at the focal point of the objective lens (Fig. 20.9).

A more serious type of optical or IR noise and interference is produced by modulated sources of visible light or IR. Lightning is one example. Incandescent, fluorescent,

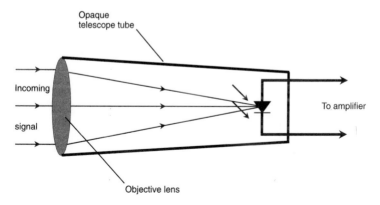

Figure 20.9 A telescope can be used to cut down on optical and IR noise and interference.

and elemental-vapor lamps are another. A scheme that can reduce "bulb hum" is illustrated in Fig. 20.10. The hum sensor is situated so it picks up little or none of the desired signal but plenty of "bulb hum." The switch and the potentiometer are adjusted by trial and error until a null or minimum occurs in the hum. This circuit is similar to the low-frequency RF noise-canceling scheme described earlier in this chapter.

Acoustic systems

In acoustic wireless systems and devices, noise and interference occur because of physical vibration in the molecules of gases such as air, liquids such as water, and solids such as concrete or steel.

In an ultrasonic communications system, noise within the human hearing range has little effect, unless the noise has harmonic content. Some noise results from air movement and the movements of objects (such as tree leaves) caused by wind. Insects and various other small animals sometimes emit sonic and ultrasonic waves.

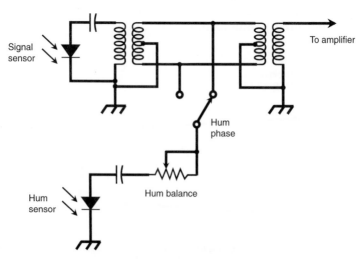

Figure 20.10 A hum-cancellation scheme for modulated-light communications.

In a computer *speech recognition* system, acoustic interference can present a major problem. Slamming doors, raised voices from an adjacent desk or cubicle, ringing telephones, and shuffling papers can be interpreted by the system as sounds it should translate.

In ultrasonic intrusion-detection or motion-detection systems, acoustic noise rarely causes a problem in itself, although if the noise is loud enough to physically vibrate objects in the vicinity, an alarm can be falsely triggered. These systems typically work by detecting a change in the phase of an acoustic wave having a short wavelength. Ultrasonic systems can be fooled if an acoustic wave of the same, or nearly the same, frequency is emitted by some other device.

Sonar, used for underwater depth-finding and/or the location of objects, can sometimes be fooled by obstructions such as fish, sand or mud in the water, or an abrupt change in temperature or salinity. False "echoes" can also be produced by deliberately transmitting signals at the same wavelength and pulse frequency as the sonar.

Measurement and Monitoring Systems

Electronic measurement and monitoring is commonly done for parameters such as current, voltage, power, frequency, and the motions of objects.

The Ammeter

An *ammeter* is a device for measuring electric current. Some ammeters can measure currents up to many amperes; others can measure currents down to a fraction of a microampere. Some ammeters work only with dc; others can measure ac.

Analog versus digital

The basic analog ammeter employs a *D'Arsonval movement.* The meter needle is attached to a current-carrying coil, and the coil is surrounded by a magnet. The needle/coil assembly swings on a spring bearing. Current in the coil produces a magnetic field. This field generates a variable force that deflects the needle in direct proportion to the current.

The sensitivity of a D'Arsonval meter depends on the strength of the magnet and the number of turns in the coil.

The stronger the magnet, and the larger the number of turns in the coil, the less current that is required to produce a given needle deflection.

A *digital ammeter* contains no moving parts; it is entirely electronic. Most digital meters provide a direct numeric readout. Direct-reading digital meters eliminate interpolation error. Digital meters are difficult to use in situations where frequent adjustments must be made that affect the meter indications or when the current fluctuates constantly and rapidly.

AC ammeter

The conventional ammeter is designed to measure dc. If ac is applied to such a meter, the reading will be zero because the average current is zero. To measure ac, the current must be rectified. The meter scale must be calibrated in a laboratory. Figure 21.1 is a simple functional diagram of an ac ammeter.

Ammeter shunt

If a resistor of the correct value is connected in parallel with an ammeter, its full-scale deflection can be increased. This is commonly done by orders of magnitude, that is, by powers of 10. The resistor must be capable of carrying the current without burning out. This is an *ammeter shunt*. Shunts are used when it is necessary to measure very large currents, such as hundreds of amperes. Shunts also allow a

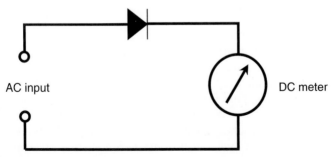

Figure 21.1 A simple ac ammeter.

microammeter or milliammeter to be used as a *multimeter* with several current ranges.

Thermal ammeter

When current flows through a resistive wire, the wire heats up. The extent of heating is proportional to the square of the current. Heating is directly proportional to the wire resistance if the current remains constant. By choosing the right metal or alloy, making the wire a certain length and diameter, employing a sensitive thermometer, and putting the whole assembly inside a thermally insulating package, a *thermal ammeter,* also called a *hot-wire ammeter,* can be made. This instrument can measure ac as well as dc, because the extent of heating does not depend on the direction of current flow.

Another type of thermal ammeter can be made by placing two different metals in physical contact with each other. If the correct metals are chosen, the junction heats up when current flows through it. This constitutes a *thermocouple ammeter.* As with the hot-wire meter, a thermometer can be used to measure the extent of the heating.

Basic voltmeter

If a large resistor is connected in series with a microammeter, the meter gives a reading directly proportional to the voltage across the meter/resistor combination. Using a microammeter and a large resistance in series, a voltmeter can be devised that will draw almost no current. A *voltmeter* can be made to have various full-scale ranges by switching different resistors in series with the microammeter (Fig. 21.2).

Basic ohmmeter

An *ohmmeter* is used for measuring dc resistances. A set of internal fixed resistors varies the measurement range. The resistors must have close tolerances for optimum metering accuracy. A schematic diagram of a simple ohmmeter is shown in Fig. 21.3.

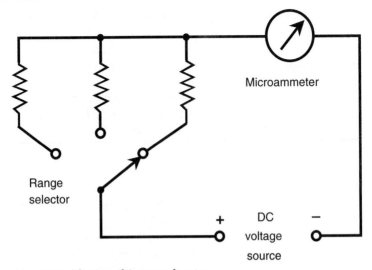

Figure 21.2 A basic multirange voltmeter.

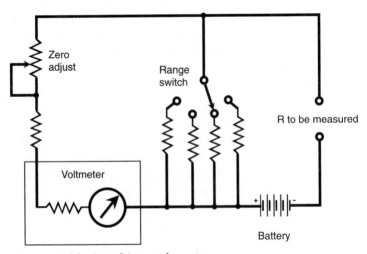

Figure 21.3 A basic multirange ohmmeter.

Most ohmmeters have several ranges, labeled according to the magnitude of the resistances in terms of the scale indication. The scale is calibrated from 0 to "infinity." Usually, 0 Ω (representing maximum current) is at the extreme right-hand end of the scale, and "infinity" ohms (representing zero current) is at the extreme left-hand end. When an ohmmeter is used, the circuit under test must not be carrying current, because such current will cause false readings.

Volt-ohm-milliammeter

In electronics labs, a common piece of test equipment is the *volt-ohm-milliammeter* (VOM). It can measure voltage, resistance, or current. Commercially available VOMs have limits in the values they can measure. The maximum voltage is around 1000 V; larger voltages require special leads and heavily insulated wires, as well as other safety precautions. The maximum current that a common VOM can measure is a few amperes. The highest measurable resistance is several tens of megohms.

Some VOMs use active amplifiers to enhance the voltmeter function. A good voltmeter disturbs circuits under test as little as possible; this requires that the meter have extremely high internal resistance. This can be achieved by sampling a current that is too small for a meter to directly indicate and amplifying it for measurement (Fig. 21.4).

Basic wattmeter

A *wattmeter* is employed to measure electrical power. When reactance is not a factor, such as in a dc or utility ac circuit, the power P, in watts, is the product of the voltage E, in volts, and the current I, in amperes:

$$P = EI$$

In these applications, a voltmeter can be connected in parallel with a circuit to get a reading of the voltage across it, and an ammeter can be connected in series to get a reading of the current through it. The voltage and current are multiplied to obtain the power.

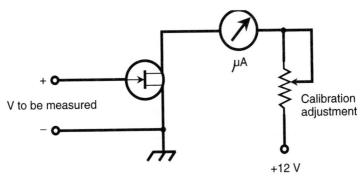

Figure 21.4 A voltmeter can employ an active amplifier to maximize the internal resistance.

Specialized Meters

The following are examples of special-purpose meters often encountered in electronic devices, systems, and test equipment.

Bar-graph meter

A *bar-graph meter* consists of several lamps or LED or liquid-crystal-display (LCD) elements. The components are arranged in a row, so the meter has an incrementally graduated scale. The drawings of Fig. 21.5 show a bar-graph meter (Fig. 21.5A) and an analog meter (Fig. 21.5B), both connected to a source that has an amplitude of 75 units. It is difficult to interpolate the reading given by the bar-graph meter, but the analog meter allows some interpolation. Because they are entirely electronic, bar-graph meters can withstand more physical abuse than analog meters. Bar-graph meters are less costly than electromechanical meters.

Directional wattmeter

A *directional wattmeter* facilitates measurement of RF transmitter output power and can also give an indication of how well an antenna is matched to a transmission line.

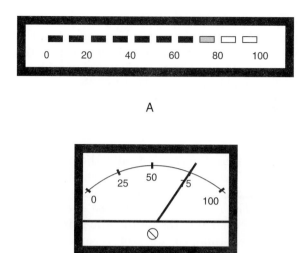

Figure 21.5 (A) A bar-graph meter; (B) an analog meter. Both meters show a reading of 75 units.

Directional wattmeters fall into two categories. The first, simpler type has a graduated scale calibrated in watts and sometimes also in milliwatts and/or kilowatts. Multiple scales are switch-selectable. The meter reads *forward power* or *reflected power* depending on the position of a switch or rotatable internal element. The second type of directional wattmeter, known as a *crossed-needle meter*, has two needles in a single enclosure, with a calibrated scale for each needle. One needle/scale indicates forward power, and the other needle/scale indicates reflected power. A third scale is calibrated for the point where the two needles cross. This scale indicates the SWR.

Volume unit meter

In audio high-fidelity equipment, meters are used to measure loudness. These are calibrated in *decibels*. In audio applications, 1 dB is approximately equal to the smallest increase or decrease in sound level that a listener can detect when a change is anticipated. Audio loudness is expressed in *volume units* (VUs), and the meter that indicates it is called a *VU meter*. The scale has a zero marker with a red line to the right and a black line to the left and is calibrated in decibels above and below zero (Fig. 21.6). Such a meter can alternatively be calibrated in rms watts.

S meter

In a wireless receiver, an *S meter* indicates the amplitude of incoming signals. Two common types are shown in Fig. 21.7. The signal for driving the meter is usually obtained from the IF chain. The most common method of obtaining this signal is by measuring the automatic-gain-control (AGC) voltage. The standard unit of received signal strength is the *S unit*. One S unit represents a change in signal voltage of 6 dB (a 2:1 ratio) across the receiver antenna terminals, assuming constant impedance. A signal level of 50 μV is considered to represent nine S units (S-9). The S-unit scale in a typical S meter ranges from 1 to 9 or 0 to 9. Most S meters are also marked off in decibels above S-9.

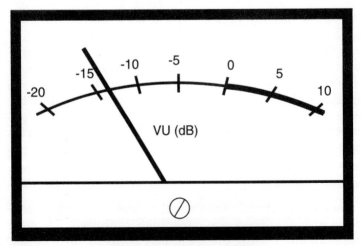

Figure 21.6 An audio VU meter, calibrated in decibels.

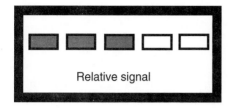

A

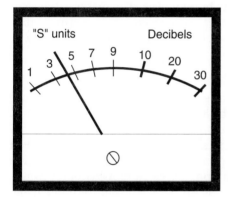

B

Figure 21.7 (A) A bar-graph S meter; (B) an analog S meter.

Applause meter

Sound levels can be measured by means of a device called an *applause meter*. The instrument gets its name from the fact that it is used to measure applause and vocal disturbances (such as laughter) in auditoriums and theaters. Figure 21.8 is a schematic diagram of a simple applause meter. The scale can be uncalibrated, in which case it will give relative but not absolute indications; or it can be marked off in decibels and connected to the output of a precision amplifier with a microphone of known sensitivity.

Illumination meter

A basic *illumination meter,* also called a *light meter,* can be made by connecting a microammeter to a photovoltaic cell (Fig. 21.9). Sophisticated illumination meters use dc amplifiers to enhance sensitivity and to allow for several ranges of readings. Solar cells are not sensitive to light at exactly the same wavelengths as human eyes. This effect can be neutralized by placing a color filter in front of a solar cell so that the cell becomes sensitive to the same wavelengths, in the same proportions, as human eyes. The meter can be calibrated in units such as *lumens* or *candela.* Sometimes, illumination meters are used to measure IR or UV intensity.

Pen recorder

A meter movement can be equipped with a marker to produce a graphic hard copy of the level of some quantity with respect to time. This is a *pen recorder.* A sheet of paper with a calibrated scale is attached to a rotating drum. The drum, driven by a clock motor, turns at a slow rate, such as one revolution per hour (h) or one revolution every 24 h. A computer can be programmed to function as an electronic pen recorder using specialized transducers and software.

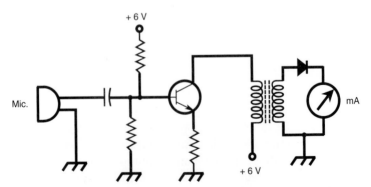

Figure 21.8 An applause meter using a bipolar-transistor amplifier.

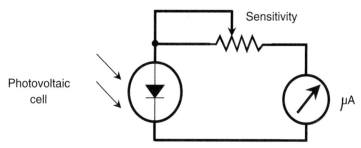

Figure 21.9 A simple illumination meter.

Oscilloscope

The *oscilloscope* is a graphic meter that measures and records quantities that vary rapidly, at frequencies ranging from a few hertz up to hundreds of megahertz. It creates a "graph" by throwing a beam of electrons at a phosphor screen. The "scope" displays the instantaneous signal voltage as a function of time. A CRT, similar to a TV picture tube, is employed. The scope reveals not only the pk and pk-pk levels of a signal but can display the signal waveform if the sweep frequency is high enough (Fig. 21.10). An oscilloscope can, in some cases, be used to approximately infer the frequency of a signal.

Frequency counter

A *frequency counter* is a digital instrument that measures and displays the frequency of a signal. It literally counts the signal cycles or pulses that occur within a set interval of time. A *gate* defines the starting and ending time for each counting period. The longer the *gate time,* the more accurate the measurement, if all other factors remain constant. A crystal oscillator is used as the *clock* whose frequency can be synchronized with an accepted time standard such as WWV and WWVH broadcasts from the National Bureau of Standards. Frequency counters are commonly available that allow accurate measurements of signal frequencies into the gigahertz range, with displays that show eight or ten significant digits.

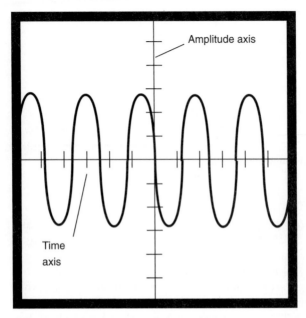

Figure 21.10 An oscilloscope displays instantaneous signal amplitude as a function of time.

Spectrum analyzer

A *spectrum analyzer* graphically displays signal amplitude as a function of frequency. Figure 21.11 shows a typical spectrum-analyzer display. In this example, each vertical division represents 10 dB and each horizontal division represents 10 kHz. These parameters can be changed by the user. The amplitude display can be set for a linear scale (say, 0.1 V per division) if desired, rather than the more common logarithmic (decibel) scale.

A radio receiver can be adapted for spectral analysis by connecting a spectrum analyzer into the IF chain. The result is a *panoramic receiver*. The frequency to which the receiver is tuned appears at the center of the horizontal scale. The frequency increment per horizontal division can be set for spectral analysis of individual signals (for example, 500 Hz per division), or it might be set for observation of a specific range of frequencies (for example, 10 kHz per division).

Monitoring Systems

Most *monitoring systems* consist of radio or IR transmitters and receivers. Some have mechanical hardware operated via signals reaching the receiver.

Baby monitor

A short-range, AM or FM radio transmitter and receiver can be used to listen at a distance to the sounds in an infant's room. The transmitter contains a sensitive microphone, a whip antenna, and a power supply. The receiver is battery powered and portable. It has an inductively loaded, short antenna similar to the antennas on cordless telephone sets. The receiver can pick up signals from the transmitter at distances of up to about 200 feet (ft). The RF signals pass easily through walls, ceilings, and floors.

The so-called baby monitor is subject to interference from other units that might be operating nearby on the same

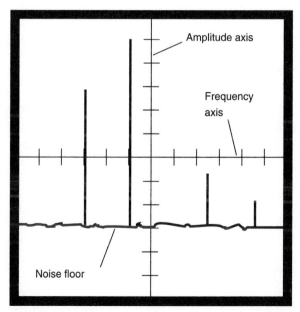

Figure 21.11 A spectrum analyzer displays signal amplitude as a function of frequency.

channel. Sometimes high-power radio transmitters in the vicinity produce *radio-frequency interference* (RFI). There are several channels to choose from. If interference occurs, another channel can be selected. Communications privacy and security are not a concern.

Smoke detector

Smoke changes the characteristics of the air. It is often accompanied by changes in the relative amounts of gases. Fire burns away oxygen and produces other gases, such as carbon dioxide, carbon monoxide, and/or sulfur dioxide. The smoke itself consists of solid particles. A *smoke detector* senses changes in the dielectric constant and/or the ionization potential of the air. Two electrically charged plates are spaced a fixed distance apart (Fig. 21.12). These two plates form an air-dielectric capacitor. A source of dc is connected to the plates. Normally, the plates retain a constant charge, and the current in the circuit is zero. If the properties of the air change, the capacitance varies, causing a small electric current to flow in the circuit for a short time. This current can be detected, and the resulting signals can actuate an alarm. The signals might also actuate a simple robotic system, such as a water sprinkler to extinguish fires.

Quality control

An application of wireless technology, the *laser,* can be found in *quality control* (QC) checking of bottles for height as they move along an assembly line. A laser/robot combination can find and remove bottles that are not of the correct height. The laser device is a pair of electric eyes. The principle is shown in Fig. 21.13. If a bottle is too short, both laser beams reach the photodetectors. If a bottle is too tall, neither laser beam reaches the photodetectors. In either of these situations, a robot arm/gripper picks the faulty bottle off the line and discards it. Only when a bottle is within a narrow range of heights (the acceptable range) will the top laser reach its photodetector while the bottom laser is blocked. Then the bottle is allowed to pass.

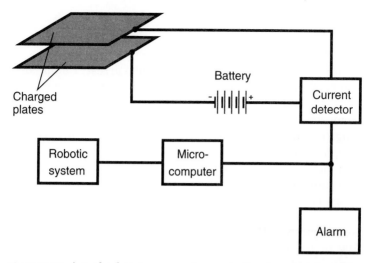

Figure 21.12 A smoke detector senses changes in the characteristics of the air between a pair of electrodes.

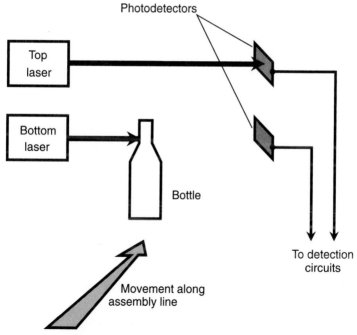

Figure 21.13 A pair of lasers in a QC application.

Tracking people

Suppose a person is sentenced to house arrest. Compliance can be monitored by having the person carry a conventional beeper (pager). An officer can page the person at random times; the person must then call the officer within a certain length of time. The call can be traced, and the location of the telephone verified.

A more secure method of ascertaining that a person is at a certain place, at a certain time, is by means of a short-range radio transmitter and receiver. The person wears the transmitting unit. Receiving units are placed at the home, in the car, and the place of work. The transmitter range is 100 to 200 ft. Receiver signals are sent to a central monitoring point. The signals are encoded so the monitoring personnel (or computers) know whether the person is at home, in the car, or at work. Any deviation from normal patterns can be detected.

Radiolocation provides another way to keep track of people. A *transponder* can be carried or worn by the person to be tracked; continuous signals can be sent to the unit asking for a position fix, and the unit can respond via a wireless network such as the cellular telephone system.

Electronic bug

An *electronic bug* is a tiny radio transmitter that can be hidden in a room, placed in a shirt pocket, or planted in a car. The antenna is a length of wire. A receiver can be located nearby. The device must operate at a low RF power level (a few milliwatts) to conserve battery energy. If the transmitter is near a wireless repeater that connects into a larger system, eavesdropping can be done anywhere within the coverage of the wireless system. With the advent of LEO satellite systems, it is theoretically possible to bug a room on the other side of the world.

Electronic bugs can be detected rather easily, although reliable detection equipment is expensive. The presence of a suspicious RF field gives away the existence of the transmitter.

Eye-in-hand system

To help a robot gripper (hand) find its way, a camera can be placed in the mechanism. The camera must be equipped for work at close range, from about 3 ft down to a fraction of an inch. The positioning error must be as small as possible. To be sure that the camera gets a good image, a lamp is included in the gripper along with the camera (Fig. 21.14). This *eye-in-hand system* can be used to precisely measure how close the gripper is to whatever object it is seeking. It can also make positive identification of the object.

The eye-in-hand system uses a *servo*. The robot is equipped with, or has access to, a controller (computer) that processes the data from the camera and sends instructions back to the gripper. Although most eye-in-hand systems use visible light for guidance and manipulation, it is also possible to use IR. This can be useful when it is necessary for the robot hand to sense differences in temperature.

Flying eyeball

In environments hostile to humans, robots find many uses, from manufacturing to exploration. One such device, especially useful underwater, has been called a *flying eyeball*. A

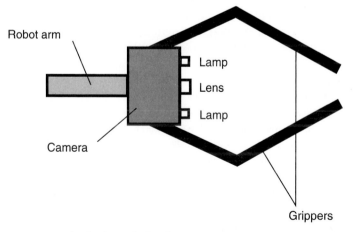

Figure 21.14 A robotic eye-in-hand system.

cable, containing the robot in a special launcher housing, is dropped from a boat. When the launcher gets to the desired depth, it lets out the robot, which is connected to the launcher by a tether. The tether and the drop cable convey data back to the boat. In some cases, the tether can be eliminated, and a wireless link can be used to convey data from the robot to the launcher. The link is usually in the IR or visible red portion of the spectrum. The robot contains a video camera and one or more lamps to illuminate the underwater environment. It also has a set of thrusters, or propellers, that let it move around according to control commands sent via the link between the boat and the robot. Human operators on board the boat watch the images and guide the robot.

Telepresence

Telepresence is an advanced form of remote monitoring and control. The robot operator gets a sense of being on location, even if the *telechir* (remotely controlled robot) and the operator are miles apart. Control and feedback are done via telemetry sent over wires, optical fibers, radio waves, or, in some cases, IR or visible line-of-sight links.

In an ideal telepresence system, the telechir is an autonomous robot that resembles a human being. The more humanoid the robot, the more realistic the telepresence. The control station consists of a suit that the operator wears or a chair in which the operator sits with various manipulators and displays. Sensors give feelings of pressure, vision, and sound. The operator can wear a helmet with a viewing screen that shows whatever the robot camera sees. When the operator's head turns, the robot's head, with its vision system, follows. *Binocular machine vision* gives a sense of depth. *Binaural machine hearing* allows realistic perception of sounds.

A block diagram of a telepresence system is shown in Fig. 21.15. At the telechir end of the circuit, the transducer converts sounds, images, and mechanical resistance into electrical impulses to be sent to the operator and also converts electrical impulses from the operator into mechanical motion. At the operator end, the transducer converts mechanical motion into electrical impulses to be sent to the telechir and converts electrical impulses from the telechir into sounds, images, and mechanical resistance.

Property Protection

Visible, IR, and ultrasonic devices are extensively used in home and business security systems. Most devices detect motion; some detect body heat.

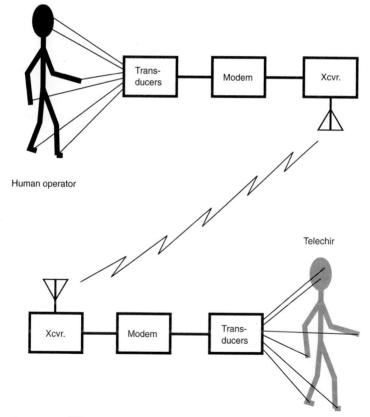

Figure 21.15 Telepresence combines remote monitoring and control.

Electric eye

The simplest device for detecting an unwanted visitor is an *electric eye*. Narrow beams of IR or visible light are shone across all reasonable points of entry, such as doorways and window openings. A photodetector receives energy from the beam. If, for any reason, the photodetector stops receiving the beam, an alarm is actuated. A person breaking into a property cannot avoid interrupting at least one beam, especially if large openings have several electric eyes spaced at suitable intervals.

IR motion detector

A common intrusion alarm device employs an *IR motion detector.* Two or three wide-angle IR pulses are transmitted at regular intervals; these pulses cover most of the room in which the device is installed. A receiving transducer picks up the returned IR energy, normally reflected from the walls, the floor, the ceiling, and the furniture. The intensity of the received pulses is noted by a microprocessor. If anything in the room changes position, there is a change in the intensity of the received energy. The microprocessor will notice this change and will trigger the alarm (Fig. 21.16). These devices consume very little power in regular operation, so batteries can serve as the power source.

Radiant heat detector

Infrared devices can detect changes in the indoor environment via direct sensing of the IR energy (often called *radiant heat*) emanating from objects. Humans, and all warm-blooded animals, emit IR. So, of course, does fire. A simple IR sensor, in conjunction with a microprocessor, can detect rapid or large increases in the amount of radiant heat in a room. The time threshold can be set so that gradual or small changes, such as might be caused by sunshine, do not trigger the alarm, but significant changes, such as a person entering the room, will trigger it. The temperature-change (increment) threshold can be set so that a small animal will not actuate

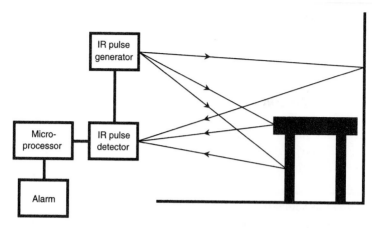

Figure 21.16 An IR intrusion alarm system.

the alarm, but a full-grown person will. This type of device, like the IR motion detector, can operate from batteries.

The main limitation of radiant-heat detectors is the fact that they can be fooled. False alarms are a risk; the sun might suddenly shine directly on the sensor and trigger the alarm. It is also possible that a person clad in a winter parka, boots, hood, and face mask, just entering from a subzero outdoor environment, might fail to set off the alarm. For this reason, radiant-heat sensors are used more often as fire-alarm actuators than as intrusion detectors.

Ultrasonic motion detector

Motion in a room can be detected by sensing the changes in the relative phase of acoustic waves. An *ultrasonic motion detector* employs a set of transducers that emit acoustic waves at frequencies above the range of human hearing (higher than 20 kHz). Another set of transducers picks up the reflected acoustic waves, whose wavelength is a fraction of an inch. If anything in the room changes position, the relative phase of the waves, as received by the various acoustic pickups, will change. This data is sent to a microprocessor, which can trigger an alarm and/or notify the police.

22

Physical Data

This chapter contains definitions of basic units encountered in electronics and related disciplines. Conversion tables are included, and common constants are denoted.

The SI System

The *Standard International (SI) System of Units,* formerly called the *meter/kilogram/second (mks) system,* defines seven quantities manifested in nature.

Displacement

One *meter* (1 m) is equivalent to 1.65076373×10^6 wavelengths in a vacuum of the radiation corresponding to the transition between the two levels of the krypton-86 atom. It was originally defined as 10^{-7} of the way from the north geographic pole to the equator as measured over the surface of the earth. Displacement is represented in equations by the lowercase letters d or s.

Mass

One *kilogram* (1 kg) is the mass of 1000 cm³ (1.000×10^3 cm³) of pure liquid water at the temperature of its greatest density (approximately 281 K). Mass is represented in equations by the lowercase letter m.

Time

One *second* (1 s) is $1/86{,}400 = 1.1574 \times 10^{-5}$ part of a solar day. Also defined as the time required for a beam of visible light to propagate over a distance of 2.99792×10^8 m in a vacuum. Time is represented in equations by the lowercase letter t.

Temperature

One *degree Kelvin* (1 K) is 3.66086×10^{-3} part of the difference between absolute zero (the absence of all heat) and the freezing point of pure water at standard atmospheric temperature and pressure. Temperature is represented in equations by the uppercase letter T.

Electric current

One *ampere* (1 A) represents the movement of 6.24×10^{18} charge carriers (usually electrons) past a specific fixed point in an electrical conductor over a time span of 1 s. Current is represented in equations by the uppercase letter I.

Luminous intensity

One *candela* (1 cd) represents the radiation from a surface area of 1.667×10^{-6} m^2 of a blackbody at the solidification temperature of pure platinum. Luminous intensity is represented in equations by the uppercase letters B, F, I, or L.

Material quantity

One *mole* (1 mol) is the number of atoms in precisely 0.012 kg of carbon-12. This is approximately 6.022169×10^{23}, also known as *Avogadro's number*. Material quantity is represented in equations by the uppercase letter N.

Electrical Units

Electrical units are defined for many quantities and phenomena. Standard units, derived from SI basic units, are defined here. For conversions to and from other units representing these properties, refer to the tables later in this chapter.

Unit electric charge

The *unit electric charge* is the charge contained in a single electron. This charge is also contained in a hole (electron absence within an atom), a proton, a positron, and an antiproton. Charge quantity in terms of unit electric charges is represented in equations by the lowercase letter *e*.

Electric charge quantity

The standard unit of charge quantity is the *coulomb* (C), which represents the total charge contained in 6.24×10^{18} electrons. Charge quantity is represented in equations by the uppercase Q or lowercase q.

Energy

The standard SI unit of energy is the *joule* (J). Mathematically, it is expressed in terms of unit mass multiplied by unit distance squared per unit time squared:

$$1 \text{ J} = 1 \text{ kg} \cdot \frac{\text{m}^2}{\text{s}^2}$$

Energy is represented in equations by the uppercase letter E. Occasionally it is represented by the uppercase H, T, or V.

Electromotive force

The standard unit of electromotive force (EMF), also called electric potential or potential difference, is the *volt* (V). It is equivalent to 1 J/C. Electromotive force is represented in equations by the uppercase E or V.

Resistance

The standard unit of resistance is the *ohm* (Ω). It is the resistance that results in 1 A of electric current with an applied EMF of 1 V. Resistance is represented in equations by the uppercase letter R.

Resistivity

The standard unit of resistivity is the *ohm-meter* ($\Omega \cdot m$). If a length of material measuring 1 m carries 1 A of current when a potential difference of 1 V is applied, it has a resistivity of $1\Omega \cdot m$. Resistivity is represented in equations by the lowercase Greek letter ρ.

Conductance

The standard unit of conductance is the *siemens* (S), formerly called the *mho*. Mathematically, conductance is the reciprocal of resistance. Conductance is represented in equations by the uppercase letter G. If R is the resistance of a component in ohms, and G is the conductance of the component in siemens,

$$G = \frac{1}{R}$$

and

$$R = \frac{1}{G}$$

Conductivity

The standard unit of conductivity is the *siemens per meter* (S/m). If a length of material measuring 1 m carries 1 A of current when a potential difference of 1 V is applied, it has a conductivity of 1 S/m. Conductivity is represented in equations by the lowercase Greek letter σ.

Power

The standard unit of power is the *watt* (W), equivalent to 1 J/s. Power is represented in equations by the uppercase P or W. In electrical and electronic circuits containing no reactance, if P is the power in watts, E is the voltage in volts, I is the current in amperes, and R is the resistance in ohms, the following holds:

$$P = EI = I^2 R = \frac{E^2}{R}$$

Period

The standard unit of ac cycle period is the *second* (s). This is a large unit in practice; typical signals have periods on the order of thousandths, millionths, billionths, or trillionths of a second. Period is represented in equations by the uppercase letter T.

Frequency

The standard unit of frequency is the *hertz* (Hz), formerly called the *cycle per second* (cps). This is a small unit in practice; typical signals have frequencies on the order of thousands, millions, billions, or trillions of hertz. Frequency, which is the mathematical reciprocal of period, is represented in equations by the lowercase letter f. If T is the period of a wave disturbance in seconds, the frequency f in hertz is given by

$$f = \frac{1}{T}$$

Capacitance

The standard unit of capacitance is the *farad* (F), which is equal to 1 C/V. This is a large unit in practice. In electronic circuits, most values of capacitance are on the order of millionths, billionths, or trillionths of a farad. Capacitance is represented in equations by the uppercase letter C.

Inductance

The standard unit of inductance is the *henry* (H), which is equal to 1 V·s/A. This is a large unit in practice. In electronic circuits, most values of inductance are on the order of thousandths or millionths of a henry. Inductance is represented in equations by the uppercase letter L.

Reactance

The standard unit of reactance is the *ohm* (Ω). Reactance is represented in equations by the uppercase letter X. It can be positive (inductive) and symbolized by X_L, or negative (capacitive) and symbolized by X_C. Reactance is dependent on frequency. Let f represent frequency in hertz, L represent inductance in henrys, and C represent capacitance in farads. Then the following formulas hold:

$$X_L = 2\pi fL$$

$$X_C = \frac{1}{2\pi fC}$$

Complex impedance

In the determination of complex impedance, there are two components: resistance (R) and reactance (X). The reactive component is multiplied by the unit imaginary number known as the *j operator*. Mathematically, j is equal to the positive square root of -1, so the following formulas hold:

$$j = (-1)^{1/2}$$
$$j^2 = j \cdot j = -1$$
$$j^3 = j^2 \cdot j = -1 \cdot j = -j$$
$$j^4 = j^3 \cdot j = -j \cdot j = 1$$

For powers of j beyond 4, the cycle repeats, so in general, for integers $n > 4$:

$$j^n = j^{(n-4)}$$

Let Z represent complex impedance, R represent resistance, and X represent reactance (either inductive or capacitive). Then,

$$Z = R + jX$$

Absolute-value impedance

Complex impedance can be represented as a vector in a rectangular coordinate plane, where resistance is plotted on the abscissa (horizontal axis) and reactance is plotted on the ordinate (vertical axis). The length of this vector is called the *absolute-value impedance,* symbolized by the uppercase Z and expressed in ohms. This impedance is usually discussed only when $jX = 0$, that is, when the impedance is a pure resistance $(Z = R)$. In the broad sense, if Z is the absolute-value impedance,

$$Z = (R^2 + X^2)^{1/2}$$

There are, in theory, infinitely many combinations of R and X that can result in a given absolute-value impedance Z.

Electric field strength

The standard unit of electric field strength is the *volt per meter* (V/m). An electric field of 1 V/m is represented by a potential difference of 1 V existing between two points displaced by 1 m. Electric field strength is represented in equations by the uppercase letter E.

Electromagnetic field strength

The standard unit of electromagnetic (EM) field strength is the *watt per square meter* (W/m^2). An EM field of 1 W/m^2 is represented by 1 W of power impinging perpendicularly on a flat surface whose area is 1 m².

Electric susceptibility

The standard unit of electric susceptibility is the *coulomb per volt-meter,* abbreviated C/(V·m). This quantity is represented in equations by the lowercase Greek letter η.

Permittivity

The standard unit of permittivity is the *farad per meter* (F/m). Permittivity is represented in equations by the lowercase Greek letter ε.

Charge-carrier mobility

The standard unit of charge-carrier mobility, also called carrier mobility or simply mobility, is the *meter squared per volt-second,* abbreviated m²/(V·s). Mobility is represented in equations by the lowercase Greek letter μ.

Magnetic Units

Magnetic units, derived from SI basic units, are defined below. For conversions to and from other units representing these properties, refer to the tables later in this chapter.

Magnetic flux

The standard unit of magnetic flux is the *weber* (Wb), defined as 1 V·s. This is a large unit in practice, equivalent to 1 A·H, represented by a constant, direct current of 1 A flowing through a coil having an inductance of 1 H. Magnetic flux is represented in equations by the uppercase Greek letter Φ.

Magnetic flux density

The standard unit of magnetic flux density is the *tesla* (T), equivalent to 1 Wb/m². Sometimes, magnetic flux density is spoken of in terms of the number of lines of flux per unit area; this is an imprecise terminology.

Magnetic field intensity

The standard unit of magnetic field intensity is the *oersted* (Oe), equivalent to 79.6 A/m. Magnetic field intensity is represented in equations by the uppercase letter H.

Magnetic pole strength

The standard unit of magnetic pole strength is the *ampere-meter* (A·m). Pole strength is represented in equations by the lowercase p or uppercase P.

Magnetomotive force

The standard unit of magnetomotive force is the *ampere-turn* (A·T), produced by a constant, direct current of 1 A flowing in a single-turn, air-core coil. Magnetomotive force is independent of the radius of the coil.

Miscellaneous Units

The following units are occasionally used in electronics. For conversions to and from other units, refer to the tables later in this chapter.

Area

The standard unit of area is the *square meter* or *meter squared* (m^2). Area is represented in equations by the uppercase letter A.

Volume

The standard unit of volume is the *cubic meter* or *meter cubed* (m^3). Volume is represented in equations by the uppercase letter V.

Plane angular measure

The standard unit of angular measure is the *radian* (rad). It is the angle subtended by an arc on a circle, whose length, as measured on the circle, is equal to the radius of the circle. Angles are represented in equations by lowercase Greek letters, usually ϕ or θ.

Solid angular measure

The standard unit of solid angular measure is the *steradian* (sr). A solid angle of 1 sr is represented by a cone with its apex at the center of a sphere, intersecting the surface of the sphere in a circle such that, within the circle, the enclosed area on the sphere is equal to the square of the radius of the sphere.

Velocity

The standard unit of linear speed is the *meter per second* (m/s). The unit of velocity requires two specifications: speed and direction. Direction is indicated in radians clockwise from geographic north on the earth's surface and counterclockwise from the positive x axis in the coordinate xy-plane. In three dimensions, direction can be specified in rectangular, spherical, or cylindrical coordinates. Speed and velocity are represented in equations by the lowercase letter v.

Angular velocity

The standard unit of angular velocity is the *radian per second* (rad/s). Angular velocity is represented in equations by the lowercase Greek letter ω.

Acceleration

The standard unit of acceleration is the *meter per second per second,* or *meter per second squared* (m/s^2). Linear acceleration is represented in equations by the lowercase letter a.

Angular acceleration

The standard unit of angular acceleration is the *radian per second per second,* or *radian per second squared* (rad/s^2). Angular acceleration is represented in equations by the lowercase Greek letter α.

Force

The standard unit of force is the *newton* (N). It is the impetus required to cause the linear acceleration of a 1-kg mass at a rate of 1 m/s^2. Force is represented in equations by the uppercase letter F.

Prefix Multipliers

Any unit can be expressed in larger or smaller units that are multiples or fractions of the fundamental unit. These multipliers and divisors are given standard values and prefix names. *Decimal (power-of-10) prefix multipliers* represent orders of magnitude in base 10 (the decimal number system); they are used in analog electronics and general science. *Binary (power-of-2) prefix multipliers* represent orders of magnitude in base 2 (the binary number system); they are used in digital electronics and computer science. Table 22.1 lists prefix names and multiplication factors for both schemes.

Alternative Unit Systems

The SI System of Units is the most widely accepted system. However, other schemes are sometimes encountered. The most common among these are the *centimeter / gram / second (cgs) system* and the *foot / pound / second (English) system.*

SI Unit Conversions

Table 22.2 is a conversion database for basic SI units to and from various other units. The first column lists units to be converted; the second column lists units to be derived. The

TABLE 22.1   **Prefix Multipliers and their Abbreviations**

Designator	Symbol	Decimal	Binary
Yocto-	y	10^{-24}	2^{-80}
Zepto-	z	10^{-21}	2^{-70}
Atto-	a	10^{-18}	2^{-60}
Femto-	f	10^{-15}	2^{-50}
Pico-	p	10^{-12}	2^{-40}
Nano-	n	10^{-9}	2^{-30}
Micro-	μ or mm	10^{-6}	2^{-20}
Milli-	m	10^{-3}	2^{-10}
Centi-	c	10^{-2}	
Deci-	d	10^{-1}	
(None)		10^{0}	2^{0}
Deka-	da or D	10^{1}	
Hecto-	h	10^{2}	
Kilo-	K or k	10^{3}	2^{10}
Mega-	M	10^{6}	2^{20}
Giga-	G	10^{9}	2^{30}
Tera-	T	10^{12}	2^{40}
Peta-	P	10^{15}	2^{50}
Exa-	E	10^{18}	2^{60}
Zetta-	Z	10^{21}	2^{70}
Yotta-	Y	10^{24}	2^{80}

third column lists numbers by which units in the first column must be multiplied to obtain units in the second column. The fourth column lists numbers by which units in the second column must be multiplied to obtain units in the first column.

Electrical Unit Conversions

Table 22.3 is a conversion database for electrical SI units to and from various other units. The first column lists units to be converted; the second column lists units to be derived. The third column lists numbers by which units in the first column must be multiplied to obtain units in the second column. The fourth column lists numbers by which units in the second column must be multiplied to obtain units in the first column.

TABLE 22.2 SI Unit Conversions

When no coefficient is given, the coefficient is meant to be precisely equal to 1.

To convert	To	Multiply by	Conversely, multiply by
Meters	Angstroms	10^{10}	10^{-10}
Meters	Nanometers	10^9	10^{-9}
Meters	Micrometers	10^6	10^{-6}
Meters	Millimeters	10^3	10^{-3}
Meters	Centimeters	10^2	10^{-2}
Meters	Inches	39.37	0.02540
Meters	Feet	3.281	0.3048
Meters	Yards	1.094	0.9144
Meters	Kilometers	10^{-3}	10^3
Meters	Statute miles	6.214×10^{-4}	1.609×10^3
Meters	Nautical miles	5.397×10^{-4}	1.853×10^3
Meters	Light seconds	3.336×10^{-9}	2.998×10^8
Meters	Astronomical units	6.685×10^{-12}	1.496×10^{11}
Meters	Light years	1.057×10^{-16}	9.461×10^{15}
Meters	Parsecs	3.241×10^{-17}	3.085×10^{16}
Kilograms	Atomic mass units	6.022×10^{26}	1.661×10^{-27}
Kilograms	Nanograms	10^{12}	10^{-12}
Kilograms	Micrograms	10^9	10^{-9}
Kilograms	Milligrams	10^6	10^{-6}
Kilograms	Grams	10^3	10^{-3}
Kilograms	Ounces	35.28	0.02834
Kilograms	Pounds	2.205	0.4535
Kilograms	English tons	1.103×10^{-3}	907.0
Seconds	Minutes	0.01667	60.00
Seconds	Hours	2.778×10^{-4}	3.600×10^3
Seconds	Days	1.157×10^{-5}	8.640×10^4
Seconds	Years	3.169×10^{-8}	3.156×10^7
Seconds	Centuries	3.169×10^{-10}	3.156×10^9
Seconds	Millennia	3.169×10^{-11}	3.156×10^{10}
Degrees Kelvin	Degrees Celsius	Subtract 273	Add 273
Degrees Kelvin	Degrees Fahrenheit	Multiply by 1.80, then subtract 459	Multiply by 0.556, then add 255
Degrees Kelvin	Degrees Rankine	1.80	0.556
Amperes	Carriers per second	6.24×10^{18}	1.60×10^{-19}
Amperes	Statamperes	2.998×10^9	3.336×10^{-10}
Amperes	Nanoamperes	10^9	10^{-9}
Amperes	Microamperes	10^6	10^{-6}
Amperes	Abamperes	0.10000	10.000
Amperes	Milliamperes	10^3	10^{-3}
Candela	Microwatts per steradian	1.464×10^3	6.831×10^{-4}
Candela	Milliwatts per steradian	1.464	0.6831

TABLE 22.2   SI Unit Conversions (Continued)

When no coefficient is given, the coefficient is meant to be precisely equal to 1.

To convert	To convert	Multiply by	Conversely, multiply by
Candela	Lumens per steradian	Identical; no conversion	Identical; no conversion
Candela	Watts per steradian	1.464×10^{-3}	683.1
Moles	Coulombs	9.65×10^4	1.04×10^{-5}

TABLE 22.3   Electrical Unit Conversions

When no coefficient is given, the coefficient is meant to be precisely equal to 1.

To convert	To	Multiply by	Conversely, multiply by
Unit electric charges	Coulombs	1.60×10^{-19}	6.24×10^{18}
Unit electric charges	Abcoulombs	1.60×10^{-20}	6.24×10^{19}
Unit electric charges	Statcoulombs	4.80×10^{-10}	2.08×10^9
Coulombs	Unit electric charges	6.24×10^{18}	1.60×10^{-19}
Coulombs	Statcoulombs	2.998×10^9	3.336×10^{-10}
Coulombs	Abcoulombs	0.1000	10.000
Joules	Electronvolts	6.242×10^{18}	1.602×10^{-19}
Joules	Ergs	10^7	10^{-7}
Joules	Calories	0.2389	4.1859
Joules	British thermal units	9.478×10^{-4}	1.055×10^3
Joules	Watt-hours	2.778×10^{-4}	3.600×10^3
Joules	Kilowatt-hours	2.778×10^{-7}	3.600×10^6
Volts	Abvolts	10^8	10^{-8}
Volts	Microvolts	10^6	10^{-6}
Volts	Millivolts	10^3	10^{-3}
Volts	Statvolts	3.336×10^{-3}	299.8
Volts	Kilovolts	10^{-3}	10^3
Volts	Megavolts	10^{-6}	10^6
Ohms	Abohms	10^9	10^{-9}
Ohms	Megohms	10^{-6}	10^6
Ohms	Kilohms	10^{-3}	10^3
Ohms	Statohms	1.113×10^{-12}	8.988×10^{11}

TABLE 22.3   Electrical Unit Conversions (Continued)

When no coefficient is given, the coefficient is meant to be precisely equal to 1.

To convert	To	Multiply by	Conversely, multiply by
Siemens	Statsiemens	8.988×10^{11}	1.113×10^{-12}
Siemens	Microsiemens	10^6	10^{-6}
Siemens	Millisiemens	10^3	10^{-3}
Siemens	Absiemens	10^{-9}	10^9
Watts	Picowatts	10^{12}	10^{-12}
Watts	Nanowatts	10^9	10^{-9}
Watts	Microwatts	10^6	10^{-6}
Watts	Milliwatts	10^3	10^{-3}
Watts	British thermal units per hour	3.412	0.2931
Watts	Horsepower	1.341×10^{-3}	745.7
Watts	Kilowatts	10^{-3}	10^3
Watts	Megawatts	10^{-6}	10^6
Watts	Gigawatts	10^{-9}	10^9
Hertz	Degrees per second	360.0	0.002778
Hertz	Radians per second	6.283	0.1592
Hertz	Kilohertz	10^{-3}	10^3
Hertz	Megahertz	10^{-6}	10^6
Hertz	Gigahertz	10^{-9}	10^9
Hertz	Terahertz	10^{-12}	10^{12}
Farads	Picofarads	10^{12}	10^{-12}
Farads	Statfarads	8.898×10^{11}	1.113×10^{-12}
Farads	Nanofarads	10^9	10^{-9}
Farads	Microfarads	10^6	10^{-6}
Farads	Abfarads	10^{-9}	10^9
Henrys	Nanohenrys	10^9	10^{-9}
Henrys	Abhenrys	10^9	10^{-9}
Henrys	Microhenrys	10^6	10^{-6}
Henrys	Millihenrys	10^3	10^{-3}
Henrys	Stathenrys	1.113×10^{-12}	8.898×10^{11}
Volts per meter	Picovolts per meter	10^{12}	10^{-12}
Volts per meter	Nanovolts per meter	10^9	10^{-9}
Volts per meter	Microvolts per meter	10^6	10^{-6}
Volts per meter	Millivolts per meter	10^3	10^{-3}
Volts per meter	Volts per foot	3.281	0.3048
Watts per square meter	Picowatts per square meter	10^{12}	10^{-12}
Watts per square meter	Nanowatts per square meter	10^9	10^{-9}

TABLE 22.3   Electrical Unit Conversions (Continued)

When no coefficient is given, the coefficient is meant to be precisely equal to 1.

To convert	To	Multiply by	Conversely, multiply by
Watts per square meter	Microwatts per square meter	10^6	10^{-6}
Watts per square meter	Milliwatts per square meter	10^3	10^{-3}
Watts per square meter	Watts per square foot	0.09294	10.76
Watts per square meter	Watts per square inch	6.452×10^{-4}	1.550×10^3
Watts per square meter	Watts per square centimeter	10^{-4}	10^4
Watts per square meter	Watts per square millimeter	10^{-6}	10^6

Magnetic Unit Conversions

Table 22.4 is a conversion database for magnetic SI units to and from various other units. The first column lists units to be converted; the second column lists units to be derived. The third column lists numbers by which units in the first column must be multiplied to obtain units in the second column. The fourth column lists numbers by which units in the second column must be multiplied to obtain units in the first column.

Miscellaneous Unit Conversions

Table 22.5 is a conversion database for miscellaneous SI units to and from various other units. The first column lists units to be converted; the second column lists units to be derived. The third column lists numbers by which units in the first column must be multiplied to obtain units in the second column. The fourth column lists numbers by which units in the second column must be multiplied to obtain units in the first column.

TABLE 22.4   Magnetic Unit Conversions

When no coefficient is given, the coefficient is meant to be precisely equal to 1.

To convert	To	Multiply by	Conversely, multiply by
Webers	Maxwells	10^8	10^{-8}
Webers	Ampere-microhenrys	10^6	10^{-6}
Webers	Ampere-millihenrys	10^3	10^{-3}
Webers	Unit poles	1.257×10^{-7}	7.956×10^6
Teslas	Maxwells per square meter	10^8	10^{-8}
Teslas	Gauss	10^4	10^{-4}
Teslas	Maxwells per square centimeter	10^4	10^{-4}
Teslas	Maxwells per square millimeter	10^2	10^{-2}
Teslas	Webers per square centimeter	10^{-4}	10^4
Teslas	Webers per square millimeter	10^{-6}	10^6
Oersteds	Microamperes per meter	7.956×10^7	1.257×10^{-8}
Oersteds	Milliamperes per meter	7.956×10^4	1.257×10^{-5}
Oersteds	Amperes per meter	79.56	0.01257
Ampere-turns	Microampere-turns	10^6	10^{-6}
Ampere-turns	Milliampere-turns	10^3	10^{-3}
Ampere-turns	Gilberts	1.256	0.7956

TABLE 22.5   Miscellaneous Unit Conversions

When no coefficient is given, the coefficient is meant to be precisely equal to 1.

To convert	To	Multiply by	Conversely, multiply by
Square meters	Square angstroms	10^{20}	10^{-20}
Square meters	Square nanometers	10^{18}	10^{-18}
Square meters	Square microns	10^{12}	10^{-12}
Square meters	Square millimeters	10^6	10^{-6}
Square meters	Square centimeters	10^4	10^{-4}
Square meters	Square inches	1.550×10^3	6.452×10^{-4}
Square meters	Square feet	10.76	0.09294
Square meters	Acres	2.471×10^{-4}	4.047×10^3
Square meters	Hectares	10^{-4}	10^4
Square meters	Square kilometers	10^{-6}	10^6
Square meters	Square statute miles	3.863×10^{-7}	2.589×10^6

TABLE 22.5 Miscellaneous Unit Conversions (Continued)

To convert	To	Multiply by	Conversely, multiply by
Square meters	Square nautical miles	2.910×10^{-7}	3.434×10^6
Square meters	Square light years	1.117×10^{-17}	8.951×10^{16}
Square meters	Square parsecs	1.051×10^{-33}	9.517×10^{32}
Cubic meters	Cubic angstroms	10^{30}	10^{-30}
Cubic meters	Cubic nanometers	10^{27}	10^{-27}
Cubic meters	Cubic microns	10^{18}	10^{-18}
Cubic meters	Cubic millimeters	10^9	10^{-9}
Cubic meters	Cubic centimeters	10^6	10^{-6}
Cubic meters	Milliliters	10^6	10^{-6}
Cubic meters	Liters	10^3	10^{-3}
Cubic meters	U.S. gallons	264.2	3.785×10^{-3}
Cubic meters	Cubic inches	6.102×10^4	1.639×10^{-5}
Cubic meters	Cubic feet	35.32	0.02831
Cubic meters	Cubic kilometers	10^{-9}	10^9
Cubic meters	Cubic statute miles	2.399×10^{-10}	4.166×10^9
Cubic meters	Cubic nautical miles	1.572×10^{-10}	6.362×10^9
Cubic meters	Cubic light seconds	3.711×10^{-26}	2.695×10^{25}
Cubic meters	Cubic astronomical units	2.987×10^{-34}	3.348×10^{33}
Cubic meters	Cubic light years	1.181×10^{-48}	8.469×10^{47}
Cubic meters	Cubic parsecs	3.406×10^{-50}	2.936×10^{49}
Radians	Degrees	57.30	0.01745
Meters per second	Inches per second	39.37	0.02540
Meters per second	Kilometers per hour	3.600	0.2778
Meters per second	Feet per second	3.281	0.3048
Meters per second	Statute miles per hour	2.237	0.4470
Meters per second	Knots	1.942	0.5149
Meters per second	Kilometers per minute	0.06000	16.67
Meters per second	Kilometers per second	10^{-3}	10^3
Radians per second	Degrees per second	57.30	0.01745
Radians per second	Revolutions per second	0.1592	6.283
Radians per second	Revolutions per minute	2.653×10^{-3}	377.0
Meters per second per second	Inches per second per second	39.37	0.02540

TABLE 22.5   Miscellaneous Unit Conversions (Continued)

To convert	To	Multiply by	Conversely, multiply by
Meters per second per second	Feet per second per second	3.281	0.3048
Radians per second per second	Degrees per second per second	57.30	0.01745
Radians per second per second	Revolutions per second per second	0.1592	6.283
Radians per second per second	Revolutions per minute per second	2.653×10^{-3}	377.0
Newtons	Dynes	10^5	10^{-5}
Newtons	Ounces	3.597	0.2780
Newtons	Pounds	0.2248	4.448

TABLE 22.6   Physical, Electrical, and Chemical Constants

Quantity or phenomenon	Value	Symbol
Mass of sun	1.989×10^{30} kg	m_{sun}
Mass of earth	5.974×10^{24} kg	m_{earth}
Avogadro's number	6.022169×10^{23}	N
Mass of moon	7.348×10^{22} kg	m_{moon}
Mean radius of sun	6.970×10^8 m	r_{sun}
Speed of electromagnetic-field propagation in free space	2.99792×10^8 m/s	c
Faraday constant	9.649×10^7 C/kmol	F
Mean radius of earth	6.371×10^6 m	r_{earth}
Mean orbital speed of earth	2.978×10^4 m/s	
Base of natural logarithms	2.718282	e or ε
Mean radius of moon	1.738×10^6 m	r_{moon}
Characteristic impedance of free space	376.7 Ω	Z_0
Speed of sound in dry air at standard atmospheric temperature and pressure	344 m/s	
Gravitational acceleration at sea level	9.8067 m/s^2	g
Gas constant	8.3145 J/$^\circ$K·mol	R_0

TABLE 22.6   Physical, Electrical, and Chemical Constants (Continued)

Quantity or phenomenon	Value	Symbol
Wien's constant	0.0029 m·K	σ_w
Second radiation constant	0.01439 m·K	c_2
Permeability of free space	1.257×10^{-6} H/m	μ_0
Stefan-Boltzmann constant	5.6697×10^{-8} W/m^2·°K	σ
Gravitational constant	6.673×10^{-11} N·m^2/kg^2	G
Permittivity of free space	8.85×10^{-12} F/m	ε_0
Boltzmann's constant	1.3807×10^{-23} J/°K	k
First radiation constant	4.993×10^{-24} J·m	c_1
Mass of alpha particle	6.64×10^{-27} kg	m_α
Mass of neutron	1.675×10^{-27} kg	m_n
Mass of proton	1.673×10^{-27} kg	m_p
Mass of electron	9.109×10^{-31} kg	m_e
Planck's constant	6.6261×10^{-34} J·s	h

Constants

Table 22.6 lists common physical, electrical, and chemical constants. Expressed units can be converted to other units by referring to Tables 22.2 through 22.5.

23

Mathematical Data

In electronics, a variety of symbols are used to represent units, constants, variables, mathematical operations, and logical operations. This chapter contains tables of symbols and their most common uses and/or meanings. Scientific notation, coordinate systems, and other concepts are discussed.

Greek Alphabet

Table 23.1 lists uppercase Greek letters, character names (as written in English), and usages. Table 23.2 lists lowercase Greek letters, character names, and usages. Uppercase or lowercase Greek letters are sometimes italicized. In these tables, characters are not italicized.

General Symbols

Table 23.3 lists symbols used to depict operations, relations, and specifications in mathematics relevant to physical sciences and engineering. Italics are sometimes used for alphabetic characters; other symbols are rarely italicized.

TABLE 23.1 Uppercase Greek Alphabet

Symbol	Character name	Common representations
A	Alpha	
B	Beta	Magnetic flux density
Γ	Gamma	Gamma match; general index set; curve; contour
Δ	Delta	Delta match; three-phase ac circuit with no common ground; increment; difference quotient; difference sequence; laplacian
E	Epsilon	Voltage; energy
Z	Zeta	Impedance
H	Eta	Efficiency
Θ	Theta	Order
I	Iota	Current
K	Kappa	Magnetic susceptibility; degrees Kelvin
Λ	Lambda	General index set
M	Mu	Mutual inductance
N	Nu	Avogadro's number (6.022169×10^{23})
Ξ	Xi	
O	Omicron	Order
Π	Pi	Product; infinite product; homotopy
P	Rho	Power
Σ	Sigma	Summation; series; infinite series
T	Tau	Time constant; temperature
Υ	Upsilon	
Φ	Phi	Magnetic flux; Frattini subgroup
X	Chi	Reactance
Ψ	Psi	Dielectric flux
Ω	Omega	Ohms; volume of a body

TABLE 23.2   Lowercase Greek Alphabet

Symbol	Character name	Common representations
α	Alpha	Current gain of bipolar transistor in common-base configuration; alpha particle; angular acceleration; angle; direction angle; transcendental number; scalar coefficient
β	Beta	Current gain of bipolar transistor in common-emitter configuration; magnetic flux density; beta particle; angle; direction angle; transcendental number; scalar coefficient
γ	Gamma	Gamma radiation; electrical conductivity; Euler's constant; gravity; direction angle; scalar coefficient; permutation; cycle
δ	Delta	Derivative; variation of a quantity; point evaluation; support function; metric function; distance function; variation of an integral; laplacian
ε	Epsilon	Electric permittivity; natural logarithm base (approximately2.71828); eccentricity; signature
ζ	Zeta	Impedance, coefficient; coordinate variable in a transformation
η	Eta	Electric susceptibility; hysteresis coefficient; efficiency; coordinate variable in a transformation
θ	Theta	Angle; phase angle; angle in polar coordinates; angle in cylindrical coordinates; angle in spherical coordinates; parameter; homomorphism
ι	Iota	Definite description (in predicate logic)
κ	Kappa	Dielectric constant; coefficient of coupling; curvature
λ	Lambda	Wavelength; Wien Displacement Law constant; ratio; Lebesgue measure; eigenvalue

TABLE 23.2   Lowercase Greek Alphabet (Continued)

Symbol	Character name	Common representations
μ	Mu	Micro-; magnetic permeability; amplification factor; charge carrier mobility; mean; statistical parameter
ν	Nu	Frequency; reluctivity; statistical parameter; natural epimorphism
ξ	Xi	Coordinate variable in a transformation
o	Omicron	Order
π	Pi	Ratio of circle circumference to diameter (approximately 3.14159); radian; permutation
ρ	Rho	Electrical resistivity; variable representing an angle; curvature; correlation; metric; density
σ	Sigma	Electrical conductivity; Stefan-Boltzmann constant; standard deviation; variance; mathematical partition; permutation; topology
τ	Tau	Time-phase displacement; torsion; mathematical partition; topology
υ	Upsilon	
ϕ or φ	Phi	Angle; phase angle; dielectric flux; angle in spherical coordinates; Euler phi function; mapping; predicate
χ	Chi	Magnetic susceptibility; characteristic function; chromatic number; configuration of a body
ψ	Psi	Angle; mapping; predicate; chart
ω	Omega	Angular velocity; period; modulus of continuity

Subscripts and Superscripts

Subscripts modify the meanings of units, constants, and variables. A subscript is placed to the right of the main character (without spacing), is set in smaller type than the main character, and is set below the base line. Numeric subscripts are generally not italicized; alphabetic subscripts are sometimes italicized. Examples of subscripted quantities are

TABLE 23.3   General Mathematical Symbols and Their Common Meanings

For meanings of Greek letters, refer to Tables 23.1 and 23.2.

Symbol	Character name	Common representations
.	Decimal or radix point	Separates integral part of number from fractional part
®	Universal qualifier	Read as for all
#	Pound sign	Number; pounds
∃	Existential qualifier	Read as there exists or for some
%	Percent sign	Read as parts per hundred or percent
0/00	Per mil sign	Read as parts per thousand or permil
&	Ampersand	Logical AND operation
@	At sign	Read as at the rate of or at the cost of
( )	Parentheses	Encloses elements defining coordinates of a point; encloses elements of a set of ordered numbers; encloses bounds of an open interval
[ ]	Brackets	Encloses a group of terms that includes one or more groups in parentheses; encloses elements of an equivalence class; encloses bounds of a closed interval
{ }	Braces	Encloses a group of terms that includes one or more groups in brackets; encloses elements comprising a set
[ ) or ( ]	Half-brackets	Encloses bounds of a half open interval
] [	Inside-out brackets	Encloses bounds of an open interval
$\left(\ \right)$ or $\left[\ \right]$	Parentheses or brackets (enlarged)	Encloses elements of a matrix
*	Asterisk	Multiplication; logical AND operation
×	Cross	Multiplication; logical AND operation; vector (cross) product of two vectors
Π	Uppercase Greek letter pi (enlarged)	Product of many values
·	Dot	Multiplication; logical AND operation; scalar (dot) product of two vectors
+	Plus sign	Addition; logical OR operation
Σ	Uppercase Greek letter sigma (enlarged)	Summation of many values

TABLE 23.3 General Mathematical Symbols and Their Common Meanings (Continued)

For meanings of Greek letters, refer to Tables 23.1 and 23.2.

Symbol	Character name	Common representations
,	Comma	Separates large numbers by thousands; separates elements defining coordinates of a point; separates elements of a set of ordered numbers; separates bounds of an interval
−	Minus sign	Subtraction; logical NOT symbol
±	Plus/minus sign	Read as plus or minus; defines the extent to which a value can deviate from the nominal value
/	Slash or slant	Division; ratio; proportion; separates parts of a Web site uniform resource locator (URL)
÷		Division
:	Colon	Ratio; separates hours from minutes; separates minutes from seconds
::	Double colon	Mean
!	Exclamation mark	Factorial
≤	Inequality sign	Read as is less than or equal to
<	Inequality sign	Read as is less than
<<	Inequality sign	Read as is much less than
=	Equal sign	Read as is equal to; logical equivalence
≥	Inequality sign	Read as is greater than or equal to
>	Inequality sign	Read as is greater than
>>	Inequality sign	Read as is much greater than
≅	Congruence sign	Read as is congruent with
≠	Unequal sign	Read as is not equal to
≡	Equivalence sign	Read as is logically equivalent to
≈	Approximation sign	Read as is approximately equal to
∝		Read as is proportional to
~	Squiggle	Read as is similar to
...	Triple dot	Read as and so on or and beyond
│	Vertical line	Read as is exactly divisible by

TABLE 23.3   General Mathematical Symbols and Their Common Meanings (Continued)

For meanings of Greek letters, refer to Tables 23.1 and 23.2.

Symbol	Character name	Common representations
| |	Vertical lines	Absolute value of quantity between lines; length of vector quantity denoted between lines; distance between two points; cardinality of number; modulus
|	Vertical line (elongated)	Denotes limits of evaluation for a function
| |	Vertical lines (elongated)	Determinant of matrix whose elements are enumerated between lines
ℵ	Uppercase Hebrew letter aleph	Transfinite cardinal number; Continuum Hypothesis
∩	Intersection sign	Set-intersection operation
∪	Union sign	Set-union operation
∅	Null sign	Set containing no elements (empty set or null set)
∈		Read as is an element of
∉		Read as is not an element of
⊂		Read as is a proper subset of
⊃	Implication sign	Read as logically implies
⊆		Read as is a subset of
⊄		Read as is not a proper subset of
∠	Angle sign	Angle; angle measure
⊥		Read as is perpendicular to
∇	Del or nabla	Vector differential operator
√	Radical or surd	Root; square root
⇔ or ↔	Double arrow	Read as if and only if or is logically equivalent to
⇒	Right arrow	Logical implication
∴	Three dots	Read as therefore
→	Right arrow	Logical implication; convergence; mapping
↑	Upward arrow	Read as above or increasing

TABLE 23.3   General Mathematical Symbols and Their Common Meanings (Continued)

For meanings of Greek letters, refer to Tables 23.1 and 23.2.

Symbol	Character name	Common representations
$\downarrow$	Downward arrow	Read as below or decreasing
∂		Partial derivative; jacobian; surface of a body
$\int$		Integral
$\iint$		Double integral
$\int_E$		Riemann integral
$\int_\Gamma$		Contour integral
$\iint_S$		Surface integral
$\iiint$		Triple integral
$\circ$	Degree sign (superscript)	Degree of angle; degree of temperature
∞	Infinity sign	Infinity; an arbitrarily large number; an arbitrarily great distance away

Z_0 Read as Z sub nought; stands for characteristic impedance

R_{out} Read as R sub out; stands for output resistance

x_3 Read as x sub 3; represents a variable

Superscripts represent exponents (the taking of the base quantity or variable to the indicated power). Superscripts are usually numerals, but they are sometimes alphabetic characters. Italicized, lowercase English letters from the second half of the alphabet (n through z) are generally used to represent variable exponents. A superscript is placed to the right of the main character (without spacing), is set in smaller type than the main character, and is set above the base line. Examples of superscripted quantities are

2^3 Read as two cubed; represents $2 \times 2 \times 2$

e^x Read as e to the xth; represents the exponential function of x

$y^{1/2}$ Read as y to the one-half; represents the square root of y

Scientific Notation

Scientific notation is used to represent extreme numerical values. It also facilitates arithmetic operations among numbers ranging over many orders of magnitude. A numeral in scientific notation is written in the form

$$m.n \times 10^z$$

where m (to the left of the radix point) is a number from the set $\{1, 2, 3, 4, 5, 6, 7, 8, 9\}$, n (to the right of the radix point) is a nonnegative integer, and z (the power of 10) can be any integer. Some examples of numbers written in scientific notation are

$$2.56 \times 10^6$$

$$8.0773 \times 10^{-18}$$

$$1.000 \times 10^0$$

In some countries, scientific notation requires that $m = 0$. In this rarely used form, the above numbers appear as

$$0.256 \times 10^7$$

$$0.80773 \times 10^{-17}$$

$$0.1000 \times 10^1$$

The multiplication sign can be denoted in various ways. Instead of the common cross symbol (x or $\times$), an asterisk (*) can be used, so the above expressions become

$$2.56 \ * \ 10^6$$

$$8.0773 \ * \ 10^{-18}$$

$$1.000 \ * \ 10^0$$

Another alternative is to use the dot ($\cdot$), so the expressions appear as

$$2.56 \cdot 10^6$$

$$8.0773 \cdot 10^{-18}$$

$$1.000 \cdot 10^0$$

Sometimes it is necessary to express numbers in scientific notation using plain text. This is the case, for example, when transmitting information within the body of an e-mail message (rather than as an attachment). Some electronic calculators and computers use this system. The uppercase letter E indicates that the quantity immediately following is an exponent. In this format, the above expressions are written as

$$2.56E6$$

$$8.0773E{-}18$$

$$1.000E0$$

Sometimes the exponent is written with two numerals and includes a plus or minus sign, so the above expressions appear as

$$2.56E + 06$$

$$8.0773E{-}18$$

$$1.000E + 00$$

Another alternative is to use an asterisk to indicate multiplication and the symbol ^ to indicate a superscript, so the expressions appear as

$$2.56 \,*\, 10\text{^}6$$

$$8.0773 \,*\, 10\text{^}{-}18$$

$$1.000 \,*\, 10\text{^}0$$

In all of these examples, the numerical values represented are identical. Respectively, if written out in full, they are:

$$2{,}560{,}000$$

0.0000000000000000080773

1.000

In printed literature, it is common practice to use scientific notation only when z (the power of 10) is fairly large or small. If $-2 \leq z \leq 2$, numbers are written out in full as a rule, and the power of 10 is not shown. If $z = -3$ or $z = 3$, numbers are sometimes written out in full and are sometimes depicted in scientific notation. If $z \leq -4$ or $z \leq 4$, values are expressed in scientific notation as a rule. Calculators set to display quantities in scientific notation will usually show the power of 10 for all numbers, even those for which the power of 10 is zero.

Addition and subtraction of numbers is best done by writing numbers out in full, if possible. Thus, for example,

$$(3.045 \times 10^2) + (6.853 \times 10^3) = 304.5 + 6{,}853 = 7157.5$$
$$= 7.1575 \times 10^3$$

When numbers are multiplied or divided in scientific notation, the decimal numbers (to the left of the multiplication symbol) are multiplied or divided by each other. Then the powers of 10 are added (for multiplication) or subtracted (for division). Finally, the product or quotient is reduced to standard form. An example is

$$(3.045 \times 10^2)(6.853 \times 10^3) = 20.867385 \times 10^5 = 2.0867385 \times 10^6$$

Significant Figures

In scientific notation, the term *significant figures* refers to the number of numerals in the decimal portion of an expression that can be relied upon to portray the quantity to a known degree of accuracy. For example, 3.83×10^{-25} has three significant digits, whereas 3.83018×10^{-25} has six significant digits and portrays the quantity to a greater degree of accuracy.

Truncation

The process of *truncation* deletes all the numerals to the right of a certain point in the decimal part of an expression.

Many, if not most, electronic calculators use this process to fit numbers within their displays. For example, the number 3.830175692803 can be shortened in steps as follows:

3.830175692803

3.83017569280

3.8301756928

3.830175692

3.83017569

3.8301756

3.830175

3.83017

3.83

3.8

3

Rounding

Rounding is a more accurate, and preferred, method of rendering numbers in shortened form. In this process, when a given digit (call it r) is deleted at the right-hand extreme of an expression, the digit q to its left (which becomes the new r after the old r is deleted) is not changed if $0 \leq r \leq 4$. If $5 \leq r \leq 9$, q is increased by 1 ("rounded up"). Some electronic calculators use rounding rather than truncation. If rounding is used, the number 3.830175692803 can be shortened in steps as follows:

3.830175692803

3.83017569280

3.8301756928

3.830175693

3.83017569

3.8301757

3.830176

3.83018

3.8302

3.830

3.83

3.8

4

In calculations

When calculations are performed using scientific notation, the number of significant figures in the result cannot be greater than the number of significant figures in the shortest expression in the calculation.

In the foregoing example showing addition, the sum, 7.1575×10^3, must be cut down to four significant figures because the addends have only four significant figures. If the resultant is truncated, it becomes 7.157×10^3. If rounded, it becomes 7.158×10^3.

In the foregoing example showing multiplication, the resultant, 2.0867385×10^6, must be cut down to four significant figures because the multiplicands have only four significant figures. If the resultant is truncated, it becomes 2.086×10^6. If rounded, it becomes 2.087×10^6.

"Downsizing" of resultants is best done at the termination of a calculation process, if that process involves more than one computation.

Theorems in Algebra

Some common theorems, also called rules or laws, in algebra are depicted in Table 23.4. These theorems apply to all real numbers, with one exception: When a variable appears as the denominator of a quotient (for example $1/x$), the expression is undefined for $x = 0$.

TABLE 23.4 **Common Theorems in Algebra**

These theorems apply to all real numbers as long as the denominator is nonzero.

Equation	Description
$x + 0 = x$	Additive identity
$x \cdot 1 = x$	Multiplicative identity
$x \cdot 0 = 0$	Multiplication by zero
$-(-x) = x$	Double negation
$x + (-x) = 0$	Additive inverse
$x(1/x) = 1$	Multiplicative inverse
$1(1/x) = x$	Reciprocal-of-a-reciprocal rule
$x + y = y + x$	Commutative law of addition
$xy = yx$	Commutative law of multiplication
$x + (y + z) = (x + y) + z$	Associative law of addition
$x(yz) = (xy)z$	Associative law of multiplication
$x(y + z) = xy + xz$	Distributive law
$(w + x)(y + z) = wy + wz + xy + xz$	Product of sums
$w/x = y/z \rightarrow wz = xy$	Cross-multiplication rule
$1/(xy) = (1/x)(1/y)$	Reciprocal of a product
$1/(x/y) = y/x$	Reciprocal of a quotient

Coordinate Systems

Relations and functions are commonly plotted in coordinate systems. These schemes show characteristics of devices and phenomena, such as antenna radiation patterns, waveforms, and spectral displays.

Cartesian plane

The most common two-dimensional coordinate system is the *cartesian plane* (Fig. 23.1), also called *rectangular coordinates* or the *xy-plane*. The independent variable is plotted along the x axis, or abscissa; the dependent variable is plotted along the y axis, or ordinate. The scales of the abscissa and

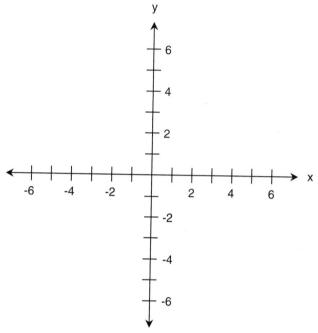

Figure 23.1 The cartesian or rectangular coordinate plane, also called the *xy*-plane.

ordinate are normally linear, although the divisions need not represent the same increments. Variations of this scheme include the *semilog graph,* in which the ordinate scale is logarithmic, and the *log-log graph,* in which both scales are logarithmic.

Polar coordinate plane

Another two-dimensional system is the *polar coordinate plane.* The independent variable is plotted as the radius r, and the dependent variable is plotted as an angle θ. Figure 23.2A shows the polar system used in mathematics and physical sciences; θ is in radians and is plotted counter-clockwise from the ray extending to the right ("east"). Figure 23.2B shows the polar system employed in wireless

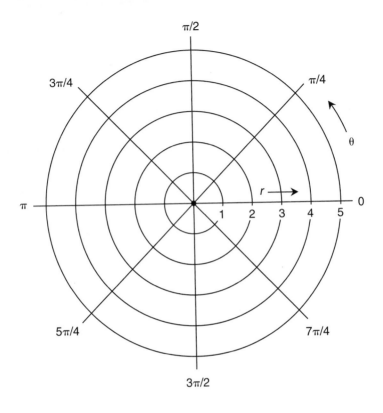

A

Figure 23.2 (A) The polar plane for mathematics and physical sciences.

communications, navigation, and location applications; θ is in degrees and is plotted clockwise from the ray extending upward ("north"). The angular scale is always linear in any polar system. The radial scale is linear in most polar graphs, but in some cases it is logarithmic.

Latitude and longitude

Latitude and *longitude* angles uniquely define the positions of points on the surface of a sphere or in the sky. The scheme for

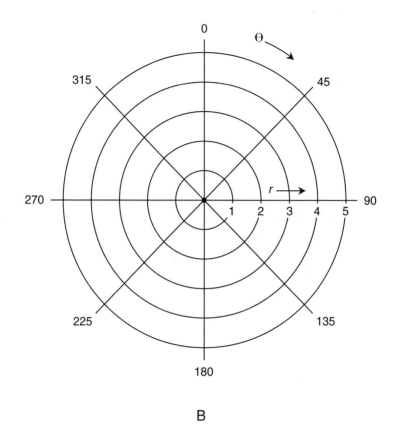

Figure 23.2 (*Continued*) (B) The polar plane for wireless communications, location, and navigation.

geographic locations on the earth is illustrated in Fig. 23.3A. The *polar axis* connects two specified points at antipodes on the sphere. These points are assigned latitude $\theta = 90°$ (north pole) and $\theta = -90°$ (south pole). The *equatorial axis* runs outward from the center of the sphere at a 90° angle to the polar axis. It is assigned longitude $\phi = 0°$. Latitude θ is measured positively (north) and negatively (south) relative to the *plane of the equator*. Longitude ϕ is measured counterclockwise (east) and clockwise (west) relative to the equatorial axis. The angles are restricted as follows:

$$-90° \leq \theta \leq 90°$$

$$-180° \leq \phi \leq 180°$$

On the earth's surface, the half-circle connecting the 0° longitude line with the poles passes through Greenwich, England, and is known as the *Greenwich meridian*. Longitude angles are defined with respect to this meridian.

Celestial coordinates

Celestial latitude and *celestial longitude* are extensions of the earth's latitude and longitude into the heavens. Figure 23.3A applies to this system. An object whose celestial latitude and longitude coordinates are (θ, ϕ) appears at the zenith in the sky from the point on the earth's surface whose latitude and longitude coordinates are (θ, ϕ).

Declination and *right ascension* define the positions of objects in the sky relative to the stars. Figure 23.3B applies to this system. Declination (θ) is identical to celestial latitude. Right ascension (ϕ) is measured eastward from the *vernal equinox* (the position of the sun in the heavens at the moment spring begins in the northern hemisphere). The angles are restricted as follows:

$$-90° \leq \theta \leq 90°$$

$$0° \leq \phi < 360°$$

Cartesian three-space

An extension of rectangular coordinates into three dimensions is *cartesian three-space* (Fig. 23.4), also called *xyz-space*. Independent variables are usually plotted along the x and y axes; the dependent variable is plotted along the z axis. The scales are normally linear, although the divisions need not represent the same increments. Variations of these schemes can employ logarithmic graduations for one, two, or all three scales.

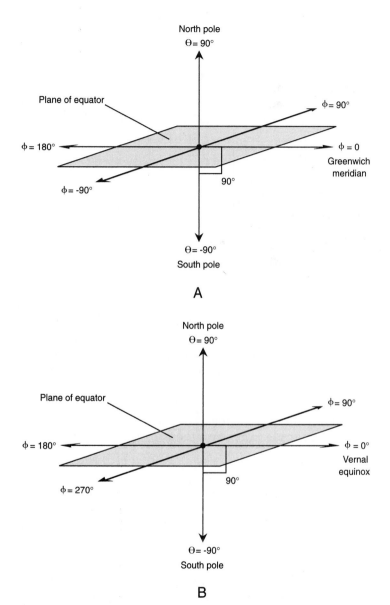

Figure 23.3 (A) Scheme for latitude and longitude; (B) scheme for declination and right ascension.

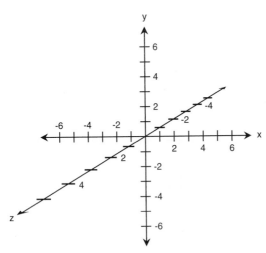

Figure 23.4   Cartesian three-space, also called *xyz*-space.

Cylindrical coordinates

Figure 23.5 shows a system of *cylindrical coordinates* for specifying the positions of points in three-space. Given a set of cartesian coordinates or *xyz*-space, an angle θ is defined in the *xy* plane and measured in radians counterclockwise from the *x* axis. Given a point *P* in space, consider its projection *P′* onto the *xy*-plane. The position of *P* is defined by the ordered triple (θ, r, z) such that

θ = angle between *P′* and the *x* axis in the *xy*-plane

r = distance (radius) from *P* to the origin

z = distance (altitude) of *P* above the *xy* plane

Spherical coordinates

Figure 23.6 shows a system of *spherical coordinates* for defining points in space. This scheme is identical to the system for declination and right ascension, with the addition of a *radius* vector *r* representing the distance of point *P*

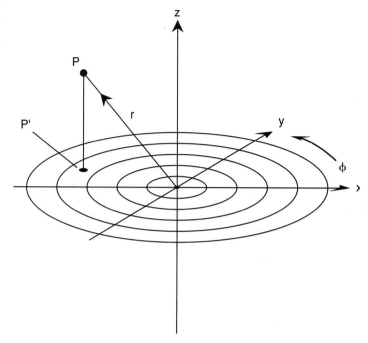

Figure 23.5 Cylindrical coordinates for defining points in three-space.

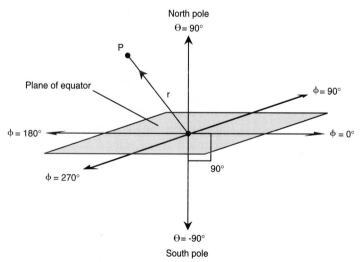

Figure 23.6 Spherical coordinates for defining points in three-space.

from the origin. The location of a point P is defined by the ordered triple (θ,ϕ,r) such that

θ = declination of P

ϕ = right ascension of P

r = distance (radius) from P to the origin

In this example, angles are specified in degrees; alternatively they can be expressed in radians. There are several variations of this system, all of which are commonly called *spherical coordinates*.

Trigonometry

There are three basic *trigonometric functions: sine, cosine,* and *tangent*. These functions typically apply to angles θ in radians such that $0 \le \theta \le 2\pi$. In formulas and equations, these functions are abbreviated $\sin \theta$, $\cos \theta$, and $\tan \theta$.

Basic functions

Consider a circle in the cartesian plane with the following equation:

$$x^2 + y^2 = 1$$

This is called the *unit circle* because its radius is one unit, and it is centered at the origin $(0,0)$. Let θ be an angle whose apex is at the origin and is measured counterclockwise from the abscissa (x axis). Suppose this angle corresponds to a ray that intersects the unit circle at some point $P = (x_0, y_0)$. Then,

$$y_0 = \sin \theta$$
$$x_0 = \cos \theta$$
$$\frac{y_0}{x_0} = \tan \theta$$

Secondary functions

Three more trigonometric functions are derived from those defined above. They are *cosecant, secant,* and *cotangent.* In formulas and equations, these functions are abbreviated as csc θ, sec θ, and cot θ. They are defined as follows:

$$\csc \theta = \frac{1}{\sin \theta} = \frac{1}{y_0}$$

$$\sec \theta = \frac{1}{\cos \theta} = \frac{1}{x_0}$$

$$\cot \theta = \frac{1}{\tan \theta} = \frac{x_0}{y_0}$$

Trigonometric identities

Theorems in trigonometry, known as *trigonometric identities,* have been demonstrated. Some common identities are listed in Table 23.5. All angles θ are in radians. Note the following standard abbreviations:

$$\sin^2 \theta = (\sin \theta)^2 = (\sin \theta)(\sin \theta)$$

$$\cos^2 \theta = (\cos \theta)^2 = (\cos \theta)(\cos \theta)$$

$$\tan^2 \theta = (\tan \theta)^2 = (\tan \theta)(\tan \theta)$$

$$\csc^2 \theta = (\csc \theta)^2 = (\csc \theta)(\csc \theta)$$

$$\sec^2 \theta = (\sec \theta)^2 = (\sec \theta)(\sec \theta)$$

$$\cot^2 \theta = (\cot \theta)^2 = (\cot \theta)(\cot \theta)$$

Logarithms

A *logarithm* is an exponent to which a constant is raised to obtain a given number. Suppose the following relationship exists among three real numbers a, m, and x:

$$a^m = x$$

TABLE 23.5   Common Trigonometric Identities

These theorems apply to all real numbers as long as the denominator is nonzero.

Equation	Description
$\sin^2 \theta + \cos^2 \theta = 1$	Theorem of Pythagoras for sine and cosine functions
$\sec^2 \theta - \tan^2 \theta = 1$	Theorem of Pythagoras for secant and tangent functions
$\sin -\theta = -\sin \theta$	Sines of negative angles
$\cos -\theta = \cos \theta$	Cosines of negative angles
$\tan -\theta = -\tan \theta$	Tangents of negative angles
$\sin (\theta + 2\pi) = \sin \theta$	Periodicity of sine function
$\cos (\theta + 2\pi) = \cos \theta$	Periodicity of cosine function
$\tan (\theta + 2\pi) = \tan \theta$	Periodicity of tangent function
$\sin 2\theta = 2 \sin \theta \cos \theta$	Double-angle formula for sine function
$\cos 2\theta =$ $1 - (2 \sin^2 \theta) = (2 \cos^2 \theta) - 1$	Double-angle formula for cosine function
$\tan 2\theta$ $= (2 \tan \theta) / (1 - \tan^2 \theta)$	Double-angle formula for tangent function
$\sin (\theta / 2)$ $= \pm [(1 - \cos \theta) / 2]^{1/2}$	Half-angle formula for sine function
$\cos (\theta / 2)$ $= \pm [(1 + \cos \theta) / 2]^{1/2}$	Half-angle formula for cosine function
$\tan (\theta / 2)$ $= (\sin \theta) / (1 + \cos \theta)$	Half-angle formula for tangent function
$\sin (\theta + \phi)$ $= \sin \theta \cos \phi + \cos \theta \sin \phi$	Sum formula for sine function
$\cos (\theta + \phi)$ $= \cos \theta \cos \phi - \sin \theta \sin \phi$	Sum formula for cosine function
$\tan (\theta + \phi)$ $= (\tan \theta + \tan \phi) /$ $(1 - \tan \theta \tan \phi)$	Sum formula for tangent function
$\sin (\theta - \phi)$ $= \sin \theta \cos \phi - \cos \theta \sin \phi$	Difference formula for sine function
$\cos (\theta - \phi)$ $= \cos \theta \cos \phi + \sin \theta \sin \phi$	Difference formula for cosine function
$\tan (\theta - \phi) = (\tan \theta - \tan \phi) /$ $(1 + \tan \theta \tan \phi)$	Difference formula for tangent function

TABLE 23.5 Common Trigonometric Identities

These theorems apply to all real numbers as long as the denominator is nonzero.

Equation	Description
$\sin\theta = \cos(\pi/2 - \theta)$	Complementary-angle rule for sine and cosine functions
$\cos\theta = \sin(\pi/2 - \theta)$	Complementary-angle rule for sine and cosine functions
$\tan\theta = \cot(\pi/2 - \theta)$	Complementary-angle rule for tangent and cotangent functions
$\cot\theta = \tan(\pi/2 - \theta)$	Complementary-angle rule for tangent and cotangent functions
$\sec\theta = \csc(\pi/2 - \theta)$	Complementary-angle rule for secant and cosecant functions
$\csc\theta = \sec(\pi/2 - \theta)$	Complementary-angle rule for secant and cosecant functions
$\sin\theta = \sin(\pi - \theta)$	Supplementary-angle rule for sine function
$\cos\theta = -\cos(\pi - \theta)$	Supplementary-angle rule for cosine function
$\tan\theta = \tan(\pi - \theta)$	Supplementary-angle rule for tangent function

Then m is the logarithm of x in base a. This expression is written

$$m = \log_a x$$

The two most common logarithm bases are $a = 10$ and $a = e \approx 2.71828$.

Base-10 logarithms are also called *common logarithms*. In equations, common logarithms are denoted by writing log. For example,

$$\log 10 = 1.000$$

Base-e logarithms are also known as *natural logarithms* or *napierian logarithms*. In equations, the natural-logarithm function is usually abbreviated ln, although it is sometimes denoted as $\log_e$. Hence, for example,

$$\ln 2.71828 = \log_e 2.71828 \approx 1.00000$$

Table 23.6 lists some formulas and relations involving logarithms.

TABLE 23.6   Common Logarithmic Identities

These theorems apply to all real numbers as long as the denominator is nonzero.

Equation	Description
$\log x = \ln x / \ln 10 \approx 0.434 \ln x$	Conversion of natural (base-e) logarithm to common (base-10) logarithm
$\ln x = \log x / \log e$ $\approx 2.303 \log x$	Conversion of common logarithm to natural logarithm
$\log xy = \log x + \log y$	Common logarithm of a product
$\ln xy = \ln x + \ln y$	Natural logarithm of a product
$\log (x/y) = -\log (y/x)$ $= \log x - \log y$	Common logarithm of a quotient
$\ln (x/y) = -\ln (y/x)$ $= \ln x - \ln y$	Natural logarithm of a quotient
$\log x^y = y \log x$	Common logarithm of a power
$\ln x^y = y \ln x$	Natural logarithm of a power
$\log 1/x = -\log x$	Common logarithm of a reciprocal
$\ln 1/x = -\ln x$	Natural logarithm of a reciprocal
$\log (x)^{1/y} = (\log x) / y$	Common logarithm of a root
$\ln (x)^{1/y} = (\ln x) / y$	Natural logarithm of a root
$\log 10^x = x$	Common logarithm of base-10 exponential function
$\ln e^x = x$	Natural logarithm of base-e exponential function

24

Chemical Data

This chapter contains information about chemical substances and their applications in electronics.

Atoms and Molecules

All matter is made up of particles known as *atoms*. Atoms combine to form *molecules*. A molecule with two or more different elemental atoms is a *compound*.

Atomic structure

The *nucleus* of an atom gives a chemical element unique characteristics. The nucleus can contain *protons* and *neutrons*. Protons carry a positive electric charge; neutrons have zero (neutral) charge. The number of protons in the nucleus of an atom is the *atomic number*. For a given element, the number of neutrons can vary. Differing numbers of neutrons result in various *isotopes* for a given element. The *atomic weight* of an element is approximately equal to the sum of the number of protons and neutrons in its atomic nucleus.

In most atoms, the nucleus is orbited by low-mass *electrons* having negative charge. At any instant in time, an electron has a 50 percent probability of existing within a closed region called an *electron shell.* Normally, shells are spheres of various radii, all centered at the nucleus. Each shell represents a certain electron energy. The greater the radius of the shell, the more energy the electron has. If an electron attains enough energy, it will escape the nucleus. It might then become a *shared electron,* wandering between the original nucleus and one or more other nuclei. Or it might become a *charge carrier,* "jumping" from one atom to another in a more or less constant direction. In the extreme case, it might become a *free electron,* not associated with any nuclei.

Ions

If an atom has more or less electrons than neutrons, that atom acquires an electrical charge. A shortage of electrons results in a positive charge; an excess of electrons gives a negative charge. A charged atom is an *ion.* When a substance contains numerous ions, the material is said to be *ionized.* Ionized materials generally conduct electric current well, even if the substance is normally not a good conductor. This is true of gases, liquids, and solids.

Compounds

The atoms of two or more different elements can join, sharing electrons. When this occurs, the result is a chemical *compound.* Compounds often behave much differently than the elements that make them up. At room temperature and pressure, both hydrogen and oxygen are gases, but water under the same conditions is a liquid. If it gets a few degrees colder, water turns solid at standard pressure. If it is sufficiently heated, water becomes a colorless, odorless gas, like its elemental constituents. Another example of a compound is iron oxide (rust). This forms when iron combines with oxygen. Iron is a dull gray solid and oxygen is a gas, but rust is a brownish powder unlike either of the elements from which it is formed.

Molecules

The natural form of an element is also known as its molecule. Oxygen tends to occur in pairs most of the time in the atmosphere. Thus, an oxygen molecule is denoted by the symbol O_2. Sometimes there are three atoms of oxygen in a molecule. This is ozone (O_3). Compounds are always molecules. For example, water is H_2O; each molecule has two hydrogen atoms and one oxygen atom. As another example, carbon dioxide is CO_2; each molecule has one carbon atom and two oxygen atoms.

Conductors

An electrical *conductor* is a substance in which the electrons can move easily among the atoms. The best conductor at room temperature is pure elemental silver. Copper and aluminum are also excellent conductors. Iron, steel, and various other metals are fair to good conductors of electricity. Some liquids are good electrical conductors; mercury is an example. Salt water is a fair conductor. Gases are, in general, poor conductors, except when ionized.

Insulators

An electrical *insulator* is a substance that is an exceptionally poor conductor. Most gases are good insulators. Glass, dry wood, paper, and plastics are other examples. Pure water is a good electrical insulator, although it conducts some current if it contains any conductive impurity. Metal oxides can be good insulators, even though the metal in pure form is a good conductor.

When an insulating material is deliberately used to separate electric charge poles, the substance is called a *dielectric*. Porcelain and glass are employed in electrical systems to keep short circuits from occurring. Metal oxides, mica, various ceramics, and air are commonly used as dielectric materials in capacitors.

Semiconductors

In a *semiconductor,* electrons flow, but not as well as in a conductor. Common semiconductors include silicon, selenium, indium, germanium, or gallium that has been *doped* by the addition of impurities. Electrical conduction in these materials is always a result of the motion of electrons. But sometimes the conduction is described in terms of *holes* (electron vacancies within atoms) rather than electrons. Holes carry positive charge, equal in magnitude to the charge on an electron. In a semiconductor, the more abundant type of charge carrier is the *majority carrier.* The less abundant one is the *minority carrier.* Semiconductors are used in diodes, transistors, and integrated circuits.

The Elements

Table 24.1 is a list of the known chemical elements in alphabetical order, including chemical symbols and atomic numbers from 1 through 112. The following lists the elements in order by atomic number.

1. *Hydrogen.* A gaseous element. The lightest and most abundant element in the universe. Used in the semiconductor manufacturing process and in glow-discharge lamps. Highly flammable; combines readily with oxygen to form water (H_2O).

2. *Helium.* A gaseous element. The second lightest and second most abundant element in the universe. Product of hydrogen fusion. Used in helium-neon (He-Ne) lasers, which produce visible red coherent light.

3. *Lithium.* An element of the alkali-metal group, and the lightest elemental metal. Used in electrochemical cells known for long life in low-current applications.

4. *Beryllium.* An elemental metal. Found in various dielectrics, alloys, and phosphors. The oxide of this element is used as an insulator.

5. *Boron.* Can exist either as a powder or as a black, hard metal. The metal is a poor electrical conductor. Used in nuclear reactors, the manufacture of electronic semiconductor devices, and various industrial applications.

6. *Carbon.* Exists in two forms: *graphite* (a black powder), which is common, and *diamond* (a clear solid), which is rare. Used in electrochemical cells, thermocouples, and noninductive resistors.

TABLE 24.1 **The Chemical Elements in Alphabetical Order by Name, Including Chemical Symbols and Atomic Numbers 1 through 112**

Element name	Chemical symbol	Atomic number
Actinium	Ac	89
Aluminum	Al	13
Americium	Am	95
Antimony	Sb	51
Argon	Ar	18
Arsenic	As	33
Astatine	At	85
Barium	Ba	56
Berkelium	Bk	97
Beryllium	Be	4
Bismuth	Bi	83
Bohrium	Bh	107
Boron	B	5
Bromine	Br	35
Cadmium	Cd	48
Calcium	Ca	20
Californium	Cf	98
Carbon	C	6
Cerium	Ce	58
Cesium	Cs	55
Chlorine	Cl	17
Chromium	Cr	24
Cobalt	Co	27
Copper	Cu	29
Curium	Cm	96
Dubnium	Db	105
Dysprosium	Dy	66
Einsteinium	Es	99
Erbium	Er	68
Europium	Eu	63
Fermium	Fm	100
Fluorine	F	9
Francium	Fr	87
Gadolinium	Gd	64
Gallium	Ga	31
Germanium	Ge	32
Gold	Au	79
Hafnium	Hf	72
Hassium	Hs	108
Helium	He	2
Holmium	Ho	67
Hydrogen	H	1
Indium	In	49
Iodine	I	53
Iridium	Ir	77
Iron	Fe	26

TABLE 24.1   **The Chemical Elements in Alphabetical Order by Name, Including Chemical Symbols and Atomic Numbers 1 through 112** (*Continued*)

Element name	Chemical symbol	Atomic number
Krypton	Kr	36
Lanthanum	La	57
Lawrencium	Lr or Lw	103
Lead	Pb	82
Lithium	Li	3
Lutetium	Lu	71
Magnesium	Mg	12
Manganese	Mn	25
Meitnerium	Mt	109
Mendelevium	Md	101
Mercury	Hg	80
Molybdenum	Mo	42
Neodymium	Nd	60
Neon	Ne	10
Neptunium	Np	93
Nickel	Ni	28
Niobium	Nb	41
Nitrogen	N	7
Nobelium	No	102
Osmium	Os	76
Oxygen	O	8
Palladium	Pd	46
Phosphorus	P	15
Platinum	Pt	78
Plutonium	Pu	94
Polonium	Po	84
Potassium	K	19
Praseodymium	Pr	59
Promethium	Pm	61
Protactinium	Pa	91
Radium	Ra	88
Radon	Rn	86
Rhenium	Re	75
Rhodium	Rh	45
Rubidium	Rb	37
Ruthenium	Ru	44
Rutherfordium	Rf	104
Samarium	Sm	62
Scandium	Sc	21
Seaborgium	Sg	106
Selenium	Se	34
Silicon	Si	14
Silver	Ag	47
Sodium	Na	11
Strontium	Sr	38

TABLE 24.1   The Chemical Elements in Alphabetical Order by Name, Including Chemical Symbols and Atomic Numbers 1 through 112 (*Continued*)

Element name	Chemical symbol	Atomic number
Sulfur	S	16
Tantalum	Ta	73
Technetium	Tc	43
Tellurium	Te	52
Terbium	Tb	65
Thallium	Tl	81
Thorium	Th	90
Thulium	Tm	69
Tin	Sn	50
Titanium	Ti	22
Tungsten	W	74
Ununbium	Uub	112
Ununnilium	Uun	110
Unununium	Uuu	111
Uranium	U	92
Vanadium	V	23
Xenon	Xe	54
Ytterbium	Yb	70
Yttrium	Y	39
Zinc	Zn	30
Zirconium	Zr	40

7. *Nitrogen.* The most abundant component of the earth's atmosphere (approximately 78 percent at the surface). Reacts to some extent with certain combinations of other elements.

8. *Oxygen.* The second most abundant component of the earth's atmosphere (approximately 21 percent at the surface). Combines readily with many other elements, particularly metals. When three oxygen atoms form a molecule, the result is *ozone.*

9. *Fluorine.* A gaseous element of the halogen family. Reacts readily with many other elements.

10. *Neon.* An inert gas present in trace amounts in the atmosphere. Used in specialized low-frequency oscillators, flip-flops, voltage regulators, indicator lamps, and displays. Also used in helium-neon (He-Ne) lasers, which produce visible red coherent light.

11. *Sodium.* A metallic element of the alkali-metal group. Used in gas-discharge lamps. Has a characteristic candle-flame-colored glow when it fluoresces. Reacts readily with various other elements.

12. *Magnesium.* A metallic element. Compounds of this element are used as phosphors for some types of CRTs. Also mixed with aluminum in antenna hardware. Reacts readily with various other elements.

13. *Aluminum.* A metallic element and a good electrical conductor. Used as chassis, in antenna hardware, and in electrical wiring. Serves as the electrode material in electrolytic capacitors. Can be doped to form a semiconductor or oxidized to form a dielectric.

14. *Silicon.* A metalloid abundant in the earth's crust. Commonly doped to form a semiconductor. Can be oxidized to form a dielectric. This element and various compounds of it are used in diodes, transistors, and ICs.

15. *Phosphorus.* A nonmetallic element of the nitrogen family. Used as a dopant in semiconductor manufacture and as an alloy constituent in some electrical and electronic components.

16. *Sulfur.* A nonmetallic element. Reacts with some other elements. Compounds of sulfur are used in rechargeable electrochemical cells and in power transformers.

17. *Chlorine.* A gas at room temperature and a member of the halogen family. Reacts readily with various other elements. Potentially dangerous; displaces oxygen in the blood if inhaled.

18. *Argon.* An inert gas at room temperature; present in small amounts in the atmosphere. Used in some types of lasers and glow lamps.

19. *Potassium.* A member of the alkali metal group. Compounds of this element are used in CRT phosphor coatings, in electroplating, and as ferroelectric substances.

20. *Calcium.* A metallic element of the alkaline-earth group. Compounds of this element are used as phosphor coatings in CRTs. Various display colors can be obtained by using different compounds of calcium.

21. *Scandium.* In pure form, it is a soft metal.

22. *Titanium.* Used in the construction of large antenna support towers. Certain compounds, especially oxides, are used as dielectric materials.

23. *Vanadium.* A metallic element. Used in the manufacture of specialized industrial alloys.

24. *Chromium.* In pure form, it is a shiny, silver-colored metal. Used as a plating for metals to improve resistance to corrosion. Compounds of this element are used in specialized recording tapes and in thermocouple devices.

25. *Manganese.* A metallic element. An alloy of manganese is used in the manufacture of permanent magnets.

26. *Iron.* In pure form, it is a dull gray metal. Used in magnetic circuits and transformer cores. Also used in thermocouples. An alloy of this element (steel) is used in wireless communications antenna structures.

27. *Cobalt.* A metallic element. Used in the manufacture of stainless steel. The isotope Co-60 is radioactive and has medical applications in radiology.

28. *Nickel.* Nickel compounds are used in rechargeable electrochemical cells and in the manufacture of certain semiconductor diodes. Alloys are used as resistance wire. Elemental nickel is employed in some electron tubes.

29. *Copper.* In pure form, it is a reddish metal. An excellent conductor of electricity and heat. Used in the manufacture of wires and cables. Oxides of this element are used in specialized diodes and photoelectric cells.

30. *Zinc.* In pure form, it is a dull gray metal. Used as the negative-electrode material in electrochemical cells and as a protective coating for metals, especially iron and steel. Certain zinc compounds are used as CRT phosphors.

31. *Gallium.* A semiconducting element. Forms a compound with arsenic (GaAs) that is used in low-noise, high-gain FETs and also in specialized diodes.

32. *Germanium.* A semiconducting metalloid. Used in specialized diodes, transistors, rectifiers, and photoelectric cells. In semiconductor components, germanium has been largely replaced by other materials in recent years.

33. *Arsenic.* A metalloid used as a dopant in the manufacture of semiconductors. Forms a compound with gallium (GaAs) that is used in low-noise, high-gain FETs and also in specialized diodes.

34. *Selenium.* A semiconducting element used in diodes, rectifiers, and photoelectric cells.

35. *Bromine.* A nonmetallic element of the halogen family. A reddish-brown liquid at room temperature. Reacts readily with various other elements.

36. *Krypton.* An inert gas present in trace amounts in the earth's atmosphere. Some common isotopes of this element are radioactive.

37. *Rubidium.* A metallic element. Reacts easily with oxygen and chlorine.

38. *Strontium.* A metallic element of the alkaline-earth group. Compounds of this element are used in the manufacture of specialized ceramic dielectrics.

39. *Yttrium.* A metallic element of the rare-earth-metals group. This element and compounds containing it are used in electrooptical devices, particularly lamps and lasers.

40. *Zirconium.* A metallic element. The oxide of this element is used as a dielectric at high temperatures. Zinc-beryllium-zirconium silicate is employed as a CRT phosphor.

41. *Niobium.* A metallic element chemically resembling tantalum. In industry, it is used in specialized welding processes.

42. *Molybdenum.* In its pure form, it is a silver-colored metal. Used in the grids and plates of certain vacuum tubes. Also used to harden steel.

43. *Technetium.* Formerly called *masurium.* In pure form it is a metal and occurs when the uranium atom is split by nuclear fission.

44. *Ruthenium.* A rare metallic element. Used in certain alloys.

45. *Rhodium.* A metallic element. Occurs in nature along with platinum. It is used in scientific work. In particular, it makes a good silvering for first-surface mirrors in optical devices and instruments.

46. *Palladium.* A metallic element of the platinum group. In nature, palladium is found with copper ore.

47. *Silver.* A metallic element. An excellent conductor of electricity and heat. Resists corrosion. Used in circuits where low resistance and/or high Q factor (selectivity) are mandatory. Also used for plating of electrical contacts.

48. *Cadmium.* A metallic element used with nickel in the manufacture of rechargeable electrochemical cells. Also employed as a protective plating. Highly toxic, similar to lead. The disposal process is regulated by law in many locales.

49. *Indium.* A metallic element used as a dopant in semiconductor processing. In nature, it is often found along with zinc. Certain compounds of this element can be used directly as semiconductors.

50. *Tin.* In pure form it is a gray metal. Mixed with lead to manufacture solder. Tin foil is used to form the plates of some fixed capacitors. Compounds of tin can be used to manufacture resistors. Tin plating can protect metals against corrosion.

51. *Antimony.* In pure form it is a blue-gray metal. Used as a dopant in the manufacture of N-type semiconductor material.

52. *Tellurium.* A rare metalloid element related to selenium. In nature it is found along with other metals such as copper. Used in the manufacture of semiconductor materials.

53. *Iodine.* A member of the halogen family. Radioactive isotopes of this element are used in medical radiology procedures.

54. *Xenon.* An inert gaseous element, present in trace amounts in the atmosphere. Used in some thyratrons, electric lamps, flash tubes, and lasers. The xenon flash tube produces brilliant, white visible output.

55. *Cesium.* A member of the alkali-metal group. The oscillations of cesium atoms have been employed as an atomic time standard. The element can be used as the light-sensitive material in phototubes and in arc lamps to produce IR output.

56. *Barium.* A metal of the alkaline-earth group. Various compounds of barium are used as dielectrics and ferroelectric materials. The oxides of barium and strontium are used as coatings of vacuum-tube cathodes to increase electron emission.

57. *Lanthanum.* An elemental metal of the rare-earth group. Used in various industrial applications.

58. *Cerium.* A metallic element of the rare-earth group. There are numerous isotopes, some radioactive. In its pure form, it is a shiny metal. Used in various industrial applications.

59. *Praseodymium.* A metallic element of the rare-earth group. Used in various industrial applications.

60. *Neodymium.* A metallic element of the rare-earth group. Used in a specialized low-to-medium-power laser along with yttrium/aluminum/garnet (YAG) crystal. The *neodymium-YAG laser* is employed in jobs where high precision is required.

61. *Promethium.* Formerly called *illinium.* An element of the rare-earth group, derived from the fission of

uranium, thorium, and plutonium. An isotope of this element is used in specialized photovoltaic cells and batteries.

62. *Samarium.* An element of the rare-earth group. Used in the manufacture of permanent magnets and in specialized alloys. Also employed in nuclear reactors.

63. *Europium.* An element of the rare-earth group. Used in color CRT displays.

64. *Gadolinium.* An element of the rare-earth group. Used in magnetic devices.

65. *Terbium.* An element of the rare-earth group. Used in CRT displays.

66. *Dysprosium.* A ferromagnetic element of the rare-earth group, derived from erbium and holmium. Used in nuclear reactors.

67. *Holmium.* An element of the rare-earth group. Forms ferromagnetic compounds. Used in nuclear reactors.

68. *Erbium.* An element of the rare-earth group. Used in the manufacture of ceramics.

69. *Thulium.* An element of the rare-earth group. Used in power supplies for x-ray-generating equipment.

70. *Ytterbium.* An element of the rare-earth group, derived from various ores including gadolinite and monazite. Used in laboratory experiments.

71. *Lutetium.* An element of the rare-earth group, silver-colored in its pure form.

72. *Hafnium.* A transition metal, silver-colored in its pure form. Readily emits electrons. Used in nuclear reactors.

73. *Tantalum.* An element of the vanadium family, gray in its pure form. Used in the manufacture of high-capacitance, close-tolerance electrolytic capacitors. Also used in the elements of vacuum tubes and in the manufacture of diodes and resistors.

74. *Tungsten.* A transition metal, silver-colored in pure form. Also called *wolfram.* Used in switch and relay

contacts, the filaments of electron tubes, and incandescent lamps. Also employed as the anodes of x-ray tubes.

75. *Rhenium.* A transition metal, silver-gray in pure form. Used in some thermocouples and in mass spectrography equipment.

76. *Osmium.* A transition metal of the platinum group, silver-colored in pure form. Used in lamp filaments and the bearings of precision analog electromechanical meters.

77. *Iridium.* A transition metal of the platinum group, white in pure form. Occurs naturally along with platinum.

78. *Platinum.* A transition metal, bright silver-white in pure form. Resists corrosion; used as a coating for electrodes and switch contacts in electronic and computer systems. Also used in specialized vacuum tubes and thermocouple-type meters.

79. *Gold.* A transition metal. Shiny, yellowish, and malleable in pure form. Resists corrosion; used as a coating for electrodes and switch contacts in electronic and computer systems. Also used in specialized semiconductor devices as electrodes or as a dopant.

80. *Mercury.* A transition metal, silver-colored and liquid at room temperature. Used in switches, relays, high-voltage rectifiers, high-vacuum pumps, lamps, barometers, thermometers, and electrochemical cells.

81. *Thallium.* A metallic element. Compounds of this element exhibit photoconductivity and are used in photoelectric cells at IR wavelengths.

82. *Lead.* A heavy, soft, malleable metal, dull gray in pure form. Exhibits relatively low melting temperature. Used in rechargeable cells and batteries and as fuse elements. Used with tin to make solder. Also employed as a shield against ionizing radiation.

83. *Bismuth.* A metallic element, white in color. Exhibits magnetoresistive properties. Used in fuses, thermocouples, and thermocouple-type meters.

84. *Polonium.* A metalloid produced from the decay of radium. Naturally found in pitchblende ore.

85. *Astatine.* Formerly called *alabamine*. A halogen produced from radioactive decay.

86. *Radon.* A radioactive, colorless, noble gas that results from the disintegration of radium. Used in medical radiation therapy. Known to contribute to lung cancer if regularly inhaled over a long period of time.

87. *Francium.* A radioactive member of the alkali-metal group. Produced artificially through radioactive disintegration of actinium.

88. *Radium.* A rare radioactive metal of the alkaline earth group, silver-gray in pure form. Used in medical radiation therapy. Occurs naturally with uranium.

89. *Actinium.* A rare radioactive metal of the rare-earth group, silver-gray in pure form.

90. *Thorium.* A rare-earth element with a silvery color. Obtained from monazite and thorite. Used in specialized alloys and compounds and in the manufacture of photoelectric cells.

91. *Protactinium.* Formerly called *protoactinium*. A rare-earth element resulting from the fission of uranium, plutonium, and thorium.

92. *Uranium.* A rare-earth element occurring in carnotite and pitchblende ore. Used as a fuel for nuclear reactors.

93. *Neptunium.* A human-made element of the rare-earth group.

94. *Plutonium.* A human-made element of the rare-earth group. Used in nuclear reactors and in the manufacture of nuclear bombs.

95. *Americium.* A human-made element of the rare-earth group. Used in some high-tech smoke detectors.

96. *Curium.* A human-made element of the rare-earth group.

97. *Berkelium.* A human-made element of the rare-earth group.

98. *Californium.* A human-made element of the rare-earth group.

99. *Einsteinium.* A human-made element of the rare-earth group.

100. *Fermium.* A human-made element of the rare-earth group.

101. *Mendelevium.* A human-made element of the rare-earth group.

102. *Nobelium.* A human-made element of the rare-earth group.

103. *Lawrencium.* A human-made element of the rare-earth group.

104. *Rutherfordium.* Also called *unnilquadium* (Unq). A human-made transition metal.

105. *Dubnium.* Also called *unnilpentium* (Unp). A human-made transition metal.

106. *Seaborgium.* Also called *unnilhexium* (Unh). A human-made transition metal.

107. *Bohrium.* Also called *unnilseptium* (Uns). A human-made transition metal.

108. *Hassium.* Also called *unniloctium* (Uno). A human-made transition metal.

109. *Meitnerium.* Also called *unnilenium* (Une). A human-made transition metal.

110. *Ununnilium.* A human-made transition metal.

111. *Unununium.* A human-made transition metal.

112. *Ununbium.* A human-made transition metal obtained from the fusion of zinc and lead.

Compounds and Mixtures

The following is a list of some chemical combinations used in electronic components, devices, and systems.

Alnico. Trade name for an alloy used in strong permanent magnets. Contains aluminum, nickel, and cobalt. Sometimes also contains copper and/or titanium.

Alumel. Trade name for an alloy used in thermocouple devices. Composed of nickel (three parts) and aluminum (one part).

Alumina. Trade name for an aluminum-oxide ceramic used in electron tube insulators and as a substrate in the fabrication of thin-film circuits.

Aluminum antimonide. Formula, AlSb. A crystalline compound useful as a semiconductor dopant.

Barium-strontium oxides. The combined oxides of barium and strontium, employed as coatings of vacuum-tube cathodes to increase electron emission at relatively low temperatures.

Barium-strontium titanate. A compound of barium, strontium, oxygen, and titanium, used as a ceramic dielectric material. Exhibits ferroelectric properties and a high dielectric constant.

Barium titanate. Formula, $BaTiO_2$. A ceramic employed as the dielectric in capacitors. Has a high dielectric constant and some ferroelectric properties.

Beryllia. Formula, BeO. Trade name for beryllium oxide; used in various forms as an insulator and structural element (as in resistor cores).

Cadmium boratek. Formula, $(CdO + B_2O_3)$: Mn. Used as a phosphor coating in CRT screens; characteristic fluorescence is green or orange.

Cadmium selenide. A compound consisting of cadmium and selenium. Exhibits photoconductive properties. Useful as a semiconductor in photoelectric cells.

Cadmium silicate. Formula, $CdO + SiO_2$. Used as a phosphor coating in CRT screens; characteristic fluorescence is orange-yellow.

Cadmium sulfide. A compound consisting of cadmium and sulfur. Exhibits photoconductive properties. Useful as a semiconductor in photoelectric cells.

Cadmium tungstate. Formula, $CdO + WO_3$. Used as a phosphor coating in CRT screens; characteristic fluorescence is blue-white.

Calcium phosphate. Formula, $Ca_3(PO_4)_2$. Used as a phosphor coating in CRT screens; characteristic fluorescence is white.

Calcium silicate. Formula, $(CaO + SiO_2)$: Mn. Used as a phosphor coating in CRT screens; characteristic fluorescence ranges from orange to green.

Calcium tungstate. Formula, $CaWO_4$. Used as a phosphor coating in CRT screens; characteristic fluorescence is blue.

Chromel. Trade name for a nickel/chromium/iron alloy that is used in the manufacture of thermocouples.

Chromium dioxide. Formula, CrO_2. Used in the manufacture of specialized thermocouples and recording tape.

Constantan. Trade name for an alloy of copper and nickel, used in thermocouples and standard resistors.

Copper oxides and sulfides. Compounds with semiconducting properties, occasionally used in the manufacture of rectifiers, meters, modulators, and photocells. Largely replaced by silicon in recent years.

Ferrite. Trade name for a ferromagnetic material consisting of iron oxide and one or more other metals. Used as core material for inductors and switching elements. Also used in CRT deflection coils and in loopstick receiving antennas at very low, low, medium, and high radio frequencies.

Gallium arsenide. Formula, GaAs. A compound of gallium and arsenic, used as a semiconductor material in low-noise diodes, varactors, and FETs.

Gallium phosphide. A compound of gallium and phosphorus, used as a semiconductor material in LEDs.

Garnet. A mineral containing silicon and various other elements, forming a hard crystalline material. Mixed with aluminum and yttrium, garnet is used in solid-state lasers.

Germanium dioxide. Formula, GeO_2. A gray or white powder obtainable from various sources; it is reduced in an atmosphere of hydrogen or helium to yield elemental germanium.

Helium/neon. Abbreviation, He-Ne. A mixture of these two gases is used in low-cost lasers for various applications. The output is in the red part of the visible spectrum.

Indium antimonide. A combination of indium and antimony, used in the manufacture of semiconductor components.

Iron oxides. Compounds consisting of iron and oxygen. The most familiar example is common rust. Used in specialized rechargeable cells and batteries.

Lead peroxide. A compound used as a constituent of the positive electrodes in lead-acid electrochemical storage cells and batteries.

Magnesium fluoride. Used as a phosphor coating on the screens of long-persistence CRTs. The fluorescence is orange.

Magnesium silicate. Used as a phosphor coating on the screens of CRTs. Fluorescence is orange-red.

Magnesium tungstate. Used as a phosphor coating on the screens of CRTs. Fluorescence is blue-white.

Magnet steel. A high-retentivity alloy of chromium, cobalt, manganese, steel, and tungsten, employed in the manufacture of permanent magnets.

Manganese dioxide. Formula, MnO_2. Mixed with powdered carbon and used as a depolarizing agent in electrochemical dry cells.

Manganin. Trade name for a low-temperature-coefficient alloy used in making wire for precision resistors. Consists of 84 percent copper, 12 percent manganese, and 4 percent nickel.

Monel. Trade name for an alloy of nickel, copper, iron, manganese, and trace amounts of various other metals.

Mercuric iodide. Formula, HgI_2. A compound whose crystals are used as detectors in high-resolution gamma-ray spectroscopy.

Mercuric oxide. A compound used is the cathodes in electrochemical mercury cells and batteries.

Mercury cadmium telluride. Formula HgCdTe. An alloy used as a semiconductor in transistors, integrated circuits, and IR detectors.

Neodymium / yttrium / aluminum / garnet. Abbreviation, neodymium-YAG. A mixture used in low-power, solid-state lasers. Employed in medical applications and other jobs where high precision is required.

Nichrome. Trade name for a nickel-chromium alloy used in the form of a wire or strip for resistors and heater elements.

Nickel / cadmium. Abbreviation, NiCd or NICAD. A mixture used in rechargeable electrochemical cells and batteries.

Nickel hydroxide. A compound used in rechargeable electrochemical power supplies. Examples are nickel/cadmium (NiCd or NICAD) and nickel/metal hydride (NiMH) batteries, used in older notebook computers.

Nickel / iron. A mixture used in a specialized rechargeable electrochemical cell in which the active positive plate material consists of nickel hydroxide, the active negative plate material is powdered iron oxide mixed with cadmium, and the electrolyte is potassium hydroxide.

Nickel oxide. A compound of nickel and oxygen, used in specialized semiconductor components, particularly diodes.

Nickel silver. An alloy of copper, nickel, and zinc, sometimes used for making resistance wire. Also called *German silver.*

Platinum / tellurium. These two metals, when placed in direct contact, form a thermocouple used in specialized metering devices.

Potassium chloride. Formula, KCl. A compound used as a phosphor coating on the screen of long-persistence CRTs. Fluorescence is magenta or white.

Potassium cyanide. Formula, KCN. A highly toxic salt used as an electrolyte in electroplating.

Potassium hydroxide. A compound used in rechargeable electrochemical cells and batteries, along with various other compounds and mixtures. An example is the nickel/cadmium (NiCd or NICAD) cell.

Proustite. Trade name for crystalline silver arsenide trisulfide. Artificial crystals of this compound are used in tunable IR emitting devices.

Silicon carbide. Formula, SiC. A compound of silicon and carbon, used as a semiconductor, an abrasive material, and a refractory substance. In industrial applications, this compound is sometimes called by its trade name, *Carborundum.*

Silicon dioxide. Formula, SiO_2. Also called *silica.* Used in IR emitting devices. In the passivation of transistors and integrated circuits, a thin layer of silicon dioxide is grown on the surface of the wafer to protect the otherwise exposed junctions.

Silicon oxides. A mixture of silicon monoxide (SiO) and silicon dioxide (SiO_2) that exhibits dielectric properties. Used in the manufacture of metal-oxide-semiconductor (MOS) devices.

Silicon steel. A high-permeability, high-resistance steel containing 2 to 3 percent silicon. Used as core material in transformers and other electromagnetic devices.

Silver solder. A solder consisting of an alloy of silver, copper, and zinc. Has a comparatively high melting temperature.

Sodium iodide. A crystalline compound that sparkles when exposed to high-speed subatomic particles or radioactivity. Useful as a detector or counter of ionizing radiation.

Sodium silicate. Also called *water glass.* A compound used as a fireproofing agent and protective coating.

Steel. An alloy of iron, carbon, and other metals, used in the construction of antenna support towers, in permanent magnets and electromagnets, and as the core material for high-tensile-strength wire.

Sulfur hexafluoride. A gas employed as a coolant and insulant in some power transformers.

Sulfuric acid. Formula, H_2SO_4. An acid consisting of hydrogen, sulfur, and oxygen. Used in a dilute solution or paste as the electrolyte in rechargeable lead-acid cells and batteries.

Tantalum nitride. A compound used in the manufacture of specialized, close-tolerance, thin-film resistors.

Thallium oxysulfide. A compound of thallium, oxygen, and sulfur, used as the light-sensitive material in photoelectric cells.

Thorium oxide. A compound mixed with tungsten to increase electron emissivity in the filaments and cathodes of electron tubes.

Tin/lead. These two elements are commonly alloyed to make solder. Usually combined in a tin-to-lead ratio of 50:50 or 60:40.

Tin oxide. A combination of tin and oxygen, useful as resistive material in the manufacture of thin-film resistors.

Titanium dioxide. Formula, TiO_2. A compound consisting of titanium and oxygen. Useful as a dielectric material.

Yttrium/aluminum/garnet. Abbreviation, YAG. A crystalline mixture used along with various elements, such as neodymium, in low-power, solid-state lasers.

Yttrium/iron/garnet. Abbreviation, YIG. A crystalline mixture used in acoustic delay lines, parametric amplifiers, and filters.

Zinc aluminate. Either of two similar compounds used as phosphor coatings in CRT screens. One form glows blue; the other form glows red.

Zinc beryllium silicate. A compound used as a phosphor coating in CRT screens. Fluorescence is yellow.

Zinc beryllium zirconium silicate. A compound used as a phosphor coating in CRT screens. Fluorescence is white.

Zinc borate. A compound used as a phosphor coating in CRT screens. Fluorescence is yellow-orange.

Zinc cadmium sulfide. Either of two similar compounds used as phosphor coatings in CRT screens. One form glows blue; the other form glows red.

Zinc germanate. A compound used as a phosphor coating in CRT screens. Fluorescence is yellow-green.

Zinc magnesium fluoride. A compound used as a phosphor coating in CRT screens. Fluorescence is orange.

Zinc orthoscilicate. Also called by the trade name *Willemite.* A compound used as a phosphor coating in CRT screens. Fluorescence is yellow-green.

Zinc oxide. A compound used as a phosphor coating in CRT screens. Fluorescence is blue-green. Also used in the manufacture of certain electronic components, such as voltage-dependent resistors (varistors).

Zinc silicate. A compound used as a phosphor coating in CRT screens. Fluorescence is blue.

Zinc sulfide. A compound used as a phosphor coating in CRT screens. Fluorescence is blue-green or yellow-green.

Zirconia. Any of various compounds containing zirconium, especially its oxide (ZrO_2), valued for high-temperature dielectric properties.

Chapter

25

Electronics Abbreviations

The following is a list of abbreviations encountered in electronics and some related technological fields. For Q-signals, 10-code signals, and phonetic-alphabet representations, see Chap. 26.

In many cases where an abbreviation is all lowercase (e.g., adu or bw), the all-uppercase counterpart (e.g., ADU or BW) is usually acceptable as well. Both the all-lowercase and all-uppercase abbreviations are shown only when both are more or less equally found.

When some characters are lowercase and others are uppercase (e.g., aH or Ah), an abbreviation should be represented only in the exact manner shown.

An item listed in **boldface** is an acronym, truncated word, or truncated set of words. If such an expression is spoken aloud, it is customarily pronounced as if it were a word. If the pronunciation is not obvious, it is given in parentheses.

a   **Atto-**; area; acceleration; acre; anode; obsolete for ab-
A   Ampere; area; gain; anode
AAC   Automatic aperture control
ab-   Prefix for multiple of 10 in cgs system
abA   Abampere

abc	Automatic bass compensation; automatic bias control; automatic brightness control; automatic brightness compensation
abC	Abcoulomb
abF	Abfarad
abH	Abhenry
abS	Absiemens
ABS	Absolute value; absolute-value function
abV	Abvolt
abW	Abwatt
abWb	Abweber
ac, AC	Alternating current; aerodynamic center; attitude control; automatic computer
a/c	Aircraft; air conditioning
ACA	Automatic circuit analyzer
acc	Automatic chrominance control; automatic color compensation; acceleration
ACIA	Asynchronous communications interface adapter
acr	Audiocassette recorder; audiocassette recording
ACS	Automatic control system
ACU	Automatic calling unit
A/D, a/d, AD	Analog-to-digital; analog-to-digital converter
adc, ADC	Analog-to-digital converter
ADF	Automatic direction finder
ADI	Alternate digit inversion
adp	Automatic data processing
ADT	Atlantic Daylight Time
adu	Automatic dialing unit
aF	Attofarad (SI); obsolete for abfarad (cgs)
AF	Audio frequency
afc, AFC	Automatic frequency control; audio-frequency choke

AFPC	Automatic frequency/phase control
AFSK	Audio-frequency-shift keying
agc, AGC	Automatic gain control
Ah	Ampere-hour
aH	Attohenry (SI); obsolete for abhenry (cgs)
AI	Artificial intelligence
AIP	American Institute of Physics
Al	Chemical symbol for aluminum
alc, ALC	Automatic level control
ALU	Arithmetic and logic unit
Am	Chemical symbol for americium
AM	Amplitude modulation; amplitude modulator
AMI	Alternate mark inversion
AMNL	Amplitude-modulation noise level
amp	Ampere; amplifier
amp-hr	Ampere-hour
amu, AMU	Atomic mass unit
anl, ANL	Automatic noise limiter; automatic noise limiting
ANSI	American National Standards Institute
ant	Antenna
apc, APC	Automatic phase control; automatic picture control
APL	Average picture level
Ar	Chemical symbol for argon
ARRL	American Radio Relay League
ARS	Amateur Radio Service; amateur radio station
As	Chemical symbol for arsenic
ASA	American Standards Association
ASCII	American Standard Code for Information Interchange (pronounced asky)
ASRA	Automatic stereo recording amplifier
AST	Atlantic Standard Time
At	Chemical symbol for astatine

ATR	Antitransmit/receive
atto-	Prefix multiplier for 10^{-18}
ATV	Amateur television
Au	Chemical symbol for gold
AU	Astronomical unit
a/v	Audio/visual
aV	Attovolt (SI); obsolete for abvolt (cgs)
avc, AVC	Automatic volume control; automatic voltage control
AWG	American Wire Gauge
az-el	Azimuth-elevation
B	Base of bipolar transistor
B	Susceptance; magnetic flux density; battery; bass; bel; chemical symbol for boron
B	Susceptance; bit; bits; base of bipolar transistor; bass; barns
B&S	Brown and Sharp (American Wire Gauge)
Ba	Chemical symbol for barium
BA	Battery
BAL	Balance
BAT	Battery; batch file
BC	Broadcast
BCD	Binary-coded decimal
BCFSK	Binary-coded frequency-shift keying
BCI	Broadcast interference
BCL	Broadcast listener; broadcast listening
BCN	Beacon
BCO	Binary-coded octal
BCST	Broadcast
BDC	Binary decimal counter
Be	Chemical symbol for beryllium
beO	Chemical symbol for beryllium oxide
BeV	Billion electron volts

BFO	Beat-frequency oscillator
BG	Birmingham Wire Gauge
B-H	Flux density and magnetic force
Bi	Chemical symbol for bismuth
biMOS	Bipolar and metal-oxide semiconductor
bip	Binary image processor
BIPM	International Bureau of Weights and Measures
bit	Binary digit
BITE	Built-in test equipment
BK	Break
BO	Beat oscillator
bp	Bandpass
bpi	Bits per inch
bps	Bits per second
Br	Chemical symbol for bromine
Btu, BTU	British thermal unit
bw	Bandwidth; black-and-white
b&w, b/w	Black-and-white
BWA	Backward-wave amplifier
BWG	Birmingham Wire Gauge
BWO	Backward-wave oscillator
BX	Armored and insulated cable
c	**Centi-**; capacitance; speed of light in vacuum; calorie; curie; candle; collector of bipolar transistor
C	Capacitance; capacitor; collector of bipolar transistor; Celsius; coulomb; chemical symbol for carbon
Ca	Chemical symbol for calcium
CAD	Computer-aided design
cal	Calorie
CAL	Conversational algebraic language
CAM	Computer-aided manufacturing; content-addressable memory

cap	Capacitance; capacitor
CAT	Computerized axial tomography
CATV	Community-antenna television
C_B	Base capacitance in bipolar transistor
Cb	Chemical symbol for columbium
CB	Citizens band
C_C	Collector capacitance in bipolar transistor
cc	Cubic centimeter; cubic centimeters
CCA	Current-controlled amplifier
CCD	Charge-coupled device
CCIS	Common-channel interface signaling
CC, c/c	Closed circuit
CCTV	Closed-circuit television
ccw	Counterclockwise
cd	Candela
Cd	Chemical symbol for cadmium
cd/m^2	Candela per square meter
CDT	Central daylight time
C_E	Emitter capacitance in bipolar transistor
Ce	Chemical symbol for cerium
centi-	Prefix multiplier for 10^{-2}
Cf	Chemical symbol for californium
C_{GK}	Grid-cathode capacitance in vacuum tube
C_{GP}	Grid-plate capacitance in vacuum tube
cgs	Centimeter-gram-second
CHIL	Current-hogging injection logic
C_i, C_{in}	Input capacitance
Ci	Curie
circ	Circuit; circular
cis	On this side of
ckt	Circuit
Cl	Chemical symbol for chlorine
Cm	Chemical symbol for curium
c.m., cir mil	Circular mil

cm^2	Square centimeter
cm^3	Cubic centimeter
C$_{max}$	Maximum capacitance
C$_{min}$	Minimum capacitance
CML	Current-mode logic
CMOS	Complementary metal-oxide semiconductor (pronounced sea-moss)
CMR	Common-mode rejection
CMRR	Common-mode rejection ratio
CMYK	Cyan/magenta/yellow/black color model
C_o, C$_{out}$	Output capacitance
Co	Cobalt
COBOL	Common business-oriented language
CODEC	Coder/decoder; compressor/decompressor
COM	Computer output on microfilm
.com	Internet domain suffix for commercial Web site
COS	Complementary symmetry
cp	Candlepower; central processor
cps	Characters per second; cycles per second
CPU	Central processing unit
Cr	Chemical symbol for chromium
CRO	Cathode-ray oscilloscope
crt, CRT	Cathode-ray tube
C_S	Source capacitance in field-effect transistor; standard capacitance
Cs	Chemical symbol for cesium
CS	Complementary symmetry
CST	Central standard time
CTL	Complementary transistor logic
Cu	Chemical symbol for copper
cu ft	Cubic foot; cubic feet
cu in	Cubic inch; cubic inches
cur	Current

CW, cw	Continuous wave; continuous-wave emission; counterclockwise
d	**Deci-**; differential; distance; density; drain of FET; dissipation; day; degree; diameter; drive; depth
D	**Deca-**; **deka-**; chemical symbol for deuterium; drain of FET; displacement; flux density; dissipation factor; dissipation; determinant; diffusion constant; diameter
da	**Deca-**; **deka-**
DA, D/A, d/a	Digital-to-analog; digital-to-analog converter
DAC	Digital-to-analog converter
DACI	Direct adjacent-channel interference
DAGC	Delayed automatic gain control
DAM	Data-addressed memory
DART	Data-analysis recording tape
DAT	Diffused-alloy transistor
DAVC	Delayed automatic volume control; delayed automatic voltage control
dB	Decibel; decibels
DB	Diffused base; double break
dBa	Adjusted decibels
dBc	Decibels relative to carrier
dBd	Decibels relative to signal from main lobes of half-wave dipole antenna
DBD	Double-base diode
dBi	Decibels relative to signal from isotropic antenna
dBj	Signal level relative to 1 mV
dBk	Signal level relative to 1 kW
dBm	Decibels relative to 1 mW
DBM	Database management
dBm0	dBm relative to zero transmission level
dBmp	dBm with psophometric weighting

dBm0p	dBm0 with psophometric weighting
dBmV	Decibels relative to 1 mV
dBr	Decibels relative to zero transmission level
dBrap	Decibels above reference acoustic power
dBrn	Decibels above reference noise
dBrnc0	Noise in dBrnc relative to zero level
dBV	Decibels relative to 1 V
dBW	Decibels relative to 1 W
dBx	Decibels above reference coupling
dc, DC	Direct current; direct-coupled; direct conversion
D/CMOS	combined **DMOS** and **CMOS** (pronounced dee-slant-seamoss)
DCM	Digital capacitance meter
DCTL	Direct-coupled transistor logic
dcu	Decimal counting unit
dcv, DCV	DC volts; dc voltage
DDA	Digital differential analyzer
DDD	Direct distance dialing
DE	Decision element
deac	Deaccentuator
deca-, deka-	Prefix multiplier for 10
deci-	Prefix multiplier for 0.1
deg	Degree; degrees
dens	Density
df, DF	Direction finder; direction finding
dg	Decigram
dia	Diameter
DIIC	Dielectric-isolated integrated circuit
DIP	Dual in-line package
dj	Diffused junction
DMA	Direct memory access; direct memory addressing
DMM, dmm	Digital multimeter

DMOS	Double-diffused metal-oxide semiconductor (pronounced dee-moss)
DNL	Differential nonlinearity
DOS	Disk operating system
dwnconv	Down converter; down conversion
dp	Double-pole
DP	Double-pole; data processing
DPDT, dpdt	Double-pole, double-throw
DPM, dpm	Digital power meter; digital panel meter; disintegrations per minute
DPST	Double-pole, single-throw
dr	dram; drams
DRO	Digital readout
DSB	Double sideband
DSBSC	Double sideband, suppressed carrier
dsc	Double silk covered
DSP	Digital signal processing; digital signal processor; double silver plated
DSR	Dynamic spatial reconstructor
DSS	Direct station selection
DSSC	Double sideband, suppressed carrier
dt	Double-throw; differential of time
DT	Double-throw; data transmission
DTL	Diode-transistor logic
DTS	Data transmission system; digital telemetry system
DU	Duty cycle
dv	Differential of velocity
DVM	Digital voltmeter
DVOM	Digital volt-ohm-milliammeter (pronounced dee-vom)
DX	Long-distance communication; radio station in a foreign country; duplex
Dy	Chemical symbol for dysprosium
dyn	Dyne
dyna-	Power

e	Voltage; emitter of bipolar transistor; electron charge; natural logarithm base (approximately 2.71828); eccentricity; erg
E	Voltage; electric field strength; emitter of bipolar transistor; **exa-**; energy
E_0	Zero reference voltage
EAM	Electronic accounting machine
E_{avg}	Average voltage
E_b	Battery voltage
E_{BB}	Plate supply voltage in vacuum tube
EBI	Equivalent background input
EBR	Electron beam recording
EBS	Electron-bombarded semiconductor
ec	Enamel-covered
E_{CC}	Grid supply voltage in vacuum tube
ECDC	Electrochemical diffused collector
ECG	Electrocardiogram; electrocardiograph
ECL	Emitter-coupled logic
ECM	Electronic countermeasures
ECO	Electron-coupled oscillator
ECTL	Emitter-coupled transistor logic
EDD	Envelope-delay distortion
EDP	Electronic data processing
EDT	Eastern daylight time; ethylene diamine tartrate
EDU	Electronic display unit
.edu	Internet domain suffix for educational institution Web site
EDVAC	Electronic discrete variable automatic computer
EE	Electrical engineering; electrical engineer
EMO_{eff}	Effective voltage
EEG	Electroencephalogram; electroencephalograph
EEPROM	Electrically erasable programmable read-only memory (pronounced ee-eeprom or double-ee-prom)
EFL	Emitter-follower logic

E_G	Grid voltage in vacuum tube; gate voltage in FET; generator voltage
E_H	Heater voltage in vacuum tube
EHF	Extremely high frequency
EHV	Extra-high voltage
E_i, E_{in}	Input voltage
EIT	Engineer in training
E_K	Cathode voltage in vacuum tube
EKG	Electrocardiogram; electrocardiograph
EL	Electroluminescent
ELD	Edge-lighted display
ELSIE	Electronic letter-sorting and indicator equipment
E_m	Maximum voltage, maximum junction field
EM	Electromagnetic; electromagnetics; efficiency modulation; electromagnetic iron; electromagnetizer; electron microscope; exposure meter; electromotive
e/m_e	Ratio of electron charge to electron mass
E_{max}	Maximum voltage
EMC	Electromagnetic compatibility
EMF, emf	Electromotive force
ENIAC	Electronic Numerical Integrator and Calculator
ENIC	Voltage negative-impedance converter
E_o, E_{out}	Output voltage
EOF	End of file
EOL	End of line
EOLM	Electrooptical light modulator
EOR	End of run
EOS	Electrooptical system
EOT	End of tape
EOTS	Electro-optical tracking system
E_p	Plate voltage in vacuum tube; peak voltage
EP, ep	Extended play

E_{pk}	Peak voltage
$E_{pk\text{-}pk}$, $E_{p\text{-}p}$, E_{pp}	Peak-to-peak voltage
EPROM	Electrically programmable read-only memory
EPU	Electronic power unit; emergency power unit
Eq, eq	Equation; equivalent
equiv	Equivalent
E_R	Voltage drop across a resistance
Er	Chemical symbol for erbium
E_{rms}	Root-mean-square voltage
ERP	Effective radiated power
E_s	Screen voltage in vacuum tube
Es	Chemical symbol for einsteinium
ESG	Electronic sweep generator
ESS	Electronic switching system
EST	Eastern daylight time
esu	Electrostatic unit; electrostatic units
E_{sup}	Suppressor voltage in vacuum tube; supply voltage
ETC	Electronic temperature control
Eu	Chemical symbol for europium
eV	Electronvolt; electronvolts
E_X	Voltage drop across a reactance; excitation energy
exa-	Prefix multiplier for 10^{18}; prefix multiplier for 2^{60} (binary)
exc	Exciter; excitation
exp	Exponential; experimental
E_Z	Voltage drop across an impedance
f	**femto-**; frequency; function; factor
F_0	Damping factor
F	Farad; farads; force; chemical symbol for fluorine; fermi; focal length; filament of vacuum tube; fuse; Faraday constant; logic false

fax	Facsimile
f_c	Carrier frequency
fc	Footcandle
FCC	Federal Communications Commission (U.S.)
f_{co}	Cutoff frequency
FDM	Frequency-division multiplex; frequency-division multiplexing
FDS	Faraday dark space
Fe	Chemical symbol for iron
FE	Ferroelectric
FE-EL	Ferroelectric-electroluminescent
femto-	Prefix multiplier for 10^{-15}
ferfi-	Prefix denoting magnetic properties
ferro-	Prefix denoting magnetic properties
FET	Field-effect transistor (usually indicates junction type)
FET VOM	Field-effect-transistor volt-ohm-milliammeter
FF	Flip-flop
FFI	Fuel-flow indicator
fig	Figure
fil	Filament
FIR	Far infrared
fL	Foot-lambert
f_m	Modulation frequency
Fm	Chemical symbol for fermium
FM	Frequency modulation
FORTRAN	Formula translation (computer language)
FOSDIC	Film optical scanning devices for input to computer
F_p	Power-loss factor
FPIS	Forward propagation by ionospheric scatter
fps	Feet per second; frames per second; foot-pound-second; foot-pound-seconds
fr	Frankline

Fr	Chemical symbol for francium
FRUGAL	**FORTRAN** rules used as a general application language
FRUSA	Flexible rolled-up solar array
FSK	Frequency-shift keying
FSM	Field-strength meter
FSR	Feedback shift register
ft	Foot; feet
ft-L	Foot-lambert
ft-lb	Foot-pound; foot-pounds
FUBAR	Fouled up beyond all recognition
g	Conductance; gate of field-effect transistor
g	Gravity; gram; generator; grid of vacuum tube
G	Conductance; **giga-**; deflection factor; perveance; gravitational constant; generator; gate of field-effect transistor
Ga	Chemical symbol for gallium
GA	Go ahead
GaAs	Chemical symbol for gallium arsenide (pronounced gas)
GaAsFET	Gallium-arsenide field-effect transistor (pronounced gas-fet)
Gd	Chemical symbol for gadolinium
GDO	Grid-dip oscillator; gate-dip oscillator
Ge	Chemical symbol for germanium
gen	Generator
GeV	Gigaelectronvolt; gigaelectronvolts
GFI	Ground-fault interrupter
g_{fs}	Forward transconductance
G/G, G-G	Ground-to-ground
GHz	Gigahertz
G_i, G_{in}, G_f, g_i, g_{in}, g_f	Input conductance
giga-	Prefix multiplier for 10^9; prefix multiplier for 2^{30} (binary)

GJD	Germanium junction diode
g_m	Transconductance
gm	Gram
gcal, gmcal	Gram-calorie
gmcm	Gram-centimeter
gmm	Gram-meter
GMT	Greenwich mean time
GND, gnd	ground
G_o, G_{out}, g_o, g_{out}	Output conductance
.gov	Internet domain suffix for government Web site
G_p, g_p	Plate conductance in vacuum tube
gpc	Germanium point-contact
GSR	Galvanic skin resistance
h	**Hecto-**; Planck's constant; hour; height
H	Magnetic field strength; magnetizing force; chemical symbol for hydrogen; unit function; horizontal; heater, henry; henrys; harmonic
hal	Halogen
HCD	Hard-copy device
HCM	Half-cycle magnetizer
HDB3	High-density bipolar-3
He	Chemical symbol for helium
hecto-	Prefix multiplier for 10^2
He-Ne	Chemical symbol for helium-neon
Hf	Chemical symbol for hafnium
HF	High frequency
Hg	Chemical symbol for mercury
HIC	Hybrid integrated circuit
HIDM	High-information delta modulation
hi-fi	High fidelity
hipot	High potential
HLL	High-level language

Ho	Chemical symbol for holmium
hor, horiz	Horizontal
HOT	Horizontal output transformer; horizontal output transistor; horizontal output tube
hp	Horsepower
h-p	High-pressure
hr	Hour; hours
HSM	High-speed memory
HTL	High-threshold logic
HV	High voltage
hy	Henry; henrys (preferred abbreviation is H)
Hz	Hertz
i	Instantaneous current; mathematician's symbol for the square root of -1; instantaneous value; intrinsic semiconductor; angle of incidence; vector parallel to abscissa; incident ray
I	Current; chemical symbol for iodine; intrinsic semiconductor; luminous intensity
I_A	Anode current in vacuum tube
I_{AC}, I_{ac}	AC component current
I_{AF}	Audio-frequency current
IAGC	Instantaneous automatic gain control
ICAS	Intermittent commercial and amateur service
ICBM	Intercontinental ballistic missile
I_{CBO}	Static reverse collector current in common-base bipolar-transistor circuit
ICBS	Interconnected business system
I_{CEO}	Static reverse collector current in common-emitter bipolar-transistor circuit
ICET	Institute for the Certification of Engineering Technicians
I_{CO}	Collector cutoff current in bipolar transistor
ICW	Interrupted continuous wave
ID	Identification; identification designator; inside diameter

i.d.	Inside diameter
I_{DC}, I_{dc}	DC component current
$I_D(\text{off})$	Drain cutoff current in FET
IDOT	Instrumentation on-line transcriber
IDP	Industrial data processing; integrated data processing; intermodulation-distortion percentage
I_{DSS}	Drain current at zero gate voltage in FET
I_E	Emitter current in bipolar transistor
IEC	Integrated electronic component
IEE	Institute of Electrical Engineers
IEEE	Institute of Electrical and Electronics Engineers
I_f	Filament current in vacuum tube
IF, i-f	Intermediate frequency
I_{FB}	IC feedback current
I_{FS}	Full-scale current
I_g	Grid current in vacuum tube
I_G	Gate current in FET
IGFET	Insulated-gate FET
I_{GSS}	Gate reverse current in FET
I_h	Heater current in vacuum tube; hold current; holding current
IHF	Inhibit flip-flop
I_i	Instantaneous current, input current
I_{in}	Input current
I_K	Cathode current in vacuum tube
I^2L	Integrated injection logic
I_m	Meter current, maximum current
IM	Intermodulation
I_{max}	Maximum current
IMPATT	Impact avalanche transit time
In	Chemical symbol for indium
in	Input; inch; inches
ind	Indicator; inductance; inductor
INV, inv	Inverter; inverse; inverse function

I/O	Input/output; input/output device; input/output port
I_o, I_{out}	Output current
I_P	Plate current in vacuum tube
I_p	Peak current
ipm	Inches per minute
ips	Inches per second
I_R	Current in a resistor
Ir	Chemical symbol for iridium
IR	Product of current and resistance; insulation resistance; infrared
IRE	Institute of Radio Engineers
I_{RF}	Radio-frequency current
I_S	Source current in FET; screen current in vacuum tube
ISCAN	Inertialess steerable communications antenna
I_{sup}	Suppressor current in vacuum tube
$I(t)$	Indicial response
I_X	Current in a reactance
I_Y	Current in an admittance
I_Z	Current in an impedance
j	j operator (engineer's symbol for the square root of -1)
J	Joule; jack; connector; emissive power
JEDEC	Joint Electron Device Engineering Council
JFET	Junction field-effect transistor (pronounced jay-fet)
JHG	Joule heat gradient
J/K	Joules per kelvin
J/(kgK)	Joules per kilogram-kelvin
J/s, J/sec	Joules per second
JSR	Jump to subroutine
k	**Kilo-**; constant (generic); dielectric constant; Boltzmann constant

K	Chemical symbol for potassium; Kelvin; constant (generic); go ahead (radiotelegraphy); cathode of vacuum tube; **kilo-**
kA	Kiloampere
Kb	Kilobit; kilobits
kc	Kilocycle; kilocycles (obsolete)
kcal	Kilocalorie; kilocalories
kCi	Kilocurie; kilocuries
kcs	1000 characters per second; kilocycles (obsolete)
KDP	Potassium dihydrogen phosphate
keV	Kiloelectronvolt; kiloelectronvolts
kg	Kilogram; kilograms
kgc	Kilogram-calorie; kilogram calories
kgm	Kilogram-meter; kilogram-meters
kg/m^3	Kilograms per cubic meter
kHz	Kilohertz
kilo-	Prefix multiplier for 10^3; prefix multiplier for 2^{10} (binary)
kJ	Kilojoule; kilojoules
km	Kilometer; kilometers
Kr	Chemical symbol for krypton
kV	Kilovolt; kilovolts
kVA	Kilovolt-ampere; kilovolt-amperes
kVAR	Reactive kilovolt-ampere; reactive kilovolt-amperes
kVARh	Reactive kilovolt-ampere-hour; reactive kilovolt-ampere-hours
kW	Kilowatt; kilowatts
kΩ	Kilohm; kilohms
l	Length; liter; low; lumen
L	Inductance
L	Lambert; lamberts; left stereo channel; mean life; low; Laplace transform
La	Chemical symbol for lanthanum

LASCR	Light-activated silicon-controlled rectifier
LASCS	Light-activated silicon-controlled switch
lb	Pound; pounds
LC	Inductance-capacitance
LC	Liquid crystal
LCD	Liquid-crystal display
LCR	Inductance-capacitance-resistance
L_d	Distributed inductance
LED	Light-emitting diode
LF	Low frequency
Li	Chemical symbol for lithium
LIY	Liquid crystal
LLL	Low-level logic
lm	Lumen; lumens
lm/ft^2	Lumens per square foot
lmhr	Lumen-hour; lumen-hours
lm/m^2	Lumens per square meter
lm/W	Lumens per watt
ln	Natural (base-e) logarithm
lo	Low
LO	Local oscillator; low
log	Base-10 logarithm
log_e	Natural (base-e) logarithm
log_{10}	Base-10 logarithm
loran	Long-range navigation
LP	Low power; long-playing; low pressure
LPB	Lighted pushbutton
lpm, l/m	Lines per minute
lpW, l/W	Lumens per watt
Lr	Chemical symbol for lawrencium (also Lw)
L + R	Sum of left and right stereo channels
L − R	Difference of left and right stereo channels
LSB	Least significant bit; lower sideband
LSC	Least significant character

LSD	Least significant digit
LSI	Large-scale integration
LSSC	Lower-sideband suppressed-carrier
LTROM	Linear transformer read-only memory (pronounced el-tee-rom)
Lu	Chemical symbol for lutetium
LUF	Lowest usable frequency
LV	Low voltage
LVDT	Linear variable differential transformer
Lw	Chemical symbol for lawrencium (also Lr)
lx	Lux
LZT	Lead zirconate-titanate
m	**Milli-**; mass; meter; meters; mile; miles; modulation coefficient
M	**Mega-**; mutual inductance; refractive modulus
m^2	Square meter; square meters
m^3	Cubic meter; cubic meters
mA	Milliampere; milliamperes
MADT	Microalloy diffused transistor
MAG	Maximum available gain
mag	Magnetic; magnification
magamp	Magnetic amplifier
MAR	Memory-address register
MAT	Microalloy transistor
max	Maximum
mb	Millibar
MBM	Magnetic bubble memory
MBO	Monostable blocking oscillator
MBS	Magnetron beam switching
mc	Millicurie; meter-candle
Mc	Megacycle; megacycles (obsolete)
MCG	Magnetocardiogram; magnetocardiograph
mCi	Millicurie

MCi	Megacurie
MCW	Modulated continuous wave
Md	Chemical symbol for mendelevium
MDI	Magnetic direction indicator
MDS	Minimum discernible signal
MDT	Mountain daylight time
m_e	Electron rest mass
meg	Megohm; megohms; megabyte; megabytes
mega-	Prefix multiplier for 10^6; prefix multiplier for 2^{20} (binary)
MESFET	Hybrid depletion/enhancement mode FET
MeV	Megaelectronvolt; megaelectronvolts
mF	Millifarad
MF	Medium frequency; midfrequency
MFSK	Multiple frequency-shift keying
Mg	Chemical symbol for magnesium
MGD	Magnetogasdynamics
MHD	Magnetohydrodynamics
Mhz	Megahertz
mi	Mile; miles
mic; mike	Microphone
MIC	Microwave integrated circuit; microphone
MICR	Magnetic ink character recognition
micro-	Prefix multiplier for 10^{-6}; extremely small
.mil	Internet domain suffix for military Web site
milli-	Prefix multiplier for 10^{-3}
min	Minimum; minute
MIR	Memory-information register
mks	Meter-kilogram-second
mL	Millilambert
ml	Milliliter; milliliters
mm	Millimeter; millimeters
MMF	Magnetomotive force

mmF	Micromicrofarad (more commonly called picofarad)
mmol	Millimole
MMV	Monostable multivibrator
m_n	Neutron rest mass
Mn	Chemical symbol for manganese
mntr	Monitor
Mo	Chemical symbol for molybdenum
MO	Master oscillator
mod	Modulator; modulation; modification; modulus
mol	Mole
MOPA	Master oscillator/power amplifier
MOS	Metal-oxide semiconductor; metal-oxide silicon
MOSFET	Metal-oxide-silicon field-effect transistor
MOSROM	Metal-oxide-silicon read-only memory
MOST	Metal-oxide-silicon transistor
MOV	Metal-oxide varistor
MPG	Microwave pulse generator
mph	Miles per hour
MPO	Maximum power output
mps	Meters per second; miles per second
MPT	Maximum power transfer
MPX	Multiplex
MR	Memory register
mrad	Milliradian
ms, msec	Millisecond
msg	Message
MSI	Medium-scale integration
MST	Mountain standard time
mtr	Meter (preferred abbreviation is m)
mu	Amplification factor; permeability; micron; micrometer; electric moment; inductivity; magnetic moment; molecular conductivity

MUF	Maximum usable frequency
MUPO	Maximum undistorted power output
MUSA	Multiple-unit steerable antenna
mV	Millivolt
MV	Megavolt; megavolts; multivibrator; medium voltage
MVA	Megavolt-ampere; megavolt amperes
mV/m	Millivolts per meter
MVP	Millivolt potentiometer
mW	Milliwatt; milliwatts
MW	Megawatt; megawatts
Mwh	Megawatt-hour; megawatt-hours
mWRTL	Milliwatt resistor-transistor logic
Mx	Maxwell; maxwells
μ	**Micro-** (10^{-6}); amplification factor; permeability; micron; micrometer; electric moment; inductivity; magnetic moment; molecular conductivity
μA	Microampere; microamperes
$\mu\mu$	**Micro-micro-** (preferred is **pico-,** abbreviated p)
μ_B	Bohr magneton
μCi	Microcurie; microcuries
μ_e	Electron magnetic moment
μF	Microfarad; microfarads
μg	Microgram; micrograms
μH	Microhenry; microhenrys
μl	Microliter; microliters
μ_n	Nuclear magneton
μ_o	Free-space permeability constant
μP	Microprocessor
$\mu\nu$	Micromho (preferred is microsiemens, abbreviated μS)
$\mu\Omega$	Microhm; microhms
$\mu\Omega$-cm	Microhm-centimeter; microhm-centimeters
μ_p	Proton magnetic moment

μs, μsec	Microsecond; microseconds
μS	Microsiemens
μV	Microvolt; microvolts
μV/m	Microvolts per meter
μW	Microwatt; microwatts
μW/cm^2	Microwatts per square centimeter
n	Number (usually an integer)
n	**Nano-** (10^{-9}); index of refraction; amount of substance
N	Chemical symbol for nitrogen; number; natural number, set of natural numbers; newton; newtons
nA	Nanoampere; nanoamperes
Na	Chemical symbol for sodium
nano-	Prefix multiplier for 10^{-9}
NAP	Nuclear auxiliary power
NAPU	Nuclear auxiliary power unit
NARTB	National Association of Radio and Television Broadcasters
NAS	National Academy of Sciences
NAVAIDS	Navigational aids
Nb	Chemical symbol for niobium
NB, N/B	Narrowband
NBFM	Narrowband frequency modulation
NBTDR	Narrowband time-domain reflectometry
NBVM	Narrowband voice modulation
NBW	Noise bandwidth
nc	No connection
NC	Normally closed; no connection; numerical control
N/C	Numerical control
NCS	Net control station
Nd	Chemical symbol for neodymium
Ne	Chemical symbol for neon

NEB	Noise equivalent bandwidth
NEC	National Electric Code
NEDA	National Electrical Distributors' Association
NEI	Noise equivalent input
NEL	National Electronics Laboratory
NELA	National Electric Light Association
NEMA	National Electrical Manufacturers' Association
NEP	Noise equivalent power
NEPD	Noise equivalent power density
NESC	National Electrical Safety Code
net	Network
Net	The Internet
NET	Noise equivalent temperature
nF	Nanofarad; nanofarads
NF	Noise figure
NFET	n-channel junction field-effect transistor (pronounced en-fet)
NFM	Narrowband frequency modulation
NG	Negative glow; no good; no go
NGT	Noise-generator tube
nH	Nanohenry; nanohenrys
Ni	Chemical symbol for nickel
NIDA	Numerically integrating differential analyzer
NIF	Noise improvement factor
NIPO	Negative input/positive output
NIR	Near infrared
NLR	Nonlinear resistance; nonlinear resistor
NLS	No-load speed
N/m^2	Newtons per square meter (pascals)
NMOS	N-channel metal-oxide semiconductor (pronounced en-moss)
NMR	Nuclear magnetic resonance; normal mode rejection
NMRI	Nuclear magnetic resonance imaging
No	Chemical symbol for nobelium

No., no.	Number
NO	Normally open
N_p, N_{pri}	Number turns in primary winding
Np	Neper; nepers; chemical symbol for neptunium
NPM	Counts per minute
NP0	Negative-positive-zero
NPS	Counts per second
NRD	Negative-resistance diode
NRZ	Nonreturn-to-zero
N_s, N_{sec}	Number of turns in secondary winding
NSPE	National Society of Professional Engineers
NTC	Negative temperature coefficient
NTSC	National Television Systems Committee
nu	Greek ν; reluctivity
nV	Nanovolt; nanovolts
nW	Nanowatt; nanowatts

O	Chemical symbol for oxygen; output
O_2	Chemical symbol for common oxygen molecule
O_3	Chemical symbol for ozone molecule
OAT	Operating ambient temperature
OBWO	O-type backward-wave oscillator
OC, o/c	Open circuit
OCR	Optical character recognition
oct	Octal
OD, o.d.	Outside diameter
Oe	Oersted, oersteds
OGL	Outgoing line
OLRT	On-line real time
OM	Optical microscope; old man (amateur radio slang)
op	Operational; operation; operate; operator
op amp	Operational amplifier
op code	Operation code

opt	Optical; optional
.org	Internet domain suffix for organization Web site
Os	Chemical symbol for osmium
osc	Oscillator
Os-Ir	Chemical symbol for osmiridium
OTA	Operational transconductance amplifier
OTL	Output-transformerless
ox	Oxygen
oz-in	Ounce-inch; ounce-inches

p	**Pico-**; peak; primary; plate of vacuum tube; pitch; per
P	Power; chemical symbol for phosphorus; plate of vacuum tube; pressure; primary; **peta-**; permeance; point
PA	Power amplifier; pulse amplifier; pulse amplitude; public address; particular average; pilotless aircraft
pA	Picoampere; picoamperes
Pa	Chemical symbol for protactinium; pascal; pascals
PACM	Pulse-amplitude code modulation
PADT	Post-alloy-diffused transistor
PAM	Pulse-amplitude modulation
PAN	Urgent message to follow
PAV	Phase-angle voltmeter
PAX	Private automatic exchange
Pb	Chemical symbol for lead
PBX	Private branch exchange
pc	Picocurie; picocuries; parsec; parsecs
PC	Personal computer; printed circuit; photocell; positive column; point contact; program counter; punched card
PCB	Printed-circuit board
PCL	Printed-circuit lamp
PCM	Pulse-code modulation; punched-card machine

Pd	Chemical symbol for palladium
PD	Pulse duration; plate dissipation; proximity detector; potential difference
PDAS	Programmable data acquisition system
PDM	Pulse-duration modulation
PDT	Pacific daylight time
PE	Potential energy; professional engineer; probable error
PEM	Photoelectromagnetic
PEP	Peak envelope power; planar epitaxial passivated
peta-	Prefix multiplier for 10^{15}; prefix multiplier for 2^{50} (binary)
pF	Picofarad; picofarads
PF, pf	Power factor
PFET	p-channel junction field-effect transistor (pronounced pee-fet)
PFM	Pulse-frequency modulation
PG	Power gain
pH	Hydrogen-ion concentration
P_i, P_{in}	Input power
PIA	Peripheral interface adapter
pico-	Prefix multiplier for 10^{-12}
PIM	Pulse-interval modulation
PIO	Parallel input/output
PIV	Peak inverse voltage
PLC	Power-line communication
PLL	Phase-locked loop
PLM	Pulse-length modulation
PLO	Phase-locked oscillator
P_m, P_{max}	Maximum power
PM	Pulse modulation; phase modulation; pulse modulator; phase modulator; permanent magnet; postmeridian
PMOS	p-channel metal-oxide semiconductor (pronounced pee-mos)

PN	Polish notation; positive-negative
PNM	Pulse-numbers modulation
P_o, P_{out}	Output power
Po	Chemical symbol for polonium
pos	Positive; position
pot	Potentiometer; dashpot; potential
P_p	Peak power; plate power
PP	Peripheral processor
ppb	Parts per billion (10^9)
ppm	Parts per million (10^6); pulses per minute
pps	Pulses per second
ppt	Parts per thousand
Pr	Chemical symbol for praseodymium
PRF	Pulse repetition frequency
pri	Primary
PROM	Programmable read-only memory
PRR	Pulse repetition rate
PRV	Peak reverse voltage
ps, psec	Picosecond; picoseconds
PS	Power supply
PSD	Phase-sensitive detector
psf	Pounds per square foot
psi	Pounds per square inch; angle; flux
psia	Pounds per square inch absolute
psig	Pounds per square inch gauge
PSK	Phase-shift keying
PSM	Pulse-spacing modulation
PST	Pacific standard time
PSVM	Phase-sensitive voltmeter
PSWR	Power standing-wave ratio
Pt	Chemical symbol for platinum
PTC	Positive temperature coefficient
PTM	Pulse-time modulation
PTO	Permeability-tuned oscillator

PTT	Press-to-talk; push-to-talk
Pu	Chemical symbol for plutonium
pV	Picovolt; picovolts
PVC	Polyvinyl chloride
pW	Picowatt; picowatts
PWM	Pulse-width modulation; plated-wire memory
pwr	Power
PZT	Lead zirconate titanate
q	Electrical quantity; charge carried by electron; value of quantum; quart
Q	Figure of merit; electrical charge; selectivity; Q band; Q output
QAVC	Quiet automatic volume control
QCW	Q-phase CW signal (television)
QFM	Quantized frequency modulation
QM	Quadrature modulation
QMQB	Quick-make, quick-break
qt	Quart
qty	Quantity
qual	Qualitative; quality
r	Roentgen; roentgens; correlation coefficient; radius
R	Resistance; resistor
R	Roger (message received); reluctance; right stereo channel; radical
$R_{(\infty)}$	Rydberg constant
RA	Random access; right ascension
R_{AC}, R_{ac}	Alternating-current resistance
RAC	Rectified alternating current
racon	Radar beacon
rad	Radiac; radian; radio; radix; radius; radical
r_e	Classical electron radius
RAM	Random-access memory

R_B	Base resistance in bipolar transistor
Rb	Chemical symbol for rubidium
R_C	Collector resistance in bipolar transistor; cold resistance
RC	Resistance-capacitance
Rc	Remote control; radio-controlled
RCL	Resistance-capacitance-inductance
RCL	Recall
RCTL	Resistor-capacitor-transistor logic
RCV, rcv	Receive
RCVR, rcvr	Receiver
R&D, R/D	Research and development
R_D	Drain resistance in FET
R_{DC}, R_{dc}	Direct-current resistance
RDF	Radio direction finder; radio direction finding
R_E	Emitter resistance in bipolar transistor
Re	Chemical symbol for rhenium
rect	Rectifier; rectification; rectified
ref	Reference; referred
R_{eff}	Effective resistance
rej	Reject; rejection
rem	Roentgen equivalent man (unit of ionizing radiation dose)
res	Resistance; resistor; resolution; research; residual
rev	Reverse; revolution
R_f	Filament resistance in vacuum tube; feedback resistance; feedback resistor
RF	Radio frequency
RFC	Radio-frequency choke
RFI	Radio-frequency interference
RFO	Radio-frequency oscillator
R_g	Grid resistance in vacuum tube
R_G	Gate resistance in FET

RGB	Red/green/blue color model
RGT	Resonant-gate transistor
Rh	Chemical symbol for rhodium
R_{HF}	High-frequency resistance
R_i, R_{in}	Input resistance
RI	Radio interference
R_K	Cathode resistance in vacuum tube
R_L	Load resistance; loss resistance
RL	Resistance-inductance
RL	Relay logic
R_{LF}	Low-frequency resistance
R_{load}	Load resistance
R_{loss}	Loss resistance
r_m	Emitter-collector transresistance in bipolar transistor
R_m	Meter resistance
RMA	Radio Manufacturers' Association
rms	Root mean square
Rn	Chemical symbol for radon
R_o, R_{out}	Output resistance
ROM	Read-only memory
R_P	Plate resistance in vacuum tube; parallel resistance; primary resistance
R_{pri}	Primary resistance
rps	Revolutions per second
RPT	Repeat; report
R_{req}	Required resistance
R_s	Secondary resistance; screen resistance in vacuum tube; series resistance
R_S	Source resistance in FET
RST	Readability/strength/tone of radiotelegraphy signal
R_t, R_T	Total resistance
R_T	Thermal resistance
RTD	Resistance-temperature detector

RTL	Resistor-transistor logic
RTTY	Radioteletype
RTZ	Return to zero
Ru	Chemical symbol for ruthenium
R_x	Unknown resistance
RY	Relay; radioteletype test signal (repeated)
RZ	Return to zero
s	Screen grid of vacuum tube; distance; displacement; standard deviation; second
S	Siemens; screen grid of vacuum tube; shell; chemical symbol for sulfur; deflection sensitivity; switch; elastance; sync; secondary; sine; entropy
SADT	Surface alloy diffused-base transistor
SAE	Shaft angle encoder; Society of Automotive Engineers
sat	Saturation; saturate; saturated
satd	Saturated
SAVOR	Signal-actuated voice recorder
Sb	Chemical symbol for antimony
SB	Sideband; simultaneous broadcast
SBC	Single-board computer
SBDT	Surface-barrier-diffused transistor
SBT	Surface-barrier transistor
sc	Sine-cosine; single crystal; science; scale
Sc	Chemical symbol for scandium
SC	Suppressed carrier; short circuit; silk covered
SCA	Subsidiary communications authorization
scc	Single cotton covered
sce	Single cotton enameled
SCEPTRON	Spectral comparative pattern recognizer
SCLC	Space-charge-limited current
sco	Subcarrier oscillator

SCR	Silicon-controlled rectifier
SCS	Silicon-controlled switch
SCT	Surface-charge transistor
SD	Standard deviation
Se	Chemical symbol for selenium
sec	Second; secondary; secant; section
sech	Hyperbolic secant
SELCAL	Selective calling
SEM	Single-electron memory; scanning electron microscope
Bser	Series; serial
SF	Safety factor; single frequency; standard frequency; stability factor
SFA	Single-frequency amplifier
SFO	Single-frequency oscillator; standard-frequency oscillator
SFR	Single-frequency receiver
SG	Screen grid of vacuum tube
SGCS	Silicon gate-controlled switch
SHF	Superhigh frequency
Si	Chemical symbol for silicon
SI	Standard International System of Units
S/I	Signal-to-intermodulation ratio; signal-to-interference ratio
SIC	Specific inductive capacity
SiC	Silicon carbide
sig	Signal
sin	Sine
SINAD	Signal-to-noise-and-distortion ratio
SIO	Serial input/output
SIP	Single inline package
SJD	Single-junction diode
SKM	Sine-cosine multiplier

SLS	Side-lobe suppression
Sm	Chemical symbol for samarium
Sn	Chemical symbol for tin
SN	Semiconductor network
S/N	Signal-to-noise ratio
SNOBOL	String-oriented symbolic language
SNR	Signal-to-noise ratio
sol	Solution; soluble
SOM	Start of message
SOP	Standard operating procedure
SOS	Distress signal (radiotelegraphy)
sp	Single-pole; specific
SP	Single-pole; single-phase; self-propelled; stack pointer
SPC	Silicon point-contact; silver-plated copper
SPDT, spdt	Single-pole, double-throw
spec	Specification; spectrum
spec an	Spectrum analyzer
specs	Specifications
SPFW	Single-phase full-wave
sp gr	Specific gravity
SPHW	Single-phase, half-wave
spkr	Speaker
SPOT	Satellite positioning and tracking
SPST, spst	Single-pole, single-throw
sq	Square
SQR	Square-rooter; square-root function
Sr	Chemical symbol for strontium
SR	Silicon rectifier; shift register; silicon rubber
S-R, S/R	Send-receive
SRAM	Static random-access memory (pronounced ess-ram)
SRF	Self-resonant frequency

SS	Solid-state; single-shot; small-signal; single-signal; same size; stainless steel
SSB	Single sideband
SSBSC	Single-sideband suppressed carrier
ssc	Single silk covered
sse	Single silk enameled
SSI	Small-scale integration
SSL	Solid-state lamp
SSSC	Single-sideband suppressed carrier
ST, st	Single-throw
sta	Station; stationary
stab	Stabilizer; stability; stabilization (pronounced stabe)
STALO	Standardized oscillator
stat-	Electrostatic
statA	Statampere; statamperes
statC	Statcoulomb; statcoulombs
statF	Statfarad; statfarads
statH	Stathenry; stathenrys
statOe	Statoersted, statoersteds
statS	Statsiemens
statV	Statvolt; statvolts
statWb	Statweber; statwebers
std	Standard
stn	Station
STO	Store; storage
STP	Standard temperature and pressure
sup	Suppressor grid in vacuum tube; suppression
SVGA	Super Video Graphics Array
sw	Switch
SW	Shortwave
SWG	Standard wire gauge
SWR	Standing-wave ratio
sym	Symmetry; symmetrical; symbol

symb	Symbol
sync	Synchronization; synchronized
t	Time; ton; tons; Celsius temperature; target; tension; technical
T	Transformer; **tera-**; thermodynamic temperature; chemical symbol for tritium; ton; tons; tesla; teslas; kinetic energy; period; logic true
Ta	Chemical symbol for tantalum
tach	Tachometer (pronounced tack)
tan	Tangent
tanh	Hyperbolic tangent
Tb	Chemical symbol for terbium
Tc	Chemical symbol for technetium
TCCO	Temperature-controlled crystal oscillator
TCL	Transistor-coupled logic
TCM	Thermocouple meter
TDM	Time-division multiplex; time-division multiplexing
TDR	Time-delay relay; time-domain reflectometry
Te	Chemical symbol for tellurium
TE	Transverse electric; trailing edge
tel	Telephone; telegraph; telegram
TEM	Transverse electromagnetic
tera-	Prefix multiplier for 10^{12}; prefix multiplier for 2^{40} (binary)
TeV	Tera-electronvolt; tera-electronvolts
TGTP	Tuned-grid tuned-plate
Th	Chemical symbol for thorium
TH	True heading
THD	Total harmonic distortion
thy	Thyratron
THz	Terahertz
Ti	Chemical symbol for titanium
Tl	Chemical symbol for thallium
T^2L	Transistor-transistor logic (also TTL)

Tm	Chemical symbol for thulium
TM	Transverse magnetic; technical manual
tot	Total; to derive a total (usually by summation)
TP	Test point; tuned plate; transaction processing
TPTG	Tuned-plate tuned-grid
t_r	Recovery time; rise time
TR, T/R	Transmit-receive
trans	Transverse, transmit; transmitter; transformer
TRF	Tuned radio frequency
trig	Trigonometry; trigonometric
ts	Tensile strength
TSS	Time-sharing system
TTL	Transistor-transistor logic (also T^2L)
TTY	Teletype; teletypewriter
TU	Terminal unit
TV	Television; terminal velocity
T/V	Temperature-to-voltage
TVI	Television interference
TVL	Television listener; television listening
TVM	Transistor voltmeter
TVO, TVOM	Transistor volt-ohmmeter
TVT	Television terminal
TW	Terawatt; terawatts; traveling wave
TWA	Traveling-wave amplifier
TWT	Traveling-wave tube
u	**Micro-** (when symbol μ is not available); unit; unified atomic mass unit
U	Chemical symbol for uranium; unit; universal set; union of sets
U_a	Unit of activity
UDOP	Ultra-high-frequency Doppler system
UEP	Underwater electric potential

UFET	Unipolar field-effect transistor (pronounced you-fet)
UHF	Ultrahigh frequency
UJT	Unijunction transistor
ULD	Ultralow distortion
ULF	Ultralow frequency
uni-	Single; one
UNIFET	Unipolar field-effect transistor
UNIPOL	Universal problem-oriented language
UNIVAC	Universal Automatic Computer
uns	Unsymmetrical; unstable
UPC	Universal product code
upconv	Up converter; up conversion
UPS	Uninterruptible power supply
USB	Upper sideband
USSC	Upper sideband suppressed carrier
UTC	Coordinated universal time
UTL	Unit transmission loss
UV	Ultraviolet; undervoltage
UVM	Universal vendor marking
v	Velocity; voltage; vector
V	Volt; volts; voltage; potential; chemical symbol for vanadium; volume; reluctivity; vertical velocity; vector
VA	Volt-ampere; volt-amperes
V/A	Volts per ampere
V_{ac}, v_{ac}	AC voltage; volt ac; volts ac
vac	Vacuum
VAC	Volt ac; volts ac; vector analog computer
val	Value
var	Variable
VAR	Volt-ampere reactive; volt-amperes reactive
V_B	Base voltage in bipolar transistor

V_{BB}	Base-voltage supply in bipolar-transistor circuit
V_C	Collector voltage in bipolar transistor
VCA	Voltage-controlled amplifier
V_{CC}	Collector-voltage supply in bipolar transistor circuit
VCCO	Voltage-controlled crystal oscillator
VCD	Variable-capacitance diode
VCG	Voltage-controlled generator
VCO	Voltage-controlled oscillator
VCR	Videocassette recorder; videocassette recording
VCSR	Voltage-controlled shift register
VCXO	Voltage-controlled crystal oscillator
V_D	Drain voltage in FET
VD	Voltage drop; vapor density
V_{dc}, v_{dc}	DC voltage; volt dc; volts dc
VDC	Volt dc; volts dc
VDU	Video display unit; visual display unit
VDCW	DC working voltage
VDR	Voltage-dependent resistor; videodisk recorder; videodisk recording
V_{drive}	Drive voltage
V_E	Emitter voltage in bipolar transistor
VE	Value engineering; volunteer examiner
V_{EE}	Emitter-voltage supply in bipolar transistor circuit
vel	Velocity
vers	Versed sine
vert	Vertical
VF	Video frequency
V_{FB}	Feedback voltage
VFO	Variable-frequency oscillator
V_g, v_g	Generator voltage
VGA	Variable-gain amplifier, video graphics array
V_{GD}	Gate-drain voltage in FET
V_{GS}	Gate-source voltage in FET
VHF	Very high frequency

VHR	Very high resistance
VHRVM	Very-high-resistance voltmeter
V_i, V_{in}	Input voltage
VI	Volume indicator; viscosity index
VLF	Very low frequency
VLR	Very low resistance; very long range
VLSI	Very large scale integration
V/m	Volts per meter
VMOS	Vertical metal-oxide semiconductor (pronounced vee-moss)
VMOSFET	Vertical metal-oxide-semiconductor field-effect transistor (pronounced vee-moss-fet)
V_o, V_{out}	Output voltage
VOA	Volt-ohm-ammeter; Voice of America
vol	Volume
VOM	Volt-ohm-milliammeter
VOR	Very-high-frequency omnirange
vox	Voice-operated transmission
V_p	Pinch-off voltage in FET; plate voltage in vacuum tube
Vpm	Volts per meter
VR	Voltage regulator; voltage regulation
V_{ref}	Reference voltage
VRR	Visual radio range
Vs	Volt-second; volt-seconds
Vs/A	Volt-second per ampere; volt-seconds per ampere
VSA	Voice-stress analyzer
VSB	Vestigial sideband
VSF	Vestigial-sideband filter
VSR	Very short range
VSWR	Voltage standing-wave ratio
vt	Vacuum tube; variable time
VTL	Variable-threshold logic

VTM	Voltage-tuned magnetron
VTO	Voltage-tuned oscillator; voltage-tunable oscillator
VTR	Videotape recorder; videotape recording
VTVM	Vacuum-tube voltmeter
VU	Volume unit; volume units
VVCD	Voltage-variable-capacitor diode
VVV	Test signal (radiotelegraphy)
VW	Volts working
w	Width
W	Watt; watts; work; chemical symbol for tungsten; energy; west; width
W3	World Wide Web
WAC	Worked all continents
WAS	Worked all states
WATS	Wide Area Telephone Service
WAZ	Worked all zones
Wb	Weber; webers
W_B	Base-region width in bipolar transistor
Wb/m^2	Webers per square meter
W_C	Collector-region width in bipolar transistor
W/cm^2	Watts per square centimeter
W_E	Emitter-region width in bipolar transistor
WE	Write enable
WG	Wire gauge
wgt	Weight
WH, Wh, w·h	Watt-hour; watt-hours
WHP	Water horsepower
WL	Wavelength
WM	Wattmeter
Wm^2	Watt square meter; watt square meters
W/(mK)	Watt per meter Kelvin; watts per meter Kelvin

w/o	Without
Wpc, W/c	Watt per candle; watts per candle
wpm	Words per minute
W/sr	Watt per steradian; watts per steradian
W/(sr·m²)	Watt per steradian per square meter; watts per steradian per square meter
wt	Weight
WT	Wireless telegraphy; watertight
WVDC	Working volt dc; working volts dc; working voltage dc
ww	Wirewound
WWW	World Wide Web
x	Number of carriers; unknown quantity
x	Abscissa; multiplication; trans-; cross-
X	Reactance; unknown quantity
X	No connection; abscissa; multiplication; trans-; cross-
XB	Crossbar
X_C	Capacitive reactance
XCVR, xcvr	Transceiver
Xe	Chemical symbol for xenon
xfmr, xformer	Transformer
XHV	Extremely high vacuum
xistor	Transistor
X_L	Inductive reactance
xmission, xmsn	Transmission
xmit, xmt	Transmit
XMTR, xmtr	Transmitter
XOR	Exclusive OR function; exclusive OR gate
xover	Crossover
xponder	Transponder
XR	Index register
xsection	Cross section

X_T, X_t	Total reactance
xtal	Crystal
xtalk	Crosstalk
y	Year; yard; ordinate
Y	Admittance; ordinate; chemical symbol for Yttrium; Young's modulus
YAG	Yttrium-aluminum-garnet
YIG	Yttrium-iron-garnet
yT	y-matrix of transistor
yV	y-matrix of vacuum tube
z	Zero; electrochemical equivalent; zone
Z	Impedance
Z	Atomic number; zenith distance
Zn	Chemical symbol for zinc; azimuth
Z_0	Characteristic impedance
Zr	Chemical symbol for zirconium

26

Miscellaneous Data

This chapter contains information about schematic symbols, communications signals, and other electronics-related technologies that do not fit neatly into any of the other chapters.

Schematic Symbols

A *schematic diagram* is a technical illustration of the interconnection of components in a circuit. Some schematic diagrams include component values and perhaps tolerances. Standard symbols are used (Table 26.1). A schematic diagram does not indicate the physical arrangement of the components on the chassis or circuit board; it shows only how the components are interconnected.

A schematic diagram might look complicated to the beginner in electronics, but it is simpler than a pictorial diagram. For a complex device such as a superheterodyne radio transceiver, a pictorial diagram is impractical, but a schematic diagram can fit on a single page.

Morse Code

The *Morse code* is a digital means of sending and receiving messages. It is a binary code because it has only two possible states: ON (key-down) and OFF (key-up).

TABLE 26.1   Symbols Used in Electronics Circuit Diagrams

Component	Symbol
Ammeter	
Amplifier general	
Amplifier, inverting	
Amplifier, operational	
AND gate	
Antenna, balanced	
Antenna, general	
Antenna, loop	
Antenna, loop, multiturn	
Battery	
Capacitor, feedthrough	
Capacitor, fixed	
Capacitor, variable	
Capacitor, variable, split-rotor	
Capacitor, variable, split-stator	

TABLE 26.1   Symbols Used in Electronics Circuit Diagrams (Continued)

Component	Symbol
Cathode, electron-tube, cold	
Cathode, electron-tube, directly heated	
Cathode, electron-tube indirectly heated	
Cavity resonator	
Cell, electrochemical	
Circuit breaker	
Coaxial cable	
Crystal, piezoelectric	
Delay line	
Diac	
Diode, field-effect	
Diode, general	
Diode, Gunn	
Diode, light-emitting	
Diode, photosensitive	

TABLE 26.1 Symbols Used in Electronics Circuit Diagrams (Continued)

Component	Symbol
Diode, PIN	
Diode, Schottky	
Diode, tunnel	
Diode, varactor	
Diode, zener	
Directional coupler	
Directional wattmeter	
Exclusive-OR gate	
Female contact, general	
Ferrite bead	
Filament, electron-tube	
Fuse	
Galvanometer	
Grid, electron-tube	
Ground, chassis	

TABLE 26.1   Symbols Used in Electronics Circuit Diagrams (Continued)

Component	Symbol
Ground, earth	
Handset	
Headset, double	
Headset, single	
Headset, stereo	
Inductor, air core	
Inductor, air core, bifilar	
Inductor, air core, tapped	
Inductor, air core, variable	
Inductor, iron core	
Inductor, iron core, bifilar	
Inductor, iron core, tapped	
Inductor iron core, variable	
Inductor, powdered-iron core	
Inductor, powdered-iron core, bifilar	

TABLE 26.1   Symbols Used in Electronics Circuit Diagrams (Continued)

Component	Symbol
Inductor, powdered-iron core, tapped	
Inductor, powdered-iron core, variable	or
Integrated circuit, general	
Jack, coaxial or phono	
Jack, phone, two-conductor	
Jack, phone, three-conductor	
Key, telegraph	
Lamp, incandescent	
Lamp, neon	
Male contact, general	
Meter, general	
Microammeter	
Microphone	

TABLE 26.1   Symbols Used in Electronics Circuit Diagrams (Continued)

Component	Symbol
Microphone, directional	
Milliammeter	mA
NAND gate	
Negative voltage connection	
NOR gate	
NOT gate	
Optoisolator	
OR gate	
Outlet, two-wire, nonpolarized	
Outlet, two-wire, polarized	
Outlet, three-wire	
Outlet, 234-V	
Plate, electron-tube	
Plug, two-wire, nonpolarized	

TABLE 26.1 Symbols Used in Electronics Circuit Diagrams (Continued)

Component	Symbol
Plug, two-wire, polarized	
Plug, three-wire	
Plug, 234-V	
Plug, coaxial or phono	
Plug, phone, two-conductor	
Plug, phone, three-conductor	
Positive voltage connection	
Potentiometer	
Probe, radio-frequency	
Rectifier, gas-filled	
Rectifier, high-vacuum	
Rectifier, semiconductor	
Rectifier, silicon-controlled	

TABLE 26.1   Symbols Used in Electronics Circuit Diagrams (Continued)

Component	Symbol
Relay, double-pole, double-throw	
Relay, double-pole, single-throw	
Relay, single-pole, double-throw	
Relay, single-pole, single-throw	
Resistor, fixed	
Resistor, preset	
Resistor, tapped	
Resonator	
Rheostat	
Saturable reactor	
Signal generator	
Solar battery	

TABLE 26.1   Symbols Used in Electronics Circuit Diagrams (Continued)

Component	Symbol
Solar cell	
Source, constant-current	
Source, constant-voltage	
Speaker	
Switch, double-pole, double-throw	
Switch, double-pole, rotary	
Switch, double-pole, single-throw	
Switch, momentary-contact	
Switch, silicon-controlled	
Switch, single-pole, rotary	
Switch, single-pole, double-throw	
Switch, single-pole, single-throw	

TABLE 26.1   Symbols Used in Electronics Circuit Diagrams (Continued)

Component	Symbol
Terminals, general, balanced	
Terminals, general, unbalanced	
Test point	TP
Thermocouple	or
Transformer, air core	
Transformer, air core, step-down	
Transformer, air core, step-up	
Transformer, air core, tapped primary	
Transformer, air core, tapped secondary	
Transformer, iron core	
Transformer, iron core, step-down	
Transformer, iron core, step-up	
Transformer, iron core, tapped primary	
Transformer, iron core, tapped secondary	

TABLE 26.1   Symbols Used in Electronics Circuit Diagrams (Continued)

Component	Symbol
Transformer, powdered-iron core	
Transformer, powdered-iron core, step-down	
Transformer, powdered-iron core, step-up	
Transformer, powdered-iron core, tapped primary	
Transformer, powdered-iron core, tapped secondary	
Transistor, bipolar, *NPN*	
Transistor, bipolar, *PNP*	
Transistor, field-effect, *N*-channel	
Transistor, field-effect, *P*-channel	
Transistor, MOS field-effect, *N*-channel	
Transistor, MOS field-effect, *P*-channel	
Transistor, photosensitive, *NPN*	
Transistor, photosensitive, *PNP*	

TABLE 26.1   Symbols Used in Electronics Circuit Diagrams (Continued)

Component	Symbol
Transistor, photosensitive, field-effect, *N*-channel	
Transistor, photosensitive, field-effect, *P*-channel	
Transistor, unijunction	
Triac	
Tube, diode	
Tube, heptode	
Tube, hexode	
Tube, pentode	
Tube, photosensitive	
Tube, tetrode	

TABLE 26.1   Symbols Used in Electronics Circuit Diagrams (Continued)

Component	Symbol
Tube, triode	
Voltmeter	V
Wattmeter	W
Waveguide, circular	
Waveguide, flexible	
Waveguide, rectangular	
Waveguide, twisted	
Wires, crossing, connected	(preferred) or (alternative)
Wires, crossing, not connected	(preferred) or (alternative)

There are two different Morse codes in use by English-speaking operators. The more commonly used code is called the *international Morse code* or *continental code* (Table 26.2).

Modern communications devices can function under weak-signal conditions that would frustrate a human operator. But when human operators are involved, the Morse code is still a reliable means of getting a message through severe interference. This is because the bandwidth of a Morse-code radio signal is extremely narrow, and it is comparatively easy for the human ear to distinguish between the background noise and a code signal. The main use of Morse is by amateur radio operators, many of whom use radiotelegraphy as a means of recreation.

Q Signals

In Morse radiotelegraphy, certain statements, phrases, or words are made often. It becomes tedious to send complete sentences, phrases, and words, especially if they are repeated. To streamline Morse operation, a set of abbreviations called *Q signals* has been devised.

Each Q signal consists of the letter Q, followed by two more letters. A Q signal followed by a question mark indicates a query; if no question mark follows the signal, or if data follows, it indicates a statement. For example, "QRM?" means "Are you experiencing interference?" and "QRM" means "I am experiencing interference." As another example, "QTH?" means "What is your location?" and "QTH ROCHESTER, MN" means "My station location is Rochester, Minnesota." Table 26.3 is a list of commonly used Q signals.

Q signals were originally designed for radiotelegraph use, but many radioteletype and radiotelephone operators also use them. The Q signals can streamline radioteletype operation in the same way that they make Morse operation more convenient. There is some question as to how beneficial (if not detrimental) the Q signals are in voice communication.

TABLE 26.2 **The International Morse Code.**

Character	Symbol
A	.-
B	-...
C	-.-.
D	-..
E	.
F	..-.
G	--.
H	....
I	..
J	.---
K	-.-
L	.-..
M	--
N	-.
O	---
P	.--.
Q	--.-
R	.-.
S	...
T	-
U	..-
V	...-
W	.--
X	-..-
Y	-.--
Z	--..
0	-----
1	.----
2	..---
3	...--
4	....-
5	.....
6	-....
7	--...
8	---..
9	----.
Period	.-.-.-
Comma	--..--
Query	..--..
Slash	-..-.
Dash	-....-
Break (pause)	-...-
Semicolon	-.-.-.
Colon	---...

TABLE 26.3   Common Q Signals and Their Meanings

Signal	Query and response
QRA	What is the name of your station? The name of my station is —.
QRB	How far from my station are you? I am — miles or — kilometers from your station.
QRD	From where are you coming, and where are you going? I am coming from —, and am going to —.
QRG	What is my frequency, or that of —? Your frequency, or that of —, is — (kHz, MHz, GHz).
QRH	Is my frequency unstable? Your frequency is unstable.
QRI	How is the tone of my signal? Your signal tone is: 1 (good), 2 (fair), 3 (poor).
QRK	How readable is my signal? Your signal is: 1 (unreadable), 2 (barely readable), 3 (readable with difficulty), 4 (readable with almost no difficulty), 5 (perfectly readable).
QRL	Are you busy? Or, Is this frequency in use? I am busy. Or, This frequency is in use.
QRM	Are you experiencing interference from other stations? I am experiencing interference from other stations.
QRN	Is your reception degraded by sferics or electrical noise? My reception is degraded by sferics or electrical noise.
QRO	Should I increase my transmitter output power? Increase your transmitter output power.
QRP	Should I reduce my transmitter output power? Reduce your transmitter output power.
QRQ	Should I send (Morse code) faster? Send (Morse code) faster.
QRS	Should I send (Morse code) more slowly? Send (Morse code) more slowly.
QRT	Shall I stop transmitting? Or, Are you going to stop transmitting? Stop transmitting. Or, I am going to stop transmitting.
QRU	Do you have information for me? I have no information for you.
QRV	Are you ready for —? I am ready for —.

TABLE 26.3   Common Q Signals and Their Meanings (Continued)

Signal	Query and response
QRW	Should I tell — that you are calling him/her/them? Tell — that I am calling him/her/them.
QRX	When will you call me again? I will call you again at —.
QRY	What is my turn in order? Your turn is number — in order.
QRZ	Who is calling me? You are being called by —.
QSA	How strong are my signals? Your signals are: 1 (almost inaudible), 2 (weak), 3 (fairly strong), 4 (strong), 5 (very strong).
QSB	Are my signals varying in strength? Your signals are varying in strength.
QSD	Are my signals mutilated? Or, is my keying bad? Your signals are mutilated. Or, your keying is bad.
QSG	Should I send more than one message? Send — messages.
QSJ	What is your charge per word? My charge per word is —.
QSK	Can you hear me between your signals? Or, Do you have full break-in capability? I can hear you between my signals. Or, I have full break-in capability.
QSL	Do you acknowledge receipt of my message? I acknowledge receipt of your message.
QSM	Should I repeat my message? Repeat your message.
QSN	Did you hear me on — (frequency, channel, or wavelength)? I heard you on — (frequency, channel, or wavelength).
QSO	Can you communicate with —? I can communicate with —.
QSP	Will you send a message to —? I will send a message to —.
QSQ	Is there a doctor there? Or, Is —there? There is a doctor here. Or, — is here.
QSU	On what frequency, channel, or wavelength should I reply? Reply on — (frequency, channel, or wavelength).

TABLE 26.3   Common Q Signals and Their Meanings (Continued)

Signal	Query and response
QSV	Shall I transmit a series of V's for test purposes? Transmit a series of V's for test purposes.
QSW	On which frequency, channel, or wavelength will you transmit? I will transmit on — (frequency, channel, or wavelength).
QSX	Will you listen for me? Or, Will you listen for —? I will listen for you. Or, I will listen for —.
QSY	Should I change frequency, channel, or wavelength? Change frequency, channel, or wavelength to —.
QSZ	Should I send each word or word group more than once? Send each word or word group more than once.
QTA	Should I cancel message number —? Cancel message number —.
QTB	Does your word count agree with mine? My word count disagrees with yours.
QTC	How many messages do you have to send? I have — messages to send.
QTE	What is my bearing relative to you? Your bearing relative to me is — (azimuth degrees).
QTH	What is your location? My location is —.
QTJ	What is the speed at which your vehicle is traveling? My vehicle is traveling at — (miles or kilometers per hour).
QTL	In what direction are you headed? I am headed toward —. Or, My heading is — (azimuth degrees).
QTN	When did you leave —? I left — at —.
QTO	Are you airborne? I am airborne.
QTP	Do you intend to land? I intend to land.
QTR	What is the correct time? The correct time is — Coordinated Universal Time (UTC).
QTX	Will you stand by for me? I will stand by for you until —.

TABLE 26.3   Common Q Signals and Their Meanings (Continued)

Signal	Query and response
QUA	Do you have information concerning —?
	I have information concerning —.
QUD	Have you received my urgent signal, or that of —?
	I have received your urgent signal, or that of —.
QUF	Have you received my distress signal, or that of —?
	I have received your distress signal, or that of —

Ten Code

To streamline two-way radiotelephone (voice) operation, a set of abbreviations called the *ten code* has been devised. This code also affords some secrecy, because eavesdroppers who do not know the code cannot fully understand the transmissions.

The ten code is used in the Citizens Radio Service and in law-enforcement communications. Various other services, such as security companies and fire departments, also employ it. The ten code is not used in digital text modes such as Morse, radioteletype, packet, or computer communications.

Each ten-code signal consists of the spoken word *ten* followed by a number generally between 1 and 100. A list of ten-code signals is given in Table 26.4.

Phonetic Alphabet

The *phonetic alphabet* is a set of 26 words, one for each letter of the English alphabet, used by radiotelephone operators for the purpose of clarifying messages under marginal conditions. The words are chosen so they are not easily confused with other words in the list (Table 26.5).

When it is necessary to spell out a word in a radiotelephone communication, the operator will say, for example, "The name here is Stan. I spell: Sierra, Tango, Alpha, November."

TABLE 26.4   Ten-Code Signals and Their Meanings

Ten-code signals used in the Citizens Radio Service.

Signal	Query and response
10-1	Are you having trouble receiving my signals? I am having trouble receiving your signals.
10-2	Are my signals good? Your signals are good.
10-3	Shall I stop transmitting? Stop transmitting.
10-4	Have you received my message completely? I have received your message completely.
10-5	Shall I relay a message to —? Relay a message to —.
10-6	Are you busy? I am busy; stand by until —.
10-7	Is your station out of service? My station is out of service.
10-8	Is your station in service? My station is in service.
10-9	Shall I repeat my message? Or, Is reception poor? Repeat your message. Or, Reception is poor.
10-10	Are you finished transmitting? I am finished transmitting.
10-11	Am I talking too fast? You are talking too fast.
10-12	Do you have visitors? I have visitors.
10-13	How are your weather and road conditions? My weather and road conditions are —.
10-14	What is the local time, or the time at —? The local time, or the time at —, is —.
10-15	Shall I pick up — at —? Pick up — at —.
10-16	Have you picked up —? I have picked up —.
10-17	Do you have urgent business? I have urgent business.

TABLE 26.4 Ten-Code Signals and Their Meanings (Continued)

Signal	Query and response
10-18	Have you any information for me? I have some information for you; it is —.
10-19	Have you no information for me? I have no information for you.
10-20	Where are you located? I am located at —.
10-21	Shall I call you on the telephone? Call me on the telephone.
10-22	Shall I report in person to —? Report in person to —.
10-23	Shall I stand by? Stand by until —.
10-24	Are you finished with your last assignment? I am finished with my last assignment.
10-25	Are you in contact with —? I am in contact with —.
10-26	Shall I disregard the information you just sent? Disregard the information I just sent.
10-27	Shall I move to channel —? Move to channel —.
10-30	Is this action legal or is it illegal? This action is illegal.
10-33	Do you have an emergency message? I have an emergency message.
10-34	Do you have trouble? I have trouble.
10-35	Do you have confidential information? I have confidential information.
10-36	Is there an accident? There is an accident at —.
10-37	Is a tow truck needed? A tow truck is needed at —.
10-38	Is an ambulance needed? An ambulance is needed at —.

TABLE 26.4   Ten-Code Signals and Their Meanings (Continued)

Signal	Query and response
10-39	Is there a convoy at —?
	There is a convoy at —.
10-41	Shall we change channels?
	Change channels.
10-60	Please give me your message number.
	My message number is —.
10-63	Is this net directed?
	This net is directed.
10-64	Do you intend to stop transmitting?
	I intend to stop transmitting.
10-65	Do you have a net message for —?
	I have a net message for —.
10-66	Do you wish to cancel your messages number —through —?
	I wish to cancel my messages number — through —.
10-67	Shall I stop transmitting to receive a message?
	Stop transmitting to receive a message.
10-68	Shall I repeat my messages number — through —?
	Repeat your messages number — through —.
10-70	Have you a message?
	I have a message.
10-71	Shall I send messages by number?
	Send messages by number.
10-79	Shall I inform — regarding a fire at —?
	Inform —regarding a fire at —.
10-84	What is your telephone number?
	My telephone number is —.
10-91	Are my signals weak?
	Your signals are weak.
10-92	Are my signals distorted?
	Your signals are distorted.
10-94	Shall I make a test transmission?
	Make a test transmission.
10-95	Shall I key my microphone without speaking?
	Key your microphone without speaking.

TABLE 26.4 Ten-Code Signals and Their Meanings (Continued)

Signal	Query and response
Ten-code signals used in law enforcement	
10-1	Are you having trouble receiving my signals? I am having trouble receiving your signals.
10-2	Are my signals good? Your signals are good.
10-3	Shall I stop transmitting? Stop transmitting.
10-4	Have you received my message in full? I have received your message in full.
10-5	Shall I relay a message to —? Relay a message to —.
10-6	Are you busy? I am busy; stand by until —.
10-7	Is your station out of service? My station is out of service.
10-8	Is your station in service? My station is in service.
10-9	Shall I repeat my message? Repeat your message.
10-10	Is there a fight in progress at your location? There is a fight in progress at my location.
10-11	Do you have a case involving a dog? I have a case involving a dog.
10-12	Shall I stand by? Or, Shall I stand by until —? Stand by. Or, Stand by until —.
10-13	How are your weather and road conditions? My weather and road conditions are —.
10-14	Have you received a report of a prowler? I have received a report of a prowler.
10-15	Is there a civil disturbance at your location? There is a civil disturbance at my location.
10-16	Is there domestic trouble at your location? There is domestic trouble at my location.

TABLE 26.4   Ten-Code Signals and Their Meanings (Continued)

Signal	Query and response
10-17	Shall I meet the person who issued the complaint? Meet the person who issued the complaint.
10-18	Shall I hurry to finish this assignment? Hurry to finish this assignment.
10-19	Shall I return to —? Return to —.
10-20	What is your location? My location is —.
10-21	Shall I call — by telephone? Call — by telephone.
10-22	Shall I ignore the previous information? Ignore the previous information.
10-23	Has — arrived at —? —has arrived at —.
10-24	Have you finished your assignment? I have finished my assignment.
10-25	Shall I report in person to —? Report in person to —.
10-26	Are you detaining a subject? I am detaining a subject.
10-27	Do you have data on driver license number —? Here is data on driver license number —.
10-28	Do you have data on vehicle registration number —? Here is data on vehicle registration number —.
10-29	Shall I check records to see if — is a wanted person? Check records to see if —is a wanted person.
10-30	Is — using a radio illegally? — is using a radio illegally.
10-31	Is there a crime in progress at your location (or at —)? There is a crime in progress at my location (or at —).
10-32	Is there a person with a gun at your location (or at —)? There is a person with a gun at my location (or at —).

TABLE 26.4   Ten-Code Signals and Their Meanings (Continued)

Signal	Query and response
10-33	Is there an emergency at your location (or at —)? There is an emergency at my location (or at —.)
10-34	Is there a riot at your location (or at —)? There is a riot at my location (or at —).
10-35	Do you have an alert concerning a major crime? I have an alert concerning a major crime.
10-36	What is the correct time? The correct time is — local (or — UTC).
10-37	Shall I investigate a suspicious vehicle? Investigate a suspicious vehicle.
10-38	Are you stopping a suspicious vehicle? I am stopping a suspicious vehicle (of type —).
10-39	Is your (or this) situation urgent? My (or this) situation is urgent, use lights and/or siren.
10-40	Shall I refrain from using my light or siren? Refrain from using your light or siren.
10-41	Are you just starting duty? I am just starting duty.
10-42	Are you finishing duty? I am finishing duty.
10-43	Do you need, or are you sending, data about —? I need, or am sending, data about —.
10-44	Do you want to leave patrol? I want to leave patrol and go to —.
10-45	Is there a dead animal at your location (or at —)? There is a dead animal at my location (or at —).
10-46	Shall I assist a motorist at my location (or at —)? Or, are you assisting a motorist at your location (or at —)? Assist a motorist at your location (or at —). Or, I am assisting a motorist at my location (or at —).
10-47	Are road repairs needed now at your location (or at —)? Road repairs are needed now at my location (or at —).
10-48	Does a traffic standard need to be fixed at your location (or at —)? A traffic standard needs to be fixed at my location (or at —).

TABLE 26.4   Ten-Code Signals and Their Meanings (Continued)

Signal	Query and response
10-49	Is a traffic light out at your location (or at —? A traffic light is out at my location (or at —).
10-50	Is there an accident at your location (or at —)? There is an accident at my location (or at —).
10-51	Is a tow truck needed at your location (or at —)? A tow truck is needed at my location (or at —).
10-52	Is an ambulance needed at your location (or at —)? An ambulance is needed at my location (or at —).
10-53	Is the road blocked at your location (or at —)? The road is blocked at my location (or at —).
10-54	Are there animals on the road at your location (or at —)? There are animals on the road at my location (or at —).
10-55	Is there a drunk driver at your location (or at —)? There is a drunk driver at my location (or at —).
10-56	Is there a drunk pedestrian at your location (or at —)? There is a drunk pedestrian at my location (or at —).
10-57	Has there been a hit-and-run accident at your location (or at —)? There has been a hit-and-run accident at my location (or at —).
10-58	Shall I direct traffic at my location (or at —)? Direct traffic at your location (or at —).
10-59	Is there a convoy at your location (or at —)? Or, does — need an escort? There is a convoy at my location (or at —). Or, —needs an escort.
10-60	Is there a squad at your location (or at —)? There is a squad at my location (or at —).
10-61	Are there personnel in your vicinity (or in the vicinity of —)? There are personnel in my vicinity (or in the vicinity of —).
10-62	Shall I reply to the message of —? Reply to the message of —.
10-63	Shall I make a written record of —? Make a written record of —.
10-64	Is this message to be delivered locally? This message is to be delivered locally.

TABLE 26.4   Ten-Code Signals and Their Meanings (Continued)

Signal	Query and response
10-65	Do you have a net message assignment?
	I have a net message assignment.
10-66	Do you want to cancel message number —?
	I want to cancel message number —.
10-67	Shall I clear for a net message?
	Clear for a net message.
10-68	Shall I disseminate data concerning —?
	Disseminate data concerning —.
10-69	Have you received my messages numbered — through —?
	I have received your messages numbered — through —.
10-70	Is there a fire at your location (or at —)?
	There is a fire at my location (or at —).
10-71	Shall I advise of details concerning the fire at my location (or at —)?
	Advise of details concerning the fire at your location (or at —).
10-72	Shall I report on the progress of the fire at my location (or at —)?
	Report on the progress of the fire at your location (or at —).
10-73	Is there a report of smoke at your location (or at —)?
	There is a report of smoke at my location (or at —).
10-74	(No query)
	Negative.
10-75	Are you in contact with —?
	I am in contact with —.
10-76	Are you going to —?
	I am going to —.
10-77	When do you estimate arrival at —?
	I estimate arrival at — at — local time (or — UTC).
10-78	Do you need help?
	I need help at this location (or at —).
10-79	Shall I notify a coroner of —?
	Notify a coroner of —.
10-82	Shall I reserve a hotel or motel room at —?
	Reserve a hotel or motel room at —.

TABLE 26.4 Ten-Code Signals and Their Meanings (Continued)

Signal	Query and response
10-85	Will you (or —) be late?
	I (or —) will be late.
10-87	Shall I pick up checks for distribution?
	Pick up checks for distribution. Or, I am picking up checks for distribution.
10-88	What is the telephone number of —?
	The telephone number of — is —.
10-90	Is there a bank alarm at your location (or at —)?
	There is a bank alarm at my location (or at —).
10-91	Am I using a radio without cause? Or, Is — using a radio without cause?
	You are using a radio without cause. Or, — is using a radio without cause.
10-93	Is there a blockade at your location (or at —)?
	There is a blockade at my location (or at —).
10-94	Is there an illegal drag race at your location (or at —)?
	There is an illegal drag race at my location (or at —).
10-96	Is there a person acting mentally ill at your location (or at —)?
	There is a person acting mentally ill at my location (or at —).
10-98	Has someone escaped from jail at your location (or at —)?
	Someone has escaped from jail at my location (or at —).
10-99	Is — wanted or stolen?
	— is wanted or stolen. Or, there is a wanted person or stolen article at —.

Phonetics should be used only when necessary. Otherwise, receiving operators might be confused or irritated by them. Phonetics should not be used to clarify common words. If conditions are so poor that voice communication is difficult, a binary text mode, such as radioteletype, Morse, or packet, should be used.

TABLE 26.5   Phonetic Alphabet as Recommended by the International Telecommunication Union (ITU)

Letter	Phonetic
A	AL-fa
B	BRAH-vo
C	CHAR-lie
D	DEL-ta
E	ECK-o
F	FOX-trot
G	GOLF
H	ho-TEL
I	IN-dia
J	Ju-li-ETTE
K	KEE-low
L	LEE-ma
M	MIKE
N	No-VEM-ber
O	OS-car
P	pa-PA
Q	Que-BECK
R	ROW-me-oh
S	see-AIR-ah
T	TANG-go
U	YOU-ni-form
V	VIC-tor
W	WHIS-key
X	X-ray
Y	YANK-key
Z	ZOO-loo

Coordinated Universal Time

Coordinated universal time (UTC) is, for most practical purposes, the time at 0° longitude, the *Greenwich meridian* or *prime meridian* that passes near London, England. Time in UTC is often referred to as *zulu* (abbreviated Z), the phonetic expression for the letter Z, which is the designator for the time at the Greenwich meridian.

When referring to time in UTC, military time is generally used. It is employed in aviation, by government agencies, and by amateur radio operators. You will encounter military time if you are serious about electronic communications in any form, including the *on-line services* used with personal computers.

In the 24-hour system, people talk in "hundreds." You will hear expressions such as "oh three hundred UTC" or "seventeen hundred zulu." These are written 0300 Z and 1700 Z and refer to 3:00 a.m. and 5:00 p.m., respectively. When speaking about the time down to the exact minute, people say things like "fifteen forty-three UTC" or "one five four three zulu." This would be written 1543 Z and means 3:43 p.m. Note that 24-hour time has no colon in between the hour and the minute, as does 12-hour time.

Table 26.6 shows UTC versus the time in various zones in the United States. *Eastern daylight time* (EDT) is 4 h behind UTC. *Eastern standard time* (EST) and *Central daylight time* (CDT) are 5 h behind UTC. *Central standard time* (CST) and *Mountain daylight time* (MDT) are 6 h behind UTC. *Mountain standard time* (MST) and *Pacific daylight time* (PDT) are 7 h behind UTC. *Pacific Standard Time* (PST) is 8 h behind UTC.

Soldering and Desoldering

Solder is a metal alloy used for securing electrical connections between conductors. There are several types of solder, intended for use with various metals and in various applications.

Types of solder

The most common variety of solder consists of tin and lead with a rosin core. Some types of solder have an acid core. In electronic devices, rosin-core solder should be used.

In *tin/lead solder*, the ratio of the constituent metals determines the temperature at which the solder will melt. The higher the ratio of tin to lead, the lower the melting temperature. For general soldering purposes, 50:50 solder can be used. For heat-sensitive components, 60:40 solder is better because it melts at a lower temperature. Tin-lead solder is suitable for use with most metals except aluminum.

TABLE 26.6   Coordinated Universal Time (UTC) with Conversion for the Various Time Zones within the United States

UTC	EDT	EST/CDT	CST/MDT	MST/PDT	PST
0000	2000*	1900*	1800*	1700*	1600*
0100	2100*	2000*	1900*	1800*	1700*
0200	2200*	2100*	2000*	1900*	1800*
0300	2300*	2200*	2100*	2000*	1900*
0400	0000	2300*	2200*	2100*	2000*
0500	0100	0000	2300*	2200*	2100*
0600	0200	0100	0000	2300*	2200*
0700	0300	0200	0100	0000	2300*
0800	0400	0300	0200	0100	0000
0900	0500	0400	0300	0200	0100
1000	0600	0500	0400	0300	0200
1100	0700	0600	0500	0400	0300
1200	0800	0700	0600	0500	0400
1300	0900	0800	0700	0600	0500
1400	1000	0900	0800	0700	0600
1500	1100	1000	0900	0800	0700
1600	1200	1100	1000	0900	0800
1700	1300	1200	1100	1000	0900
1800	1400	1300	1200	1100	1000
1900	1500	1400	1300	1200	1100
2000	1600	1500	1400	1300	1200
2100	1700	1600	1500	1400	1300
2200	1800	1700	1600	1500	1400
2300	1900	1800	1700	1600	1500
2400	2000	1900	1800	1700	1600

*Previous day from UTC.

For soldering to this metal, *aluminum solder* is available. It melts at a higher temperature than tin-lead solder and requires the use of a blowtorch or other high-heat device for application.

In high-current circuits, *silver solder* is recommended because it can withstand the high temperatures that can be produced when large currents flow through components and connections. A blowtorch is usually required for working with silver solder. The solder must be applied in a well-ventilated area because it produces dangerous fumes when heated.

Table 26.7 is a summary of common types of solder, their characteristics, and their principal uses.

TABLE 26.7   Common Types of Solder Used in Electronics

Solder type	Melting point, °F, °C	Common uses
Tin/lead 50/50	430, 220	Electronics
Rosin core		
Tin/lead 60/40	370, 190	Electronics
Rosin core		Low heat
Tin/lead 63/37	360, 180	Electronics
Rosin core		Low heat
Silver	600, 320	Electronics
		High heat
		High current
Tin/lead 50/50	430, 220	Sheet metal
Acid core		bonding

Soldering instruments

A *soldering gun* is a quick-heating soldering tool. It is called a "gun" because of its shape. A trigger-operated switch is pressed with the finger, allowing the element to heat up within a few seconds. Soldering guns are convenient in the assembly and repair of some kinds of electronic equipment. They are available in various wattage ratings for different electronic applications.

A *soldering iron* consists of a heating element and a handle. The iron requires up to several minutes to fully heat after it is turned on; the larger the iron, the longer the warm-up time. The smallest irons are rated at a few watts and are used in miniaturized electronic equipment. The largest irons draw hundreds of watts and are used for outdoor wire splicing and sheet-metal bonding.

For sheet-metal bonding and outdoor wire splicing, a blowtorch can be used. Its main advantage is portability; it does not require electricity to function, so an extension cord is not necessary. It can supply sufficient heat in blustery weather when a soldering gun or iron will not work well.

All soldering equipment must be used with care because the heat is sufficient to start fires. Soldering equipment must not be used in the presence of volatile liquids such as alcohol or gasoline.

Printed circuits

Most *printed-circuit soldering* is done from the noncomponent (foil) side of the circuit board. The component lead is inserted through the appropriate hole, and the soldering iron is placed so that it heats both the foil and the component lead (Fig. 26.1A). If the component is heat-sensitive, a needle-nosed pliers should be used to grip the lead on the component (nonfoil) side of the board while heat is applied. The solder is allowed to flow onto the foil and the component lead after the joint becomes hot enough to melt the solder. Heating takes a couple of seconds. The solder should completely cover the foil dot or square in which the component lead is centered. Excessive solder should not be used. After the joint has cooled, the component lead should be snipped off, flush with the solder, using a diagonal cutter.

If the circuit board is double-sided (foil on both sides), it will usually have *plated-through holes,* and the soldering procedure described above will be adequate. However, if the holes are not plated through, solder must be applied, as described above, to the foil and component lead on both sides of the circuit board.

Some printed-circuit components are mounted on the foil side of the board. In such cases, the component lead and the circuit board foil are first coated, or "tinned," with a thin layer of solder. The component lead is then placed flat against the foil, and the iron is placed in contact with the component lead (Fig. 26.1B). The heat melts the solder by conduction.

Point to point

Most *point-to-point wiring* is accomplished with *tie strips,* where one or more wires terminate.

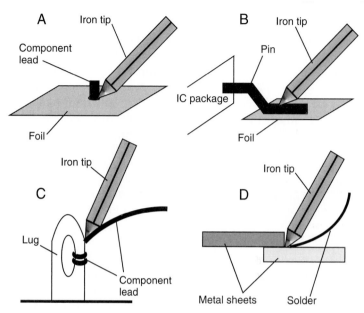

Figure 26.1 (A and B) Soldering to printed-circuit boards; (C) soldering a component lead to a tie strip; (D), bonding sheet metal.

In *tie-strip wiring,* the lug should first be coated with a thin layer of solder. The actual soldering should not be done until all the wires are attached to the lug. Wires are wrapped two or three times around the lug using a needle-nosed pliers. Excess wire is cut off using a diagonal cutter. When all wires have been attached to the lug, the soldering instrument is held against each wire "coil," one at a time, and the connection is allowed to heat up until the solder flows freely in between the wire turns, adhering to both the wire and the lug (Fig. 26.1). Enough solder should be used so that the connection is completely coated, but solder should not be allowed to ball up or drip from the connection.

If a heat-sensitive component is wired to a tie strip, a needle-nosed pliers can be used to conduct heat away from the component. The pliers should be clamped on the component lead between the connection and the body of the component.

The pliers should not be removed until the soldered joint has cooled down almost to room temperature. Water should not be applied in an attempt to hasten the cooling.

If sufficient heat has not been applied to a solder connection, a *cold solder joint* will result. A properly soldered connection has a shiny, clean appearance. A cold joint looks dull or rough. Many electronic equipment failures occur because of cold solder joints, which can exhibit high resistance and/or intermittent conduction. If a cold joint is found, as much of the solder as possible should be removed using wire braid. The surfaces should be cleaned, and then the connection should be resoldered.

Sheet metal

When soldering sheet metal, rosin-core solder should be used if possible, but sometimes the rosin does not allow a good enough mechanical bond. In such cases, acid-core solder can be used. Special solder is available for use with aluminum, which does not readily adhere to most other types of solder.

The sheet-metal surfaces should be sanded where bonding is contemplated, using fine emery paper. Then the surfaces should be cleaned with a noncorrosive, grease-free solvent such as isopropyl alcohol. A high-wattage soldering iron or blowtorch should be used to heat the metal while both sheets are "tinned" with a thin layer of solder. Then the sheets should be secured in place. Sufficient heat should be applied so the solder will flow freely. Some additional solder should be applied on each side of the bond, working gradually along the length of the bond from one side to the other (Fig. 26.1D). The bond will require some time to cool, and it should be kept free from stress until it has cooled completely. Water or other fluids should not be used in an attempt to speed up the cooling process.

Removing solder

When replacing a faulty component in an electronic circuit, it is usually necessary to desolder some connections.

With most printed-circuit boards, *desoldering* consists of the application of heat with a soldering iron and the conduction of solder away from the connection by means of wire braid. The connection and the braid must both be heated to a temperature sufficient to melt the solder. Excessive heat should be avoided so that only the desired connection is desoldered and the circuit board and nearby components are not damaged.

Many sophisticated desoldering devices are available. One popular device employs an air-suction nozzle, which swallows the solder by vacuum action as a soldering iron heats the connection. This apparatus is especially useful when a large number of connections must be desoldered, because it works fast. It is also useful in desoldering tiny connections, where there is little room for error. For large connections, such as wire splices and solder-welded joints, it is sometimes better to remove the entire connection than to try to desolder it.

Robot Generations

Some researchers have analyzed the evolution of robots, denoting progress according to *robot generations.* One of the first engineers to make formal mention of robot generations was the Japanese roboticist Eiji Nakano.

First generation

According to Nakano, *first-generation robots* were simple mechanical arms. These machines had the ability to make precise motions at high speed, many times, for a long time. Such robots are still used in industry today. First-generation robots can work in groups if their actions are synchronized. The operation of these machines must be constantly watched because if they get out of alignment and are allowed to keep working, the result can be a series of faulty production units.

Second generation

A *second-generation robot* has some *artificial intelligence (AI).* Such devices include *pressure sensors, proximity sen-*

sors, tactile sensors, and *machine vision.* A *robot controller* (computer) processes the data from these sensors and adjusts the operation of the robot accordingly. These devices came into common use around 1980. Second-generation robots can stay synchronized with each other, without having to be overseen constantly by a human operator.

Third generation

Two major avenues are developing for third-generation robot technology. *Autonomous robots* can work on their own. Such a machine contains its own controller and can do things largely without supervision. A good example is a *personal robot.*

There are some situations in which autonomous robots do not work well. In these cases, a number of simple robots, all under the control of one central computer, are used. They work like ants in an anthill or like bees in a hive. These machines are called *insect robots.*

Fourth generation and beyond

In his original paper, Nakano did not write about anything past the third generation of robots. But we might speculate about *fourth-generation robots:* machines of a sort yet to be seriously worked on. An example is a fleet of robots that reproduce and evolve or that have the ability to reason or that can replace human beings in many capacities. Beyond this, we might say that a *fifth-generation robot* is something about which nothing has been said, written, or filmed. This category is constantly retreating as researchers come up with new ideas.

Table 26.8 is a summary of robot generations, their general capabilities, and the approximate time frames of their implementation.

Wetware

Wetware is an expression for the linking of computers with the human brain. The term was originally coined by science fiction writers, but wetware is on its way to becoming reality.

TABLE 26.8 Robot Generations

Generation	Time first used	Capabilities
First	Before 1980	Mainly mechanical; stationary; good precision; high speed; physical ruggedness; use of servomechanisms; no external sensors; no artificial intelligence
Second	1980–1990	Tactile sensors; vision systems; position sensors; pressure sensors; microcomputer control; programmable
Third	1990–present	Mobile; autonomous; insectlike; artificial intelligence; speech recognition; speech synthesis; navigation systems; teleoperated
Fourth	Future	Design not yet begun; able to reproduce? Able to evolve? Artificially alive? As smart as a human? True sense of humor?
Fifth	?	Not yet discussed; capabilities unknown

Brain waves

The concept behind wetware is simple: Connect a computer to an *electroencephalograph* (EEG) and try to get the computer to respond in a controlled manner to variations in these waves. Certain mental and physical states are accompanied by *brain waves* that have various frequencies and shapes. Table 26.9 lists some of the commonly recognized brain wave types, their usual frequency ranges in hertz, and the mind/body conditions they accompany. Even if a device or system cannot infer its operator's intentions directly, it can be programmed to recognize waveforms and frequencies.

Applications

Here are some uses that have been suggested for wetware. Some are bizarre by today's standards. But technology can turn the ridiculous into the routine.

TABLE 26.9   **Brain Wave Types and Forms of Human Behavior with Which They Are Associated**

Type	Frequency, Hz	Behavior
Alpha	8 to 13	Awake, alert, relaxed
Mu	8 to 13	Nerve and muscle activity (feeling and movement)
Beta	13 to 30	Intense mental concentration
Delta	0.5 to 5	Deep sleep
Theta	4 to 7	Light sleep with dreams

1. The most promising use for wetware is in the control of *prostheses,* or artificial limbs. It would be necessary to learn a new way of giving mental commands, but the human mind/brain is good at adapting itself to challenges of this sort. It might also be possible to use wetware to control electronic nerve and muscle stimulators so that people with paralyzed limbs could use them again.

2. Complex mechanical equipment might be operated by wetware remote control. A military general could sit in a dark room, eyes fixed on a monitor screen, watching the movements of robot soldiers, tanks, ships, and aircraft in distant combat zones. The enemy would be another general, in some distant chamber, watching a different screen portraying the same theaters of battle.

3. Instead of having one's thoughts control the movements of muscles, artificial limbs, and machines, suppose that signals could be processed and sent into the brain? A pair of camera tubes might become a set of eyes; a pair of microphones could become a set of ears. *Tactile sensors* could allow a person to feel heat, cold, texture, and pressure. The data would be sent to electrodes on the scalp or implanted in the brain and/or spinal cord. Such technology might be employed for enhanced *telepresence,* in which robots take the place of human beings in dangerous or inaccessible places.

4. Some musicians have expressed interest in the idea of converting brain waves into sound. This has been tried using computers and electronic music synthesizers, with mixed results. Mind-controlled paint and draw programs could be employed to generate "brain-wave art." It has also been suggested that a mind-reading computer could help authors put their thoughts into words. As of this writing, this level of wetware technology appears unlikely to be reached for several decades.

5. *Computer games* are a realizable contemporary application for wetware. You might compete with friends to see who can achieve the finest degree of control over an animated display, such as an aircraft landing or a cartoon character's movements. The "virtual war" example above could be adapted to personal computers, changing the scenario from combat to baseball, football, basketball, or almost any other competitive but nondestructive sport.

Bibliography

Crowhurst, N., and Gibilisco, S., *Mastering Technical Mathematics,* 2d ed. (New York: McGraw-Hill, 1999).

Dorf, R., *Electrical Engineering Handbook,* 2d ed. (Boca Raton, Fla.: CRC Press, 1997).

Gibilisco, S., *Handbook of Radio and Wireless Technology* (New York: McGraw-Hill, 1999).

Gibilisco, S., *TAB Encyclopedia of Electronics for Technicians and Hobbyists* (New York: McGraw-Hill, 1997).

Gibilisco, S., *Teach Yourself Electricity and Electronics,* 2d ed. (New York: McGraw-Hill, 1997).

Van Valkenburg, M., *Reference Data for Engineers: Radio, Electronics, Computer and Communications* (Indianapolis, Ind.: Howard W. Sams, 1998).

Veley, V., *The Benchtop Electronics Reference Manual* (New York: McGraw-Hill, 1994).

INDEX

ABOUT THE AUTHOR

Stan Gibilisco has authored or coauthored dozens of nonfiction books about electronics and science. He first attracted attention with *Understanding Einstein's Theories of Relativity* (TAB Books, 1983). His *Encyclopedia of Electronics* (TAB Professional and Reference Books, 1985) and *Encyclopedia of Personal Computing* (McGraw-Hill, 1996) were annotated by the American Library Association as among the best reference volumes published in those years. Gibilisco's work has gained reading audiences in the Far East, Europe, and South America. He maintains a Web site at http://members.aol.com/stangib.